Encyclopedia of Fermented Fresh Milk Products

Encyclopedia of Fermented Fresh Milk Products

An International Inventory of Fermented Milk, Cream, Buttermilk, Whey, and Related Products

Joseph A. Kurmann
Agricultural Institute Grangeneuve
Switzerland

Jeremija Lj. Rašić
Food Research Institute Novi Sad
Yugoslavia

Manfred Kroger
The Pennsylvania State University
USA

An avi Book
Published by Van Nostrand Reinhold
New York

An AVI Book
(AVI is an imprint of Van Nostrand Reinhold)

Library of Congress Catalog Card Number 91-30210
ISBN 0-442-00869-4

Printed in the United States of America.

Van Nostrand Reinhold
115 Fifth Avenue
New York, New York 10003

Chapman and Hall
2-6 Boundary Row
London, SE1 8HN, England

Thomas Nelson Australia
102 Dodds Street
South Melbourne 3205
Victoria, Australia

Nelson Canada
1120 Birchmount Road
Scarborough, Ontario MIK 5G4, Canada

16 15 14 13 12 11 10 9 8 7 6 5 4 3 2 1

Library of Congress Cataloging-in-Publication Data

Kurmann, Joseph A., 1931-
Encyclopedia of fermented fresh milk products: an international inventory of fermented milk, cream, buttermilk, whey, and related products/Joseph A. Kurmann, Jeremija Lj. Rašić, Manfred Kroger.
p. cm.
Includes bibliographical references and appendixes.
ISBN 0-442-00869-4
1. Cultured milk—Encyclopedias. 2. Fermented milk—Encyclopedias. I. Rašić, Jeremija Lj., 1922- . II. Kroger, Manfred. III. Title.
SF275.C84K87 1992
637'.1—dc20 91-30210

Contents

Foreword

Fermented milks were available to humankind thousands of years ago, but recently they have grown immensely in importance and diversity. Despite this, no existing text devoted entirely to a modern worldwide viewpoint presents the historical, scientific, and technical nature of fermented milk and its products. This great need has been fully satisfied by the publication of this *Encyclopedia of Fermented Fresh Milk Products.*

Three highly respected authors, Kurmann, Rašić, and Kroger, from different countries with important fermented milk products industries, have combined their many talents and great breadth of international experience to produce an admirable volume that accurately portrays fermented milk products in detail and relates them to history, manufacture, characterization, health significance, and usage. This encyclopedia should be valuable to food scientists and dairy technologists, nutritionists, public health personnel, regulatory officials, educators, students, and historians. Even consumers in quest of complete nutritional descriptions of these important basic foods will benefit from reading this volume.

Fermented milks and their products should increase in importance, and the information available to the reader in this encyclopedia will accelerate such an advance. Much is known about the nutritional benefits of fermented milks, two of which are low levels of fat and cholesterol, but much more must be learned about their therapeutic effects and about the development of new products.

The authors should be congratulated on their contribution destined for great success and encouraged to periodically issue new editions to accommodate the changing times and knowledge.

Frank V. Kosikowski
Emeritus Professor of Food Science
Cornell University
Ithaca, N.Y., USA

Preface

This book is the world's first encyclopedia and inventory of fermented milk, cream, buttermilk, and whey products. It contains information on traditional and nontraditional fermented fresh milk products—those manufactured today and in earlier times—as well as proposed products not yet commercially manufactured. Related products, including nonfermented ones such as whey-drained fermented milks, products heat-treated after fermentation, "sweet" milks, acidified milk products, and pharmaceutical products are also included. Fresh cheeses are not included.

Many years of worldwide work in the field of fermented fresh milk products and the opportunity to read original papers in many of the world's languages have made this book possible. We wish to express our gratitude to the many persons and organizations who have contributed to this book. Their names are included in the acknowledgments.

The authors are, and readers should be, aware that the information compiled in this encyclopedia is in no way complete. This work is the first step in collecting current and historic data about fermented milk products known around the world. The task will continue. Therefore, the authors would appreciate receiving suggestions, constructive criticism, and items of information on products not mentioned in this edition or products not adequately covered. Please mail them to Dr. M. Kroger, Borland Laboratory, The Pennsylvania State University, University Park, PA 16802, USA.

We hope this book will prove valuable to engineers, technologists, dieticians, and those engaged in the development of new products, as well as nutritionists, students, researchers, and others interested in food. We also hope it will lead to increased consumption of fermented fresh milk products and continued or renewed studies on all those lesser known fermented milk products.

Acknowledgments

S. A. Abou-Donia, Prof., Dr., Department of Agricultural Industries, Faculty of Agriculture, Alexandria University, Alexandria, Egypt.
J. M. Davies, Director, Commonwealth Bureau of Dairy Science and Technology, Shinfield, Reading, England.
R. Forsèn, Prof., Dr., National Public Health Institute, P1 310, SF-90101 Oulu 10, Finland.
A. A. Frimer, Prof., Dr., Department of Chemistry, Bar-Ilan University, Ramat Gan, Israel.
F. Görner, Prof., Dr., Chair of Technical Microbiology and Biochemistry, Slovak Technical University, Bratislava, Czechoslovakia.
W. Hauert, Dr., Ittigen, Bern, Switzerland.
E. Hopkin, Secretary General, International Dairy Federation, Square Vergote 41, B-1040 Brussels, Belgium.
F. Ketting, Prof., Dr., Hungarian Dairy Research Institute, Budapest, Hungary.
H.-J. Klupsch, Molkereitechnisches Laboratorium H.-J. Klupsch, Osterflierich, D-4700 Hamm, Germany.
A. Kolodkin, Prof., Dr., University Irkutsk, USSR.
N. S. Koroleva, Prof., Dr., All-Union Dairy Research Institute (VNIMI), Moscow, USSR.
F. V. Kosikowski, Prof., Dr., Cornell University, Ithaca, NY, USA.
L. Laotaye, Direction CMPA, Jdjaména, Chad.
E. Lipinska, Prof., Dr., Dairy Research Institute, Warsaw, Poland.
J. F. Mostert, Dr., Director, Animal and Dairy Science Research Institute, Irene 1675, South Africa.
G. Oliver, Prof., Dr., University Tucuman, Jujuy 463, Tucuman, Argentina.
I. Rosenthal, Dr., Volcani Research Center, Bet-Dagan, Israel.
J. R. Salji, Dr., 45 Purbeck Close, Bedford MK41 9LX, England.
K. M. Shahani, Prof. Dr., University of Nebraska, Lincoln, NE, USA.
Chr. Steffen, Dr., Director of Swiss Federal Institute for Dairy Research, Liebefeld-Bern, Switzerland.
A. P. Wolfschoon, Prof., Dr., Instituto de Laticinios "Candido Tostes," Juiz de Fora, Minas Gerais, Brazil.
H. Yaygin, Prof. Dr., E. u. Ziraat Fakültesi, Süt Teknologjisi Anabilim Dali, 35100 Bornova/Izmir, Turkey.

(continued)

Special thanks for important support to:

Agricultural Institute, Grangeneuve, 1725 Posieux (Fribourg), Switzerland.
Food Research Institute, Novi Sad, Yugoslavia.
The Pennsylvania State University, Department of Food Science, and Pennsylvania Agricultural Experiment Station, University Park, PA, USA.

Encyclopedia of Fermented Fresh Milk Products

Introduction

HOW TO USE THIS ENCYCLOPEDIA

The following pages provide brief explanations for the user and draw attention to recurring commonalities.

Format of Product Descriptions

For all entries, whenever applicable, the following order of detailed descriptive information will be adhered to:

Name of the Product or Preparation. The following facts are indicated:

English term or translation (En), followed by French (Fr) and German (De) equivalents; different spellings or regional synonyms
Etymology, if known
Character of the name (e.g., trade name)

Short Description of the Product. A sketch of important characteristics is provided, including

Countries or areas of origin
Traditional products, those made before the industrial era with nondefined cultures, and nontraditional products, made approximately since 1900, with a defined microflora
Type of milk or raw material
Utilization or use
Organoleptic properties

Some articles do not contain a short description but cover a list of products manufactured in a particular country or constitute a review.

Details of the Product Descriptions. Whenever possible the following details are provided for each entry:

History
Microbiology
Manufacture
Food value (for foods) or therapeutic value (for pharmaceutical preparations). Chemical composition and energy value for 100 grams of product.
Related products
References (see below). Frequently cited books and sources only. Little-known products have usually been given several references.

Spelling Convention of Professional Terms

An effort was made to adhere to American English orthography. In many cases, however, specific authoritative sources had to be obeyed, such as

International Dairy Federation (IDF). 1983. *Dictionary of Dairy Terminology,* English, French, German and Spanish. Amsterdam: Elsevier.
Sneath, P. H. A. et al. 1986. *Bergey's Manual of Systematic Bacteriology.* vol. 2. Baltimore: Williams and Wilkins.

Abbreviations for Units, Symbols, and Terms

approx.	approximately	kg	kilogram
°C	degree Celsius	kJ	kilojoule
cfu	colony-forming unit(s)	l	liter
°D	degree Dornic	lb	pound
DM	dry matter	max.	maximum
DSA	Dairy Science Abstracts	min.	minimum
		ml	milliliter
Eh	oxidation-reduction potential	mol	mole
		pH	hydrogen ion concentration
°F	degree Fahrenheit		
g	gram	(R)	registered trademark
h	hour	ref.	reference
IU	international unit	sec	second
J	joule	°SH	degree Soxhlet-Henkel
kcal	kilocalorie		
SNF	solids-nonfat	wk	week

sp., spp.	species	yr	year
temp.	temperature	%	per cent
°Th	degree Thörner	>	greater than
TS	total solids	≥	greater or equal to; not less than
UHT	ultrahigh temperature		
UV	ultraviolet	<	less than
vol.	volume	≤	less than or equal to; not greater than
v/v	volume/volume		
w/v	weight/volume		

References to Frequently Cited Books and Sources

References are provided with each entry or article. In the interest of saving space and to avoid much repetition, the references listed below are identified throughout the book by author name(s) and year of the publication only. Less frequently used sources are cited in full with individual entries. This encyclopedia is by no means a complete bibliography. Such a compilation would encompass more than 10,000 references.

Campbell-Platt, G. 1987. *Fermented Foods in the World.* London: Butterworths.
Davis, J. G. 1963. *A Dictionary of Dairying.* London: Leonhard Hill, Ltd.
Demeter, K. J. 1941. Dairy bacteriology, 1: Microorganisms in milk. In *Handbook of Agricultural Bacteriology* (in German) vol. 1, ed. F. Löhnis. Berlin-Zehlendorf: Gebrüder Borntraeger.
FIL-IDF. 1983. *Dictionary of Dairy Terminology.* Amsterdam: Elsevier.
Fleischmann, W., and H. Weigmann. 1932. *Textbook of Dairy Science* (in German). Berlin: Paul Parey.
Földes, L. 1969. *Animal Science and Pastoral Activities* (in German). Budapest: Akadémiai Kiado.
Forsén, R. 1966. Long milk (in German). *Finnish Journal of Dairy Science* **36**(1): 1-76.
Kurmann, J. A. 1986. Yoghurt made from ewe's and goat's milk. In *Production and Utilization of Ewe's and Goat's Milk,* IDF Bulletin No. 2, pp. 153-166. Brussels: International Dairy Federation.
Martiny, B. 1907. *Glossary of Dairying Throughout the World. Terms Used in the Milk Industry and Related Animal Science Areas* (in German). Leipzig: M. Heinsius Nachfolger.
Maurizio, A. 1933. *History of Beverages* (in German). Berlin: Paul Parey.
Müller, U. 1981. Studies on Quality Criteria of Buttermilk (in German). Dissertation. Giessen: Justus—Liebig—University.

Obermann, H. 1985. Fermented milks. In *Microbiology of Fermented Foods,* ed. B. J. B. Wood, pp. 167-195. Barking: Elsevier.

Ränk, G. 1969. *Fermented Milks and Cheeses Made by Asian Pastoral Peoples* (in German). Helsinki: Suomalais-Ugrilainen Seura.

Rašić, J. Lj., and J. A. Kurmann. 1978. *Yoghurt. Scientific Grounds, Technology, Manufacture and Preparations.* Copenhagen: Technical Dairy Publishing House.

Rašić, J. Lj., and J. A. Kurmann. 1983. *Bifidobacteria and Their Role.* Basel: Birkhäuser Verlag.

Schulz, M. E. 1965. *Dairy Dictionary* (in German). Kempten/Allgäu: Volkswirtschaftlicher Verlag.

Steinkraus, K. H. 1983. *Handbook of Indigenous Fermented Foods.* New York: Marcel Dekker.

Takamiya, T. 1978. Historical Studies on the Diet and Food Preparation of the Pastoral Peoples of Central Asia (in German). Dissertation. Munich: Ludwig-Maximilian-University.

Zeuner, F. E. 1967. *History of Domesticated Animals* (in German). Munich: BLV Bayrischer Landwirtschaftsverlag.

BASIC DEFINITIONS

In order to best use this encyclopedia, the reader should understand a number of pertinent concepts, many of which are well defined and originate in the international literature.

This section contains basic definitions as they relate to the raw material (milk) utilized, the fermentation process, and the manufactured product.

The Raw Material, Milk, and Starter Cultures

Milk is understood to be the whole secreted product of female mammals, obtained from healthy animals through complete, uninterrupted milking and unmodified in its composition. It is irrelevant whether the milk is meant to be directly consumed or further processed. The unqualified term, milk, refers to the milk of cows, since that is the most widely used animal milk destined for human consumption. Whenever the milk of another species is referred to it is appropriately identified, for example, goat milk. Milk for human consumption is also obtained from sheep (ewes), horses (mares), camels, donkeys, buffalo, zebus, yaks, and reindeer in addition to dairy cows and goats.

Fluids, secretions, juices, extracts, and preparations that do not originate from the mammary glands of animals (or humans) should not be called milk. For example, so-called soymilk (soya milk, soybean milk) should instead be designated "soy-protein extract" beverage. Likewise, "coconut milk" should be called coconut juice, in order to avoid confusion in the marketplace and consumer delusion.

Source: Authors' interpretation of Swiss Food Regulations (Article 39), 1985, Bern: Eidgenössische Drucksachen und Materialzentrale.

For all traditional products, the composition and activity of their defined starter cultures must correspond to those of the main starter organisms that have always been contained in such products, even when prepared with nondefined (empirical) cultures.

The Fermentation Process

In the context of this book, fermenting is synonymous with culturing. Culturing implies the deliberate use of microorganisms to achieve specific ends. Unintentional and uncontrolled growth of microorganisms, such as in microbial spoilage (bacteria, yeasts, molds), is termed *wild fermentation.*

For a fresh milk product to be designated as fermented (or cultured) it should satisfy the following conditions:

The use and propagation of selected nonpathogenic, nontoxigenic microorganisms

A minimum of 0.6% total acidity, expressed as lactic acid, after fermentation (corresponding to a pH of approximately 4.6–4.7)

A viable cell count of at least 10^8–10^9 per ml of product after fermentation. The cells should correspond to those microbial species used in the starter culture, declared on the package label, or expected by the consumer. At the time of consumption the viable cell count should be at least 10^7 per ml.

Manufactured Products

Fermented Fresh Milk Products

The term *fermented fresh milk products* is not commonly used in English-speaking countries. In this encyclopedia the term denotes any milk or milk product that has undergone some degree of fermentation but has not been subjected to further ripening. Included in these products are milks, creams, natural buttermilk (from cultured cream), cultured buttermilk (usually made from skim milk), fermented natural sweet buttermilk (when butter is made from noncultured, nonripened cream), fermented whey, and various pasty or concentrated or otherwise modified milks.

Source: Authors' definition.

Traditional Fermented Fresh Milk Products

Traditional fermented fresh milk products are all those with a substantial historic record. They have been deliberately made for many centuries as a means of preserving milk for later consumption and preventing spoilage (wild fermentation). Yoghurt and kefir are the best-known examples. Only products containing live microorganisms can be regarded as traditional fermented milk products. There are two main types: (1) products prepared with a defined culture and (2) products with a nondefined or empirical culture.

Source: Authors' definition.

Nontraditional Fermented Fresh Milk Products

Nontraditional fermented fresh milk products are all those products that have been developed as a result of deliberate modern scientific progress, mainly during the twentieth century. Acidophilus milk and bifidus milk are representative examples. Nontraditional products may also be imitations of traditional products. Such is the case whenever there are derivations from the original, traditional microflora. As a result, it is not quite correct (and should not be allowed) to call such a product by its original name. Examples would be a yoghurt that is made without the traditional yoghurt microorganisms or kefir prepared without kefir grains and traditional kefir microorganisms.

Source: Authors' definition.

Fermented Milks

Fermented milks are prepared from milk and/or milk products fermented by the action of specific microorganisms, which shall be viable, active, and abundant in the finished product at the time of sale for consumption. After fermentation, heat treatment and whey removal shall not be allowed. The milk and milk products used may be homogenized or not and must be pasteurized. Fermented milk may contain optional additions added either before or after fermentation. As fresh products, they have a limited shelf life and should preferably be kept refrigerated throughout distribution.

Source: International Dairy Federation (IDF). 1969. International Standard (Compositional Standard for Fermented Milks) FIL-IDF 47, in revision. Brussels: International Dairy Federation.

Cultured Milk Products

The term *cultured milk product* is used in the United States to designate fermented milks that are prepared by using starter cultures and controlled fermentation.

Source: Authors' definition.

Sour Milk

Sour milk is noncultured milk fermented by indigenous lactic acid bacteria present in that milk. This term is also used mainly in German areas to name products obtained from pasteurized milk fermented with selected starter cultures (e.g., *Dickmilch* which is soured, clabbered, or clotted milk). This term is retained in certain countries despite the recent introduction and use of starter cultures for controlled fermentation and acid production.

Source: Kosikowski, F. V. 1984. Buttermilk and related fermented milks. *IDF Bulletin 179 (Fermented Milks),* pp. 116-120. (Brussels: International Dairy Federation.)

Whey-Drained Products

There is some confusion concerning whey-drained fermented milks. In principle, whey removal should not be allowed after fermentation. But there are many traditional products with whey removed (see Appendix F, whey-drained products). Also, no term has been found for designating fermented milks with the whey removed. We propose the designation of "whey-drained fermented milks." This category is situated between fermented milks and fresh cheeses but is more closely related to fermented milks.

Source: Authors' definition.

Heat-Treated Products

Heat-treated products are all those products that have been heat-treated after culturing in order to increase shelf life. The result of the heat treatment is that most or all bacteria generated by the fermentation are killed. The package label of such products when sold should indicate that the product has been "heat-treated after culturing."

Source: Authors' definition.

Acidified Milk Products

Acidified milk products are imitation fermented products obtained by acidification of milk, cream, or buttermilk with food-grade acidulants such as lactic acid or citric acid. In the case of cream, an appropriate stabilizer or stabilizer/emulsifier system and a flavoring such as starter distillate, diacetyl, or a specifically tailor-made "sour cream flavor" are also added. These products lack certain nutritional benefits that only fermented products can have. Among the more successful products are an acidified baby milk formula (e.g., Pelargon produced in Switzerland) and a sweet confection called *rasogolla* or *rasgolla* made in eastern India (Bengal) from an acid-coagulated milk curd called *chhana* or *channa.*

Sources:

Ballarin, O. 1971. The scientific rationale for the use of acidified and fermented milk in feeding infants and young children, *FAO/WHO/UNICEF PAG Doc. 1.14/19* pp. 1-16. Rome: Food and Agriculture Organization of the United Nations.

Fox, P. F. 1978. Direct acidification of dairy products: Review. *Dairy Science Abstracts* **40**:727-732.

Sangwan, R. B. and M. M. Sharma. 1984. *Literature on Indian Dairy Products.* New Delhi: Metropolitan Book Co. Ltd.

"Sweet" Milks

Pasteurized and other types of heat-treated milk may be inoculated with concentrated cultures of such bacteria as *L. acidophilus* or bifidobacteria (*B. bifidum, B. longum,* or *B. breve*). There is no subsequent incubation and such milks should not be considered fermented. The milk becomes merely a carrier for the bacteria. The term *sweet* is common consumer and industry usage: It only signifies that no fermentation has taken place and that no acid has been formed. There are no added sugars or other sweeteners in these products. They have a limited shelf life and should be kept refrigerated throughout distribution.

Source: F. V. Kosikowski, personal communication; authors' definition.

Pharmaceutical Products

Pharmaceutical products are of interest here because some exist that use fermented milk microorganisms in freeze-dried preparations or in baby food formulations.

FERMENTED FRESH MILK PRODUCTS OF THE WORLD

The conversion of milk into fermented milk products has several important advantages. It is not only a means of preserving food, it also provides improved taste and better digestibility, allows for production of a variety of foods, and, in some instances, the processes of fermentation and subsequent whey removal reduce the bulk of the starting material. The different types of traditional fermented milk products that have been developed in the course of history are defined by the geographical culture and region in which they originated.

Fermented milk product formulations can be viewed as having developed through four stages: I, with undefined microflora; II, with defined microflora; III, with microflora with selected human intestinal bacteria; and IV, with fortified preprobiotic and probiotic properties (Kurmann 1989).

Table 1 briefly surveys the history of fermented milk products, and Figure 1 shows a scheme for classifying fermented milk products and fermented milk types.

More details on the cultural history, geography, ecology of the microflora, etymology and naming, inventory, and typology of fermented milk products can be found in Kosikowski 1977, Kurmann 1984, and Kroger et al. 1989.

TABLE 1 Survey of the History of Fermented Milk Products

Stage	Time Period	Characteristics of Growth
I	Around 8000 B.C. to Middle Ages	Traditional products, homemade, using empirical cultures, e.g., soured milk, yoghurt, kumiss, kefir.
	Middle Ages to 1900	Continuation of development of specific products using empirical cultures.
II	1900-1930	Products made by using defined cultures; beginning of artisanal and industrial production. First use of pure lactic cultures in fermented products. Metchnikoff's book *On the Prolongation of Life* (Metchnikoff 1907), influenced the popularity of yoghurt and stimulated research. First use of mixed yoghurt and vegetable or fruit preparations in human nutrition around 1910 (Bircher-Rey 1953).
	1930-1970	The beginning of growing fermented milk product consumption throughout the world.
	1970 to present	Rapidly increasing fermented milk product consumption, especially yoghurt. The first scientific monograph about yoghurt (Rašić and Kurmann 1978).
III	1921	Manufacture of fermented milk products with selected strains of human intestinal bacteria. Use of *Lactobacillus acidophilus* in acidophilus milk and the first scientific monograph on the subject (Rettger and Cheplin 1921).
	1935	Use of *Lactobacillus casei,* strain Shirota, in the fermented milk *yakult.*
	1948-1968	Use of *Bifidobacterium bifidum* in fermented milk products (Mayer 1948; Schuler-Malyoth et al. 1968).
	1968 to the present	Growing consumption of fermented milk products containing acidophilus bacteria and bifidobacteria; beginning of incorporation of nutrients for in vivo stimulation of growth of these bacteria. The first scientific monograph on bifidobacteria and their use in fermented milk products (Rašić and Kurmann 1983).
	Around 2000	Further growth of industrial manufacture of products containing so-called friendly intestinal bacteria.
IV	After 2000	Manufacture of fermented products from formulated milk using selected strains of human intestinal bacteria as such or in combination with other lactic acid bacteria for health purposes.

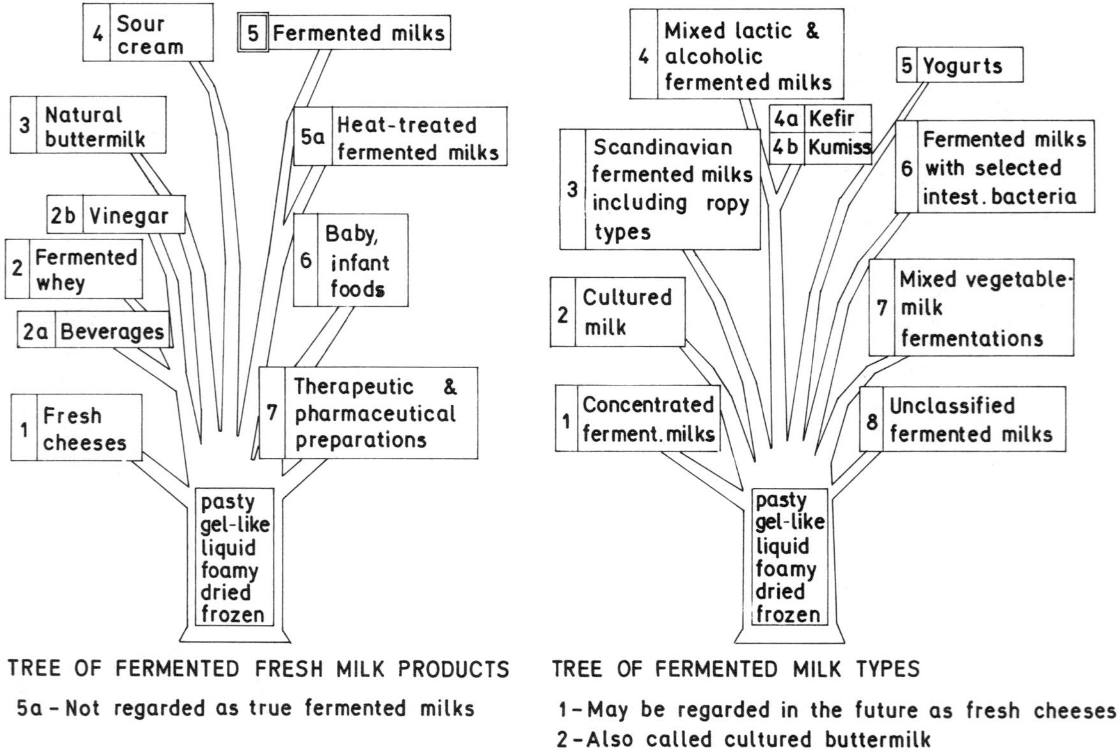

FIGURE 1. Scheme for classifying fermented milk products and fermented milk types. Source: J. A. Kurmann, The production of fermented milk in the world, *IDF Bulletin 179,* p. 22, 1984. Brussels: International Dairy Federation.

Nutritive Value of Fermented Milks

The greatest proportion of published materials about the nutritive value of fermented milks refers to yoghurt. The most important aspects include composition and digestibility, other beneficial effects, and supplementing the yoghurt flora with intestinal lactobacilli and bifidobacteria.

Other types of fermented milks may have similar nutritive value. More details on this subject are found in Rašić and Kurmann 1978; Gilliland 1979; Deeth and Tamime 1981; Rašić and Kurmann 1983; Kurmann 1986; Gurr 1987; and Rašić 1987.

Statistics

This information is based on statistical data published by the International Dairy Federation (IDF) for the years 1966-1986. More than thirty countries out of the approximately 170 in the world are members of the IDF and supply production and consumption figures. Despite varying data caused by non-reporting by some members and absence of data from nonmembers, it is the best information available. For some countries whose substantial consumption of fermented milks does not appear in the tables, for example, Bulgaria, Turkey, and Egypt, it would have been of interest to follow the development of consumption over the years.

Annual Consumption per Person

In order to facilitate reading and interpretating the annual per capita consumption figures for the different fermented milk products in the various countries, we distinguish between six different consumption levels as shown in Table 2. The same product categories are adhered to as used in IDF's reporting of the

TABLE 2 Degree and Amount of Annual Per Capita Consumption of Fermented Fresh Milk Products

		Amount Expressed as Number of Portions (Cups/Containers)		
Degree (kg)		500 g	180 g	120 g
I:	<0.1	<1	<1	<1
II:	0.1-1.4	<1-2	<1-7	<1-11
III:	1.5-4.9	3-9	8-27	12-40
IV:	5.0-9.9	10-19	28-55	41-82
V:	10.0-14.9	20-29	56-82	83-124
VI:	≥15.0	>30	>82	>124

Source: J. A. Kurmann, 1984, Consumption statistics and aspects of the production of fermented milks. *IDF Bulletin 179* (Fermented Milks) (Brussels: International Dairy Federation), p. 8-15.

statistics: (1) buttermilk, (2) yoghurt, and (3) other fermented milks apart from buttermilk and yoghurt. All figures are the best available approximations and are subject to such errors as misreportings or faulty population census figures.

The subject of buttermilk requires some clarification. Traditional (natural) buttermilk is obtained as a by-product in the manufacture of cultured cream butter. Cultured buttermilk is made from skim milk. Most reporting countries include the latter in the category of other fermented milks. Several countries, as footnoted in Table 3, include cultured buttermilk in the category of natural buttermilk. Table 3 provides an overview of the per capita consumption of buttermilk. Yoghurt and yoghurtlike products (e.g., *zabady, dahi,* and some types of *leben*) are the most widely consumed fermented milks. Table 4 shows the per capita consumption of yoghurt by country. The Netherlands and Switzerland are by far the leading countries. However, Bulgaria (which is not a member of IDF) exceeds consumption of these two substantially with about 35 kg yoghurt consumed per person per year (Kondratenko 1984). It can be assumed that in most other countries yoghurt is underconsumed and consumption could increase significantly in the future. Table 5 shows the per capita consumption figures for fermented milks (other than buttermilk and yoghurt). Over the years, only fifteen of the thirty-three reporting countries have regularly supplied data in this category, including the Scandinavian countries where there is a high consumption of a number of popular and traditional fermented milks (Stistrup 1978). Other notable countries in this category are Iceland, Israel, and Hungary. Table 6 summarizes the total consumption of all fermented milks in 1986, the last year for which such data were available. Finland and Bulgaria are by far the leading countries in the per capita consumption of fermented milks. From the annual comparison it can be seen that, in general, between 1966 and 1986 there has been a worldwide increase in the consumption of all fermented milks, especially yoghurt. It is anticipated that this trend will continue, particularly in view of increasing public recognition of the nutritional benefits of fermented milks. At the moment, fewer than ten countries enjoy a total consumption level of more than 15 kg per year (Table 6, degree of consumption VI). Consumption trends in other industrialized countries, those with a degree of consumption lower than VI, are such that they will not reach level VI in the near future. However, in certain other countries, such as China and those of Southeast Asia and South America, where fermented milk consumption is now zero or near zero, popularization of such products would greatly increase total world production figures. It is precisely in many of these areas where problems of malnutrition exist and where population increases are forecast. Development of fermented milk industries, including the recombined and reconstituted milks, in these parts of the world would bring great benefits.

Consumption figures for cultured cream (sour cream) and similar products are not included in the IDF statistics and are generally not available, except for some countries. In several, consumption of fermented cream is increasing. However, compared to the volume of consumption of total amount of fermented milks, it will remain confined to a low level. Nevertheless, in some countries, such as Hungary, USSR, Czechoslovakia, and Norway, it is respectable (Ketting 1984).

Per Capita Consumption and Total Market Volume of Fermented Milk Products

Many factors influence the production and consumption of fermented milks throughout the world including

Feed supply to support cattle breeding and milk production
Human dietary patterns
Income level
Advertising and promotion
Available fermented products
Relationship to the consumption of other dairy products
Religious and other beliefs
Distribution and sales systems

Future Development in Fermented Milk Products

Some of the most important factors that encourage the use of fermented milk products are the sensory properties of the product; the "healthy" image; increased shelf life; and the possibility of modifications such as whey removal, further ripening or aging, and use in other foods or food formulations.

The acceptance of fermented milk products and further growth in their consumption and production may depend on the following factors:

Quality improvement of the product and the introduction of various specialty products
Changes in food consumption patterns
Continued or increased interest in desirable foods
Reduced costs of production
Genetic engineering
Use in the processing of other foods
Fermented products from plant materials in combination with milk (co-fermentations)
Active marketing programs via all advertising media combined with general nutrition education for consumers and health professionals alike
Other factors, such as quality assurance programs and promotion of fermented products through research on their nutritional and health values and scientific validation thereof. More details on this subject are found in Hesseltine 1983; Rašić 1984; and Kroger et al. 1989.

TABLE 3 Per Capita Consumption of Buttermilk (in kg) in Member Countries of the IDF 1977-1988

	1977	1978	1979	1980	1981	1982	1983	1984	1985	1986	1988
Austria (AT)	0	0	2.1	2.0	1.9	2.1	2.2	1.9	2.2	2.0	2.1
Australia (AU)	—	—	—	—	—	—	—	—	—	—	—
Belgium (BE)	0	0	2.8	2.8	2.8	2.8	2.8	2.6	2.3	2.2	2.0
Canada (CA)	0	0	0.7	0.6	0.6	0.6	0.6	0.6	0.6	0.5	0.5
Switzerland (CH)	0	0	1.0	1.0	1.0	1.0	1.0	1.2	1.2	1.2	1.7
Chile (CL)	—	—	—	—	—	—	—	—	—	—	—
Czechoslovakia (CS)*	—	—	3.0	3.1	5.0	4.7	4.3	4.1	3.9	4.5	4.8
Germany F.R. (DE)	0	0	2.4	2.2	2.3	2.5	2.5	2.1	2.1	2.2	2.5
Denmark (DK)*	0	0	9.4	9.8	9.8	10.1	9.8	8.9	8.4	8.4	7.6
Spain (ES)	—	—	—	—	—	0	0	0	0	0	0
Finland (FI)	0	0	3.9	4.1	3.4	2.7	2.4	2.2	1.9	1.7	1.5
France (FR)	—	—	—	—	—	—	—	—	—	—	—
United Kingdom (GB)	—	—	—	—	—	—	—	—	—	—	—
Greece (GR)					3.1	—	—	—	—	—	—
Hungary (HU)									0.1	0.06	0.1
Ireland (IE)	0	0	0	0	0	0	0	0	0	0	0

Israel (IL)	—	—	—	—	—	—	—	—	—	—	—
India (IN)*	—	—	—	—	—	—	14.1	15.1	18.8	19.1	21.4
Iceland (IS)	—	—	—	—	—	—	—	—	0	0.6	0.1
Italy (IT)	—	—	—	—	—	—	—	0.2	—	—	—
Japan (JP)	—	—	—	—	—	—	—	—	—	—	—
Luxembourg (LU)		—	—	—	—	0	0	0	1.6	1.4	1.4
Netherlands (NL)*	9.0	8.9	9.3	9.5	9.4	10.0	9.9	8.8	8.5	9.1	9.5
Norway (NO)	—	—	—	—	—	—	—	—	—	—	—
New Zealand (NZ)	—	—	—	—	—	—	—	—	—	—	—
Poland (PL)	—	—	1.5	1.3	1.2	0.9	0.9	0.7	0.9	1.0	1.0
Sweden (SE)	0.1	0.1	0.2	0.1	0.1	0.1	0.1	0.1	0.03	0.03	0.01
USSR (SU)	—	—	—	—	—	—	0	0	—	—	—
USA (US)	0	0	2.0	1.9	1.8	1.8	1.9	2.0	2.0	2.0	1.9
South Africa (ZA)			—	—		0	—	0	0	0	0

Source: Compiled by IDF in the years 1966-1988.

Notes: Prior to 1977 data were available only from The Netherlands as follows: 1976, 11.0 kg; 1975, 10.5 kg; 1974, 9.4 kg; 1973, 10.3 kg; 1972, 9.3 kg; 1971, 9.1 kg; 1970, 8.7 kg.

Figures for 1987 are not reported because they do not differ significantly from the 1986 and 1988 figures.

A — indicates no figures; 0 indicates figures not separated from the census for the other products; blank spaces indicate the country was not a member of the IDF.

*Figures are for natural and cultured buttermilk, with cultured buttermilk predominating.

TABLE 4 Per Capita Consumption of Yoghurt (in kg) in the Member Countries of the IDF from 1966-1988

	1966	1969	1972	1975	1978	1981	1984	1986	1988
Austria (AT)	0	0	2.4	3.4	4.6	6.2	6.1	6.7	7.2
Australia (AU)	—	—	—	1.0	1.6	1.9	2.5	2.8	3.6
Belgium (BE)	1.9	2.9	4.7	5.1	4.4	4.8	4.6	6.9	6.9
Brazil (BR)			0.01	0.6	0.6				—
Canada (CA)	0.1	0.3	0.5	0.7	1.8	1.8	2.2	2.8	3.3
Switzerland (CH)	5.5	7.0	7.8	10.9	13.2	14.1	15.8	16.5	16.9
Chile (CL)	—	—	—	—	—	1.8	2.2	2.6	—
Czechoslovakia (CS)	—	—	1.0	1.3	1.7	1.8	2.1	2.6	3.2
Germany F.R. (DE)	0	3.2	4.5	4.5	6.4	6.9	7.5	8.9	10.4
Denmark (DK)	0	0	2.8	5.9	8.2	8.8	8.9	8.3	7.8
Spain (ES)	0	0	0	0	0	0	0	6.9	7.9
Finland (FI)	0	0	6.9	6.3	7.0	8.1	9.2	10.0	11.4
France (FR)	0	0	0	0	0	0	0	0	0
United Kingdom (GB)	0.4	0.6	1.0	1.6	1.9	2.9	2.7	3.5	3.9
Greece (GR)						5.5	—	—	—
Hungary (HU)						1.0	1.1	2.4	1.5

Ireland (IE)	4.2	5.1	7.7	1.0[r]	1.7	2.5	0	0	3.3
Israel (IL)	0	0	2.0	3.4	4.8	6.0	0	0	9.4
India (IN)	—	—	—	—	—	3.8	4.1	4.1	4.3
Iceland (IS)	—	—	—	1.7	2.0	5.7	6.2	6.8	8.6
Italy (IT)	—	—	—	—	1.1	1.3	0	1.9	2.4
Japan (JP)	—	—	1.1	0.8	0.7	1.3	2.7	3.1	3.8
Luxembourg (LU)	0	0	2.3	3.3	5.1	0	0	0	0
Netherlands (NL)	12.9	13.8	13.5	14.2	15.6	16.9	17.6	19.2	18.9
Norway (NO)	0	0	0.6	1.2	2.0	2.7	3.2	3.9	4.3
New Zealand (NZ)	—	—	—	—	—	—	—	—	—
Poland (PL)	0	0	0	0	0	0.1	0.2	0	0
Sweden (SE)	0	0	1.7	2.3	3.3	4.2	4.3	5.4	6.8
USSR (SU)	—	—	0.1	—	—	—	—	—	—
USA (US)	0.04	0.1	0.6	0.9	1.2	1.2	1.6	1.9	2.1
South Africa (ZA)							1.0	1.4	1.6

Source: Compiled by IDF in the years 1966-1986.

Notes: Figures for 1987 are not reported because they do not differ significantly from the 1986 and 1988 figures.
A—indicates no figures; 0 indicates figures not separated from the census for the other products; r indicates revised figure; blank spaces indicate the country was not a member of the IDF.

TABLE 5 Per Capita Consumption of Fermented Milks Other than Yoghurt and Buttermilk (in kg) in the Member Countries of the IDF from 1966-1988

	1966	1969	1972	1975	1978	1981	1984	1986	1988
Austria (AT)	0	0	3.7	3.9	2.2	2.3	2.2	2.2	2.6
Australia (AU)	—	—	—	—	—	—	—	—	—
Belgium (BE)							0.3	0.1	1.5
Brazil (BR)	0.01	—	—	—	—	—	—	—	—
Canada (CA)	—	—	—	—	—	—	—	—	—
Switzerland (CH)	—	—	—	—	—	—	—	—	—
Chile (CL)	—	—	—	—	—	—	—	—	—
Czechoslovakia (CS)	—	0.9	0.2	1.7	2.4	2.4	3.0	3.2	3.4
Germany F.R. (DE)	0	0.5	1.1	1.6	1.4	1.2	1.1	1.0	0.8
Denmark (DK)	0	0	6.8	7.1	7.8	7.7	7.8	7.3	7.0
Spain (ES)	0	0	0	0	0	0	0	—	—
Finland (FI)	0	0	30.7	29.1	27.4	28.6	29.1	27.4	27.6
France (FR)	0	0	0	0	0	0	0	0	0
United Kingdom (GB)	—	—	—	—	—	—	—	—	—
Greece (GR)	—	—	—	—	—	0.3	—	—	—

Hungary (HU)						6.9	1.2[r]	0.3	1.5
Ireland (IE)	—	—	—	—	—	—	0	0	—
Israel (IL)	—	—	9.8	10.7	10.9	9.8	9.0	8.7	12.7
India (IN)	3.2	—	—	—	—	—	—	—	—
Iceland (IS)	0.2	0.3	0.1	0.1	—	13.9	14.9	13.7	14.4
Italy (IT)	—	—	—	—	—	—	—	1.3	1.3
Japan (JP)	7.7	9.1	2.8	1.7	1.8	1.2	5.1[r]	4.6	4.2
Luxembourg (LU)	0	0	1.9	0.2	—	—	0	0	0
Netherlands (NL)	—	—	—	—	—	—	—	—	—
Norway (NO)	0	0	7.9	7.9	7.6	12.2[r]	11.9	10.7	11.0
New Zealand (NZ)	—	—	—	—	—	—	—	—	—
Poland (PL)		0	0	0	0	0.4	0.5	0	0
Sweden (SE)	0	0	14.4	17.6	19.0	19.9	22.2	21.9	22.3
USSR (SU)	—	6.2	6.6	7.2	6.8	0	0	7.4	7.9
USA (US)	—	—	—	—	—	—	—	—	—
South Africa (ZA)	—	—	—	—	—	—	0	0	2.0

Source: Compiled by IDF in the years 1966-1986.

Notes: Figures for 1987 are not reported because they do not differ significantly from the 1986 and 1988 figures.
A—indicates no figures; 0 indicates figures not separated from the census for the other products; r indicates revised figure; blank spaces indicate the country was not a member of the IDF.

TABLE 6 Overview of Degree and Amount of Annual per Capita Consumption of Fermented Fresh Milk Products in Member Countries of the IDF

Degree of Consumption*	Buttermilk	Yoghurt	Other Fermented Milks	Total of All Fermented Milks
I: <0.1 kg	Sweden 0.01			
II: 0.1-1.4 kg	Hungary 0.1 Iceland 0.1 Canada 0.5 Poland 1.0 Luxembourg 1.4		Germany 0.8 Italy 1.3	Poland 1.0
III: 1.5-4.9 kg	Finland 1.5 Switzerland 1.7 Belgium 2.0 Austria 2.1 Germany 2.5 Czechoslovakia 4.8	Hungary 1.5 S. Africa 1.6 USA 2.1 Italy 2.4 Czechoslovakia 3.2 Canada 3.3 Ireland 3.3 Australia 3.6 Japan 3.8 U. Kingdom 3.9 Norway 4.3 India 4.3	Belgium 1.5 Hungary 1.5 S. Africa 2.0 Austria 2.6 Czechoslovakia 3.4 Japan 4.2	Hungary 3.1 Ireland 3.3 S. Africa 3.6 Australia 3.6 Italy 3.7 Canada 3.8 U. Kingdom 3.9 USA 4.0

IV: 5.0–9.9 kg	Sweden 6.8 Belgium 6.9 Austria 7.2 Denmark 7.8 Spain 7.9 Iceland 8.6 Israel 9.4	Denmark 7.0 USSR 7.9	USSR 7.0 Spain 7.8 Japan 8.0 Luxembourg 8.2
V: 10.0–14.9 kg	Germany 10.4 Finland 11.4	Norway 11.0 Israel 12.7 Iceland 14.4	Belgium 10.4 Czechoslovakia 11.4 Austria 11.9 Germany 13.7
VI: ≥15.0 kg	Switzerland 16.9 Netherlands 18.9	Sweden 22.3 Finland 27.6	France 15.2 Norway 15.3 Switzerland 18.6 Israel 22.1 Denmark 22.4 Iceland 23.1 India 25.7 Netherlands 28.4 Sweden 29.1 Finland 30.3

Source: Compilation of statistics by IDF in 1988.

*See Table 2.

References

Bircher-Rey, H. 1953. *75 Yoghurt Preparations* (in German) Wengen, Switzerland: H. Bircher-Rey edition.

Deeth, H. C., and A. Y. Tamime. 1981. Yoghurt: Nutritive and therapeutic aspects. *Journal of Food Protection* **44**:78-86.

Gilliland, S. E. 1979. Beneficial interrelationships between certain microorganisms and humans: Candidate microorganisms for use as dietary adjuncts. *Journal of Food Protection* **42**(2):164-167.

Gurr, M. I. 1987. Nutritional aspects of fermented milk products. *FEMS Microbiology Reviews* **46**:337-342.

Hesseltine, C. W. 1983. The future of fermented foods. *Nutritional Review* **41**:293-301.

IDF. 1982. Consumption Statistics for Milk and Milk Products 1980, *IDF Bulletin 144.* Brussels: International Dairy Federation.

IDF. 1983. Consumption Statistics for Milk and Milk Products 1981, *IDF Bulletin 160.* Brussels: International Dairy Federation.

IDF. 1988. Consumption Statistics for Milk and Milk Products 1986, *IDF Bulletin 226.* Brussels: International Dairy Federation.

IDF. 1990. Consumption Statistics for Milk and Milk Products 1988. *IDF Bulletin 246.* Brussels: International Dairy Federation.

Ketting, F. 1984. Sour cream. unpublished manuscript. Budapest.

Kondratenko, M. S. 1984. Personal communication.

Kosikowski, F. V. 1977. *Cheese and Fermented Milk Foods,* 2nd ed. Ann Arbor, Mich: Edwards Brothers, Inc.

Kroger, M., J. A. Kurmann, and J. Lj. Rašić, 1989. Fermented milks—past, present and future. *Food Technology* **43:**92, 94-97, 99.

Kurmann, J. A. 1984. Consumption statistics and aspects of the production of fermented milks. *IDF Bulletin 179* (Fermented Milks), pp. 8-26. Brussels: International Dairy Federation.

Kurmann, J. A. 1986. Yoghurt made from ewe's and goat's milk). *IDF Bulletin 202* (Production and Utilization of Ewe's and Goat's Milk), pp. 155, 162-163. Brussels: International Dairy Federation.

Kurmann, J. A. 1989. Microorganisms of fermented milks, general aspects, bacteria, yeasts, moulds, less-known strains and microorganisms. In *Fermented Milks: Current Research.* International Congress, Palais des Congrès, Paris, France. December 16-18, 1989, pp. 3-10. London: Eurotext. (Also in French, published by John Libbey, Paris.)

Mayer, J. B. 1948. Development of a new infant food with *Lactobacillus bifidus* (in German). *Zeitschrift für Kinderheilkunde* **65**:319-345.

Metchnikoff, E. 1907. *Essais optimiste* (in French). Paris: Maloine.

Rašić, J. Lj. 1984. The future development of fermented milks. *IDF Bulletin 179* (Fermented Milks) pp. 27-32. Brussels: International Dairy Federation.

Rašić, J. Lj. 1987. Nutritive value of yogurt. *Cultured Dairy Products Journal* (Aug.), pp. 6-9.

Rašić, J. Lj., and J. A. Kurmann. 1978. *Yoghurt. Scientific Grounds, Technology, Manufacture and Preparations.* Copenhagen: Technical Dairy Publishing House.

Rašić, J. Lj., and J. A. Kurmann. 1983. *Bifidobacteria and Their Role. Microbiological,*

Nutritional-Physiological, Medical and Technological Aspects and Bibliography. Basel, Boston, Stuttgart: Birkhäuser.

Rettger, L. F., and H. A. Cheplin. 1921. *The Transformation of Intestinal Flora with Special Reference to the Implantation of Bacillus acidophilus.* New Haven: Yale University Press.

Schuler-Malyoth, R., A. Ruppert, and F. Müller, 1968. The microorganisms of the bifidus group (Lactobacillus bifidus) II. Technology of bifidus culture in the dairy plant (in German) *Milchwissenschaft* **23:**554-558.

Stistrup, K. 1978. Market research and trends. *IDF Doc. 107,* pp. 29-36. Brussels: International Dairy Federation.

Product Descriptions

A

AB-FERMENTED MILK AND MILK PRODUCTS (En); AB-Laits fermentés et produits laitiers (Fr); AB-Sauermilch und Milch-Produkte (De).

Short Description: Denmark; nontraditional products; cow's milk; set or stirred; specific flavor and prophylactic properties.

Microbiology: Concentrated and active cultures of *Lactobacillus acidophilus* and *Bifidobacterium bifidum* are used in the production of all kinds of AB-Products: only A-acidophilus or B-bifidus or AB-products or in combination with other lactic acid bacteria.

Related Products: AB-fermented milk, AB-buttermilk, AB-yoghurt, acidophilus (cultured) buttermilk, A-38 milk, A-fil milk, cultura, cultura drink.

Reference: Chr. Hansen's Laboratorium A/S. 1985. Copenhagen, Denmark.

A-FIL MILK (En); Lait A-fil (Fr); A-fil Milch (De).

Short Description: Scandinavia; nontraditional product; cow's milk; set or stirred; mild acid taste, pleasant flavor.

Microbiology: The starter culture consists of an acidophilus culture and a cream culture (*Streptococcus lactis* subsp. *lactis, S. lactis* subsp. *cremoris, Leuconostoc mesenteroides* subsp. *cremoris* and/or *S. lactis* subsp. *diacetilactis*).

Manufacture: A-fil milk is a product made from milk standardized to the desired fat content, homogenized and heat-treated. The manufacture involves the separate fermentation of milk with an acidophilus culture and the separate fermentation of milk with a cream culture and their mixing at a ratio of 15:85 after ripening.

Related Products: Acidophilus (cultured) buttermilk, A-38 milk.

Reference: Chr. Hansen's Laboratorium A/S. 1985. Copenhagen, Denmark.

ACETOBACTER-CONTAINING FERMENTED FRESH MILK PRODUCTS (En); Produits laitiers fermentés contenant Acetobacter (Fr); Acetobacter-enthaltende Sauermilchprodukte (De).

Short Description: Certain countries of Europe and Asia; traditional products; cow's milk, mare's milk, and whey; kefir, kumiss, and vinegar from whey (*Acetobacter aceti*).

Microbiology: Acetobacter is a genus of ellipsoidal to rod-shaped, aerobic, nonsporeforming bacteria growing in the presence of alcohol and securing energy by oxidizing organic compounds to organic acids, such as ethanol to acetic acid, and acetic acid to carbon dioxide. Optimum growth temperature is about 30°C (86°F).

Related Products: Fermented fresh milk products known to contain acetobacter bacteria are the following: kefir (where the organisms contribute to better viscosity), kumiss, and vinegar from whey (*Acetobacter aceti*).

ACIDIFIED BABY MILK FORMULA (En); Aliments acidifié artificiellement pour bébés (Fr); Künstlich gesäuertes Säuglingsnahrungsmittel (De).

Short Description: USA, Europe, South America (numerous countries); nontraditional baby milk; cow's milk; acidified milk product; in powdered form.

History: A product originally devised by Marriott in 1919 in the United States for feeding atrophic infants, later also for healthy children.

Manufacture: The manufacture of acidified baby milk formula involves the addition of small quantities of lactic acid, which contains more than 90%

of the L (+)-lactic acid isomer, or citric acid to the heat-treated whole cow's milk. After that a heat-treated mixture of maltodextrin, sucrose, and corn or rice starch is added to the milk followed by homogenization and spray drying.

Food Value: The use of an acidified whole milk formula in feeding babies and children is claimed to include the following advantages when compared with a nonacidified milk: improved digestibility due to the fine casein flocculation; more rapid attainment of the gastric pH; and protection against contamination during the preparation prior to feeding.

Clinical studies in Brazil have shown significant increases in weight and height and also a lower incidence of diarrhea in babies and children fed an acidified whole milk formula. Fermented milk formula may be even more effective in this respect (*see* Baby foods, fermented).

Related Product: Pelargon.

References:

Ballarin, O. 1971. The scientific rationale for the use of acidified and fermented milk in feeding infants and young children, *FAO/WHO/UNICEF PAG Doc. 1.14/19,* pp. 1-16. Rome: Food and Agriculture Organization of the United Nations.

Feer, F. 1937. Fermented milks in human nutrition and in modern medicine (in French). In *Volume Jubilaire Louis E. C. Dapples,* p. 202. Vevey: Nestlé and Anglo-Swiss Holding Company, Ltd.

ACIDIFIED MILK PRODUCTS, also called acidulated or imitation products (En); Produits laitiers acidifiés artificiellement (Fr); Künstlich gesäuerte Milchprodukte (De).

Short Description: USA and various countries; nontraditional products; cow's milk; acidified milk products.

History: In 1962, sour cream was the first chemically acidified milk product made commercially in the United States.

Manufacture: Acidified milk products are obtained by the chemical acidification of milk, as a rule with organic acids that have been approved for food use. Direct acidification is usually carried out until coagulation of milk occurs, for example, to a pH value of 4.6-4.4 or even lower (4.6-3.8) followed by heat treatment of the resulting product to prolong its shelf life.

Acidification of cream to a pH of 4.5-4.4 is carried out at a temperature of

21°C (69.8°F) or below followed by the addition of stabilizers and heat treatment of the product. Acidified sour cream does not have the diacetyl flavor of fermented cream, but it has a firm, heavy body.

Food Value: Both the texture and flavor characteristics of acidified milk products are generally disappointing. Also, they lack nutritional benefits that only fermented products can have. The benefits include:

- Better digestibility due to lower lactose content and possibly more advanced partial hydrolysis of the milk protein
- Presence of β-galactosidase, an enzyme produced by lactic cultures that contributes to better lactose tolerance
- Large numbers of viable lactic acid bacteria believed to aid the body's immunological defense system
- Presence of antimicrobial substances generated by lactic cultures that help to adjust the normal intestinal microflora
- Good taste and specific flavor
- Possibly other beneficial attributes due to components developed during the fermentation of milk.

Acidified sour cream is used as the base of various flavored dips, as a baked potato dressing, and in cooked and baked foods.

Related Product: Acidified baby milk formula.

References:

Fox, P. F. 1978. Direct acidification of dairy products. *Dairy Science Abstracts* **40:**727-732.

Little, L. 1967. Techniques for acidified dairy products. *Journal of Dairy Science* **50:**434-440.

ACIDOPHILIN (En, De); Acidophiline (Fr); means containing acidophilus bacteria.

Short Description: USSR; nontraditional product; cow's milk; set or stirred; flavor may vary from mild to acid depending on the quality of the starter culture and the processing conditions used.

Microbiology: Acidophilin used to be prepared with a *Lactobacillus acidophilus*-based mixture to which was added *Streptococcus lactis* subsp. *lactis* or *Lactobacillus delbrueckii* subsp. *bulgaricus.* Now the starter culture consists of *L. acidophilus, S. lactis* subsp. *lactis* and kefir culture. The cultures are mixed in a ratio of 1:1:1 before inoculation into the milk.

Manufacture: Acidophilin is an acidophilus product that should not be confused with the microbial inhibitor called acidophilin produced by some strains of *L. acidophilus* (Vakil and Shahani 1965). The product is made from pasteurized (90-92°C/194-197.6°F for 3 min), homogenized milk, inoculated with 6-9% of a mixed culture. The milk is incubated at 18-25°C (64.4-77.0°F) until 75-80°Th (0.67-0.72% titratable acidity) is produced, followed by cooling, packaging, and cold storage.

In order to increase the proportion of *L. acidophilus* in the product, the pasteurized milk is fermented with 5-8% of a mixed culture at 32°C (89.6°F) for 6-8 h until 75-80°Th (0.67-0.72% titratable acidity) is produced, followed by cooling and packaging.

Acidophilin is industrially made in only limited quantities possibly because of the product's variable composition and organoleptic properties, as well as the complexity of the starter culture.

Food Value: The final product has an acidity between 75 and 120°Th (0.67-1.08% titratable acidity) and contains organisms in the ratio of 97% *S. lactis* subsp. *lactis,* 2% *L. acidophilus* and 1% yeast. As reported, the antibacterial activity of this product is less than that of acidophilus milk.

References:

Bobrova, A. V. 1940. Methods of preparing acidophilin (in Russian). *Proceedings, Vologda Agricultural Institute* **I:**51-59. Cited in *Dairy Science Abstracts,* 1945, **7:**121.

Koroleva, N. S. 1975. *Technical Microbiology of Whole Milk Products* (in Russian). Moscow: Pishchevaya Promishlenost.

Sharma, N., and D. N. Gandhi. 1981. Preparation of acidophilin. I. Selection of the starter culture. *Cultured Dairy Products Journal* **16**(2):6-10.

Vakil, J. R., and K. M. Shahani. 1965. Partial purification of antibacterial activity of Lactobacillus acidophilus. *Bacteriological Proceedings,* p. 9. Cited in *Dairy Science Abstracts,* 1965. **27:**358.

ACIDOPHILUS BABY FOODS (En); Produits á l acidophilus pour nourisson (Fr); Acidophilus-Kindernahrungsmittel (De).

Short Description: USSR; nontraditional product; cow's milk; a category of dried or nondried baby foods containing *Lactobacillus acidophilus.*

History: The development of these products is based on increasing evidence for the beneficial role of *L. acidophilus* organisms in the human gastrointestinal tract and the finding by Robinson and Thomson (1952) that infants fed formulas supplemented with *L. acidophilus* grew better and had more lactobacilli in their stools than infants fed unsupplemented formulas.

Microbiology: Two specific acidophilus products, called Malutka and Malysh and made in the USSR, are prepared by using human intestinal strains of *L. acidophilus* characterized by their high proteolytic and antibiotic properties, resistance to phenol (0.5-0.6%), and ability to produce at least 70% L (+)-lactic acid in the total lactic acid.

Manufacture: The manufacture involves heat treatment and homogenization of the milk formula, then fermentation with 1-2% of a starter at 37-38°C (98.6-100.4°F) for 5-6 h until coagulation.

Food Value: The final product is reported to contain 10^7-10^9/ml viable *L. acidophilus* and to have 70°Th (0.6% titratable acidity). It is intended for feeding infants up to 3 or 6 months of age (Koroleva et al. 1982). Dried acidophilus product has been reported to contain 10^5/g viable *L. acidophilus* organisms.

Related Product: A related product is the formula called Baldyrgan. It is made by using an acidophilus culture that is employed in the manufacture of a product that is fortified with lysozyme at a rate of 20 mcg/l (Hyin et al. 1982). *See also* Baby foods, fermented.

References:

Hyin, A. A., H. A. Padalka, and V. V. Babich. 1982. An adapted cultured milk product for infants. *Proceedings, 21st International Dairy Congress* **1:**104.

Koroleva, N. S., V. E. Semenikhina, L. N. Ivanova, I. V. Oleneva, and M. B. Sundukova. 1982. Cultured milk products containing acidophilus bacteria and bifidobacteria for babies (in Russian). *Molochnaya Promȳshlennost'* **6:**17-20. Cited in *Dairy Science Abstracts,* 1983, **45:**523.

Robinson, E. L., and W. L. Thomson. 1952. Effect on weight gain of the addition of *Lactobacillus acidophilus* to the formula of newborn infants. *Journal of Pediatrics* **41**(4):395-398.

Sharmanov, T. Sh., P. V. Fedotov, S. A. Erdihanova, and A. K. Mashkeev. 1979. Method of producing "Baldyrgan" medicinal soured milk product for children. USSR Patent 695 645 (in Russian). Cited in *Dairy Science Abstracts,* 1980, **42:**583 and 765.

ACIDOPHILUS BIFIDUS YOGHURT (En); Yaourt á l'acidophilus et bifidus (Fr); Acidophilus-Bifidus Joghurt (De).

Short Description: Germany, United States, Japan and several other countries; nontraditional product; cow's milk; special yoghurt; set or stirred; firm body; mild acid flavor.

History: The manufacture of this product was developed in Germany following studies (Mülhens and Stamer 1969) that suggested the use of selected strains of acidophilus and bifidus cultures to increase the beneficial value of yoghurt.

Microbiology: The starter culture consists of a yoghurt culture (*Streptococcus thermophilus* and *Lactobacillus delbrueckii* subsp. *bulgaricus*) and cultures of *L. acidophilus* and *Bifidobacterium bifidum* or *B. longum.* The last two cultures should be human intestinal strains. These are separately cultivated and before use are microscopically controlled and selected for acid production.

Manufacture: The manufacture of this product involves standardization of milk, homogenization and pasteurization at 85°C (185°F) for 30 min or 90°C (194°F) for 5 min, tempering and inoculation with the separate cultures of yoghurt, *L. acidophilus* and *B. bifidum,* mixing well and packaging. The incubation is at 40-42°C (104-107.6°F) for 3-5 h until coagulation, followed by cooling and cold storage.

Food Value: The fresh product may contain 1-3 $\times$ 10^7/ml *L. acidophilus,* 1-3 $\times$ 10^7ml *B. bifidum* and large numbers of yoghurt organisms. Variations may occur in the numbers of viable bifidus and acidophilus bacteria depending on the strain characteristics, the original size of the inoculum, and manufacturing procedures.

References:

Mülhens, K., and H. Stamer. 1969. Supplementing the yoghurt flora with *Lactobacillus acidophilus* and *Lactobacillus bifidus* (in German). *Milchwissenschaft* **24:**25-28.
Rašić, J. Lj., and J. A. Kurmann. 1978, p. 348.

ACIDOPHILUS CREAM, CULTURED, (En); Crème acidifiée à l'acidophilus (Fr); Acidophilus-Sauerrahm (De).

Short Description: ***Czechoslovakia;*** nontraditional product; cream from cow's milk; set, firm body; mild acid to acid taste, a little sweet.

Microbiology: The starter culture consisting of *Lactobacillus acidophilus* is added and mixed in at a rate of 5%.

Manufacture: The product is made from homogenized, pasteurized cream containing 40-45% fat and usually 5% added sugar (in syrup form). The heat-treated cream is incubated with the above culture at 38-42°C (100.4-107.6°F) until 95-130°Th (0.85-1.17% titratable acidity) is produced, followed by cooling to 6-8°C (42.8-46.4°F).

Reference: Teply, M., B. Hylmar, C. Kalina, and V. Rumlova. 1968. *Kefir, Yoghurt, Acidophilus and Other Products* (in Czech). Praha: Naklad. Technicke Literatury.

Short Description: ***USSR;*** nontraditional product; cream from cow's milk and vegetable oil; baby food; set or stirred; mild acid taste and pleasant flavor.

Microbiology: The starter culture consisting of *Lactobacillus acidophilus* and aroma-producing streptococci is added and mixed in at a rate of 3-5%.

Manufacture: The product called *detskaya* is made from cream containing 20% milk fat and 10% vegetable oil. The heat-treated, homogenized cream is incubated with the above culture at 30-32°C (86-89.6°F), followed by cooling, packaging, and storage.

Food Value: The product is intended for children older than 1 year.

References:

Bolgar, I. 1955. Preparation of acidophilus smetana (in Russian) *Molochnaya Promÿshlennost'* **16**(6):30-31. Cited in *Dairy Science Abstracts,* 1955, **17:**1014-1015.
Brents, M. J., S. A. Fursova, V. M. Vorobyova, V. P. Aristova, and L. M. Nasonova. 1982. New products for children and dietetic nutrition. *Proceedings 21st International Dairy Congress* **1:**102.

ACIDOPHILUS (CULTURED) BUTTERMILK (En); Babeurre artificiel à l'acidophilus (Fr); Geschlagene Acidophilus-Buttermilch (De).

Short Description: Scandinavia and several other countries; nontraditional product; cow's milk; creamy consistency; a mild acid taste, pleasant flavor.

History: The limited consumption of acidophilus milk due to its unfavorable flavor has led to the development of a product called acidophilus buttermilk.

Microbiology: The starter culture consists of an acidophilus culture and a cream culture (*Streptococcus lactis* subsp. *lactis, S. lactis* subsp. *cremoris, Leuconostoc mesenteroides* subsp. *cremoris* and/or *S. lactis* subsp. *diacetilactis*).

Manufacture: The product is made either by a simultaneous fermentation of the standardized, heat-treated homogenized milk with an acidophilus culture and a cream culture or by the separate fermentation of milk with an acidophilus culture and a cream culture and their mixing at a desirable ratio, usually 1:9 after ripening (*see* Acidophilus milk; Cultured buttermilk).

Food Value: Intestinal strains of acidophilus culture give dietetic and therapeutic properties and intestinal strains of cream culture impart improved flavor and body to the product. The final product contains 90% lactic streptococci and aroma-producing bacteria and 10% *L. acidophilus.* Acidophilus milk and buttermilk may be mixed at a ratio other than 1:9 depending on the intended purpose thus changing the proportion of *L. acidophilus* in the final product.

Related Products: A-38 milk, A-fil milk, acidophilus milk.

References:

Anonymous. 1976. New technology of acidophilus milk manufacture (in Czech). *Prumysl Potravin* **27**(2):88-89. Cited in *Dairy Science Abstracts,* 1978, **40:**210.

Pedersen, A. H. 1973. Productive-technical conditions with regard to manufacture of cultured milk product A-38 (in Danish), p. 198. *Beredning Statens Forsogsmeijeri.* Hillerød, Denmark: Statens Forsogsmeijeri.

Rašić, J. Lj. 1987. Other products. In *Milk—The Vital Force,* ed. Organizing Committee of the 22nd International Dairy Congress, pp. 673-682. Dordrecht: D. Reidel.

ACIDOPHILUS DRINKS, PURE (En); Boissons à l'acidophilus (Fr); Acidophilus-Getränke (De).

Short Description: USSR; nontraditional beverage; cow's milk or natural buttermilk.

Microbiology: The starter culture consists of *Lactobacillus acidophilus.*

Manufacture: Low-fat or skim milk previously heat-treated is fermented with 1-2% (2-5%) of *L. acidophilus* culture at 37-38°C (98.6-100.4°F) for 18-24 h, then cooled and flavored with fruit juice or other flavorings. The product is packaged and stored at 5-10°C (41-50°F) (*see* Acidophilus milk).

Acidophilus drinks from natural buttermilk are made from sweet or acid buttermilk which is fermented, respectively, with 5% or 2.5-3% of an acidophilus culture at 40-42°C (104-107°F). Incubation to an acidity of 65-70°Th (0.58-63% titratable acidity) takes 2-3 h; it is followed by heating to 53-55°C

(127.4-131°F) for 40-45 min and by removing 50% of the original volume as whey. The product is stirred, cooled, and packaged, followed by ripening at 3-6°C (37.4-42.8°F) for 10-12 h and cold storage.

Related Product: Acidophilus natural buttermilk.

References:

Krivosheev, A. 1964. Acidophilus beverage from buttermilk (in Russian). *Molochnaya Promyshlennost'.* **25**(4):30. Cited in *Dairy Science Abstracts,* 1964, **26:**312.

Nahaisi, M. H., and R. K. Robinson. 1985. Acidophilus drinks: the potential for developing countries. *Dairy Industries International* **50**(12):16-17.

ACIDOPHILUS ICE CREAM (En); Crème glacée à l'acidophilus (Fr); Acidophilus Eiskrem (De).

Short Description: USA; nontraditional product; cow's milk; frozen product containing viable *Lactobacillus acidophilus;* smooth body; pleasant flavor.

Food Value: The feasibility of adding *L. acidophilus* to ice cream has been studied in the United States. Freezing the ice cream mix significantly reduced the numbers of *L. acidophilus* organisms, but the survival rate was still above the recommended 2 million bacteria per g after 28 days. Flavor trials with the product were favorable, indicating that ice cream with *L. acidophilus* could be a product with potential in the dairy industry.

Reference: Duthie, A. H., A. E. Duthie, K. M. Nilson, and H. V. Atherton. 1982. An ideal vehicle for Lactobacillus acidophilus. *Dairy Field* **164**(11):139-140.

ACIDOPHILUS MILK (En); Lait à l'acidophilus (Fr); Acidophilus-Milch (De).

Short Description: USA; nontraditional product; cow's milk; set or stirred; prophylactic properties; the lack of a pleasant flavor in plain fermented acidophilus milk and often extreme sourness discourage its consumption, but adding flavorings and/or sweeteners and other lactic acid bacteria improves the taste of the product.

History: Acidophilus milk was first produced in the United States and named by Rettger and Cheplin (1922) who attempted to negate Metchnikoff's theory and caused bulgaricus milk to be replaced in popularity by acidophilus milk. It is obtained by a homofermentative lactic fermentation of milk with a culture of *Lactobacillus acidophilus.* This bacterium is a part of the normal human gut microflora and possesses the ability to suppress undesirable microorganisms in the intestinal tract. Early popularity of acidophilus therapy for certain intestinal disorders reached its peak in the 1930s and then declined, but the introduction of antibiotics in World War II (1939-1945) against systemic infections and the increasing incidence of antibiotic diarrhea and other side effects brought back acidophilus therapy for restoring the normal intestinal conditions. Since then there has been a slow, steady increase in the human consumption of acidophilus milk. Acidophilus milk is produced in many countries in limited amounts. Its consumption is believed to be due more to its therapeutic and dietetic values than to its organoleptic properties.

Microbiology: The starter culture consists of pure *L. acidophilus.* Use of viable strains of *L. acidophilus* isolated from the feces of healthy human beings is recommended. Acidophilus cultures should be characterized by their antagonisms to intestinal pathogens and by their ability to pass into and survive in the intestinal tract. *L. acidophilus* grows only slowly in milk thus increasing the likelihood of growth of contaminating organisms, but yeast extract or autolysate or other stimulating nutrients may be added to promote its growth in milk.

Daily transfers of mother culture are desirable to ensure maximum activity. Sterilized skim milk is inoculated with 1% of a culture and incubated at 37-38°C (98.6-100.4°F) until 70-80°Th (0.63-0.72% titratable acidity) is produced. Concentrated cultures are often used as bulk starter inoculum to avoid subculturing in the laboratory.

The viscosity of low-fat products may be improved by using supplementary polysaccharide-producing strains of *L. acidophilus* mixed with normal strains at a desirable ratio.

Manufacture: Acidophilus milk is made from skim, partially skimmed, or whole milk. The manufacture is summarized in Fig. 2.

The limited shelf life of acidophilus milk has led to the development of freeze-dried preparations containing viable *L. acidophilus* either alone or in combination with other lactic acid bacteria.

Food Value: The final product has an acidity of about 110°Th (about 1% titratable acidity; DL-lactic acid) and contains 10^8-10^9/ml *L. acidophilus.* Maximum viability of the bacteria is maintained during storage of the packaged product at 5-10°C (41-50°F) for about one week.

Acidophilus milk is a desirable food; it is more easily digested than the

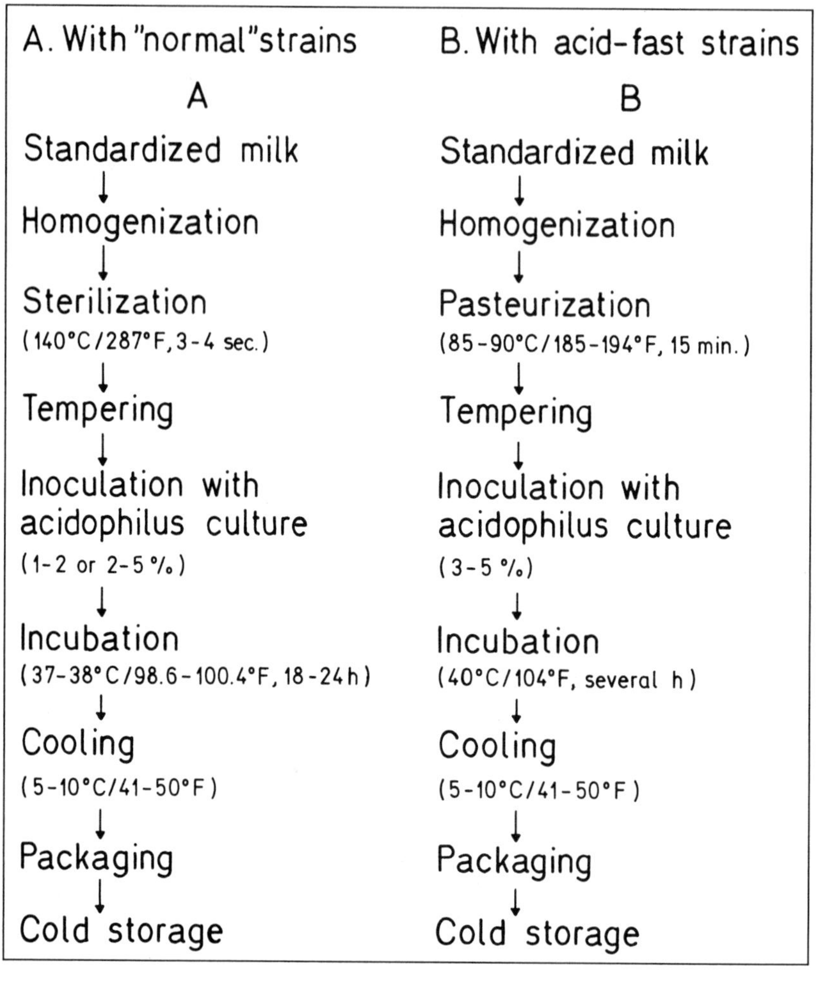

FIGURE 2. Flow scheme showing manufacturing process of acidophilus milk.

original milk because modification of the main milk constituents by the starer bacteria are believed to improve their digestibility. It is also recommended as an aid in gastrointestinal disorder therapy, and as a protective means against an imbalance in the intestinal microflora. The consumption of ½–1 l acidophilus milk per day is recommended for therapeutic purposes.

Related Products: Acidophilus (cultured) buttermilk, acidophilus yoghurt, acidophilus milk, dietetic (low-calorie acidophilus milk, vegetable oil-fortified), arla acidophilus milk (vegetable oil/protein/vitamin-fortified).

References:
Kopeloff, N. 1926. *Lactobacillus Acidophilus.* London: Bailliere, Tindall and Cox.
Koroleva, N. S. 1975. *Technical Microbiology of Whole Milk Products* (in Russian). Moscow: Pishchevaya Promishlenost.
Kurukowski, K. 1976. Acidophilus milk (in Polish). *Przeglad Mleczarski* **25**(9): 22-23.
Rettger, L. F., and H. A. Cheplin. 1922. Bacillus acidophilus and its therapeutic applications. *Archives of Internal Medicine* **29:**357-367.
Rettger, L. F., M. N. Levy, L. Weinstein, and J. E. Weiss. 1936. *Lactobacillus Acidophilus and Its Therapeutic Application.* New Haven: Yale University Press.
Sandine, W. E., K. S. Muralidhara, P. R. Elliker, and D. C. England. 1972. Lactic acid bacteria in food and health: a review with special reference to enteropathogenic *Escherichia coli* as well as certain enteric diseases and their treatment with antibiotics and lactobacilli. *Journal of Milk and Food Technology* **35**(12):691-702.

ACIDOPHILUS MILK, DIETETIC, dietetic acidophilus milk (En); Lait diététique à l'acidophilus (Fr); Diätetische Acidophilus Milch (De).

Short Description: USSR; nontraditional product; cow's milk and corn oil; dietetic product; stirred; specific flavor.

Microbiology: Starter culture consisting of *Lactobacillus acidophilus* is added and mixed in at a rate of 5%.

Manufacture: An acidophilus product made from skim milk fortified with 2% corn oil (maize oil). The refined oil, heated to 45-50°C (113-122°F), is homogenized into warm (30-35°C/86-95°F) skim milk (to obtain 10-12% fat) which is then mixed with more skim milk to a final product of 2% fat. After addition of sucrose, the mixture is pasteurized at 90-95°C (194-203°F) for 10-15 min. The incubation is at 40-42°C (104-107.6°F) for several hours, followed by cooling and packaging.

Food Value: The final product has an acidity of 90-140°Th (0.81-1.25% titratable acidity) and contains 2% vegetable fat and 7% sucrose.

Reference: Bogdanova, G. J., and E. A. Bogdanova. 1974. *New Whole Milk Products of Improved Quality* (in Russian). Moscow: Pishchevaya Promishlenost.

ACIDOPHILUS (NATURAL) BUTTERMILK (En); Babeurre à l'acidophilus (Fr); Acidophilus Buttermilch (De).

Short Description: ***Czechoslovakia;*** nontraditional beverage; sweet cow's buttermilk; stirred; mild acid taste; flavored with vanilla.

Microbiology: The starter culture consists of *Lactobacillus acidophilus* containing a supplementary polysaccharide-producing strain. It is mixed in at a rate of 5%.

Manufacture: The product is made from sweet buttermilk fortified with 5% sucrose and 0.001% vanilla. The incubation with the above-mentioned culture is at 40-42°C (104-107.6°F) until 95-130°Th (0.85-1.17% titratable acidity) is produced, followed by cooling to 6-8°C (42.8-46.4°F) and packaging.

Related Product: Acidophilus milk.

Reference: Teply, M., B. Hylmar, C. Kalina, and V. Rumlova. 1968. *Kefir, Yoghurt, Acidophilus and Other Products* (in Czech). Praha: Naklad. Technicke Literatury.

Short Description: ***USSR;*** nontraditional beverage; sweet cow's buttermilk; stirred; smooth body; pronounced flavor. Two manufacturing variations are described.

Manufacturing Process 1: Dietetic Buttermilk

Microbiology: The starter culture, consisting of 70% streptococcal culture (as used for cream culturing) and 30% acidophilus culture (polysaccharide-producing strain), is added at a rate of 1-2%.

Manufacture: Dietetic buttermilk is made from sweet buttermilk, heat-treated at 85-90°C (185-194°F) for 5-10 min. The incubation is at 28-30°C (82.4-86°F) for 12-16 h until 70-80°Th (0.63-0.72% titratable acidity) is produced, followed by cooling to 3-8°C (37.4-46.4°F) and packaging.

Food Value: The final product contains 0.5% fat, 8.0% nonfat solids, and has an acidity of 90-110°Th (0.81-0.99% titratable acidity).

Manufacturing Process 2: Product Called Ideal

Microbiology: The starter culture consisting of 70% *Streptococcus lactis* subsp. *diacetilactis* and 30% *Lactobacillus acidophilus* is added at a rate of 3-5%.

Manufacture: Ideal is made from sweet buttermilk, standardized to 1% fat by adding cream (30% fat). The blend is homogenized and pasteurized. Incubation is at 30-32°C (86-89.6°F) for 9-10 h until 80-85°Th (0.72-0.76% titratable acidity) is produced, followed by cooling and packaging.

Food Value: The final product contains 1.0% fat, 8.5% nonfat solids and 85-110°Th (0.76-0.99% titratable acidity).

References:

Bogdanova, G. J., and E. A. Bogdanova. 1974. *New Whole Milk Products of Improved Quality* (in Russian). Moscow: Pishchevaya Promishlenost.

Krivosheev, A. 1964. Acidophilus beverage from buttermilk (in Russian). *Molochnaya Promȳshlennost'* **25**(4):30. Cited in *Dairy Science Abstracts,* 1964, **26**(1873):312.

ACIDOPHILUS-ALBUMIN PASTE (En); Pâte à l'acidophilus-albumin (Fr); Acidophilus-Albumin-Paste (De).

Short Description: USSR; nontraditional product; cow's milk; whey-drained product; dietetic fermented milk product; firm body.

Microbiology: The starter culture consists of *Lactobacillus acidophilus.*

Manufacture: The product is made from acid cheese (tvorog) whey, which is pasteurized at 90°C (194°F); then the precipitated whey proteins are collected and pressed to obtain a curd with 75% moisture. The curd is mixed with an acidophilus culture and heat-treated sugar syrup (65-66% sucrose) at a suitable ratio, followed by cooling and packaging.

Food Value: The final product contains 10% whey proteins, 1.2% casein, 1.5% lactose, 0.6% lactic acid, 15% sucrose, 0.3% minerals and 71.5% moisture. It is intended for young children with slow growth rate.

ACIDOPHILUS PASTE (En); Pâte à l'acidophilus (Fr); Acidophilus-Paste (De).

Short Description: USSR; nontraditional product; cow's milk; whey-drained product; firm body; specific flavor.

History: The technology of manufacture was first described by Gibshman and Hlebnikovoy (1944) in the USSR.

Microbiology: The starter culture consisting of *Lactobacillus acidophilus* is added and mixed in at a rate of 5%.

Manufacture: Acidophilus paste is made from skim milk, partially skimmed milk, or whole milk by using two methods. In the first, milk is standardized to the required fat content or skimmed, then sterilized or pasteurized at 85-90°C (185-194°F) for 15 min. The incubation with the above culture is at 38-40°C (100.4-104°F) until 80-90°Th (0.72-0.81% titratable acidity) is produced, followed by transfer of the coagulum to sterile bags for removal of whey (at 6-8°C/42.8-46.4°F for 14-16 h) and gentle pressing to obtain the desired moisture. Alternatively, the whey may be removed from the skim milk coagulum by using a special centrifuge, followed by cooling, dressing, and packaging.

In the second method, the standardized milk or skim milk is pasteurized, then concentrated at 50-60°C (122-140°F) in a vacuum pan to 23% solids content or higher. The incubation with the above culture is at 38-40°C (100.4-104°F) for several hours until 180-200°Th (1.6-1.8% titratable acidity) is produced, followed by cooling and packaging.

Food Value: The paste is made as a nonfat product containing 80% moisture, as a 4% fat product with 65% moisture, or as an 8% fat product with 60% moisture. It may be plain or fruit flavored; maximum acidity is 200°Th (1.8% titratable acidity). The product contains $20\text{-}30 \times 10^9$ *Lactobacillus acidophilus* and has a shelf life of 10-15 days or more under refrigeration and 6-8 days at room temperature.

References:

Bogdanov, E., and A. Shchergineva. 1964. Acidophilus paste (in Russian). *Molochnaya Promȳshlennost'* **25**(8):29-30. Cited in *Dairy Science Abstracts,* 1964, **26:**518.

Gibshman, M., and N. Hlebnikovoy. 1944. Cited in M. M. Kazanski and Q. V. Tverdohleb, 1955, *Technology of Milk and Milk Products* (in Russian). Moscow: Pishchepromizdat.

Koroleva, N. S. 1975. *Technical Microbiology of Whole Milk Products* (in Russian). Moscow: Pishchevaya Promishlenost.

ACIDOPHILUS YEAST BEVERAGE (En); Boisson à l'acidophilus-levure (Fr); Acidophilus-Hefe-Getränk (De).

Short Description: USSR; nontraditional product; cow's milk whey; beverage; acidic refreshing flavor.

Microbiology: The starter culture, consisting of *Lactobacillus acidophilus* and lactose-fermenting yeasts, is added and mixed in at a rate of 5%.

Manufacture: The product is made from whey, which is pasteurized at 95-97°C (203-206.6°F) for 1 h, cooled to 35-40°C (95-104°F), centrifuged and flavored, then inoculated with a starter and bottled. The incubation is at 30-33°C (86-91.4°F) for 16-18 h until 75-100°Th (0.67-0.90% titratable acidity) is produced, followed by cooling to 6°C (42.8°F) and cold storage.

Food Value: The final product has an acidity of 75-100°Th (0.67-0.90% titratable acidity) and contains 0.4-1.0% ethanol and small quantities of carbon dioxide.

Related Product: Acidophilus yeast milk.

Reference: Bogdanova, G. J., and E. A. Bogdanova. 1974. *New Whole Milk Products of Improved Quality* (in Russian). Moscow: Pishchevaya Promishlenost.

ACIDOPHILUS YEAST MILK (En); Lait à l'acidophilus-levure (Fr); Acidophilus-Hefe-Milch (De).

Short Description: USSR; nontraditional product; cow's milk; stirred; healthy properties; sparkling yeastlike taste, which may be masked to some extent by the addition of flavoring.

History: The manufacture of this product, based on the use of *Lactobacillus acidophilus* and lactose-fermenting yeasts, has been described by Skorodumova (1956) whose intention was to develop a preparation for use in the therapy of gastrointestinal disorders and of tuberculosis.

Microbiology: The starter culture consisting of *L. acidophilus* and lactose-fermenting yeasts or sucrose-fermenting yeasts is added and mixed in at a rate of 3-5%. Reportedly, the viability of acidophilus bacteria is improved when grown together with yeasts. Selected strains are characterized by their antagonism to intestinal pathogens.

Manufacture: The product is made from partially skimmed or whole milk which is standardized, homogenized, and heat-treated. The incubation with the above starter is at 30-33°C (86-91.4°F) for 4-6 h until 70-80°Th (0.63-0.72% titratable acidity) is produced, followed by ripening at 10-17°C (50-62.6°F) for 6-10 h to assure growth of yeasts, then cooling to 8°C (46.4°F) and packaging.

When *L. acidophilus* and sucrose-fermenting yeasts (baker's yeast, brewer's yeast) are used for fermentation, 2-3% sucrose should be added to the milk before heat treatment.

Food Value: The product has an acidity of 90-110°Th (0.8-1.0% titratable acidity) and contains about 0.5% ethanol in addition to small quantities of carbon dioxide; it has a yeasty taste. Reportedly, maximum antibiotic activity is obtained in the product 3 days after completed manufacture, followed by a gradual decline.

References:

Dobrier, I. B., and A. N. Kostenko. 1955. Acidophilus-yeast milk—a new sour milk product (in Russian) *Voprosȳ Pitaniya* **14**(3):39-41. Cited in *Dairy Science Abstracts,* 1955, **17:**820.

Koroleva, N. S. 1982. Special products. *Proceedings 21st International Dairy Congress* **2:**146-151. Moscow: Mir Publishers.

Skorodumova, A. M. 1956. Antibiotic properties of "medicinal" sour milk product (in Russian) *Voprosȳ Pitaniya* **15:**32-36. Cited in *Dairy Science Abstracts,* 1956, **18:**499.

Skorodumova, A. M. 1959. Acidophilus-yeast milk—a medicinal sour-milk beverage. *Proceedings 15th International Dairy Congress* **3:**1418-1433. Brussels: International Dairy Federation.

ACIDOPHILUS YOGHURT (En); Yoghourt à l'acidophilus (Fr); Acidophilus Joghurt (De).

Short Description: Germany, United States, Scandinavia, Australia, and several other countries; nontraditional product; cow's milk; set or stirred; beneficial value in nutrition; flavor similar to that of yoghurt, usually fruit flavored.

History: The technology of acidophilus yoghurt was first described in Germany by Mülhens (1967) who also used this name for the product indicating the bacteria used in the fermentation. It is believed that human intestinal strains of acidophilus culture increase the beneficial value of the yoghurt made with them.

Microbiology: The starter culture consists of *Lactobacillus acidophilus* and a yoghurt culture (*Streptococcus thermophilus* and *L. delbrueckii* subsp. *bulgaricus*). Special attention should be given to certain characteristics of the acidophilus culture, including its capacity for acid production, antagonism to intestinal pathogens, and ability to pass into and survive in the intestinal tract.

Manufacture: Acidophilus yoghurt is a product made from a low-fat, partially skimmed, or whole milk by using two methods. In the first method, the manufacturing process is based on the separate cultivation of *L. acidophilus, S. thermophilus,* and *L. delbrueckii* subsp. *bulgaricus.* The manufacture involves standardization of milk to the required fat content, homogenization, pasteurization at 85°C (185°F) for 30 min or 90°C (194°F) for 5 min, tempering and inoculation with 0.75-1% *L. acidophilus,* 1.5-2% *S. thermophilus* and 0.75-1% *L. bulgaricus,* then mixing in. The incubation is at 43°C (109.4°F) for 3-5 h until coagulation, followed by cooling and cold storage.

In the second method, the manufacturing process involves the separate incubations of milk with an acidophilus culture and milk with a yoghurt culture, mixing them at a desirable ratio after fermentation, followed by stirring, cooling, and packaging. Reportedly, in this procedure viability of *L. acidophilus* declines rapidly during storage (Hull et al. 1984).

Food Value: The fresh product may contain $1\text{-}5 \times 10^7$/ml viable *L. acidophilus* cells in addition to large numbers of yoghurt organisms. Variations may occur in the number of *L. acidophilus* in the product depending on the strain characteristics, the rate of inoculation, and manufacturing procedures; for example, acidophilus yoghurt made in Denmark contains 90% yoghurt organisms and 10% *L. acidophilus.*

References:

Hull, R. R., A. V. Roberts, and J. J. Mayes. 1984. Survival of Lactobacillus in yoghurt. *Australian Journal of Dairy Technology* December:164-166.

Mülhens, K. 1967. The importance of fermented milk preparations (in German). *Medizin und Ernährung* **8:**11-13.

Rašić, J. Lj. 1987. Other products. In *Milk—the Vital Force,* ed. Organizing Committee of the 22nd International Dairy Congress pp. 673-682. Dordrecht: D. Reidel.

Rašić, J. Lj., and J. A. Kurmann. 1978, p. 348.

ACILMILK (En); Lait acidifié (Fr); Acilmilch (De); means acidified sterile milk.

Short Description: Germany; nontraditional beverage; cow's milk; (artificial) acidified milk product.

Microbiology: No fermentation microorganisms, sterile product, contaminants undesirable.

Manufacture: Schulz (1954) introduced a long-life milk beverage prepared with sterile (or high-heat-treated) milk that is acidified under aseptic conditions with an organic acid (or possibly also with lactic acid bacteria).

Related Product: Acidified milk products.

Reference: Schulz, M. E. 1954. *Haltbare Milch (Long-Life Milk)* (in German) Nuremberg: Verlag Hans Carl.

ACO-YOGHURT (En); Yoghourt-aco (Fr); Aco-Joghurt (De).

Short Description: Switzerland and United Kingdom; nontraditional product; cow's milk; set type of special yoghurt; consistency and flavor similar to yoghurt.

History: The product was developed by Swiss and British workers (Siegenthaler et al. 1960) to increase the dietetic and therapeutic values of yoghurt.

Microbiology: The starter culture consists of a yoghurt culture (*Streptococcus thermophilus* and *Lactobacillus delbrueckii* subsp. *bulgaricus*) and a freeze-dried culture of an intestinal strain of *L. acidophilus* (Enpac, Laccillia).

Manufacture: Aco-yoghurt is made from partially skimmed or whole milk. The manufacture involves inoculation of the heat-treated homogenized milk with a yoghurt culture and with 0.25% of a freeze-dried culture of *L. acidophilus* (containing essential nutrients and vitamins for the growth of *L. acidophilus*), mixing in and packaging. It is incubated at 40-42°C (104-107.6°F) until coagulation, followed by cooling. Maximum viability of acidophilus bacteria is obtained during storage of the final product at temperatures below 10°C (50°F) for about one week.

Food Value: The antimicrobial effect of aco-yoghurt was shown in clinical studies carried out in Switzerland (1959-1960) on hospitalized children with enteric infections: Intestinal pathogens diminished as a result of feeding with aco-yoghurt.

Reference: Siegenthaler, E., H. B. Hawley, F. Schmid, and P. Berger. 1960. Aco-yoghurt (in German). *Schweizerische Milchzeitung* **68:**537-544.

AERIN (En, Fr, De).

Short Description: USSR: nontraditional beverage for children; cow's milk.

Microbiology: The starter culture (a symbiotic mixture) consists of *Streptococcus lactis, S. lactis* subsp. *cremoris, Leuconostoc mesenteroides* subsp. *cremoris, L. mesenteroides* subsp. *dextranicum* and *Lactobacillus casei* in a ratio of 2:1:1:1/3. The selection criteria of strains are not sufficiently described.

Manufacture: After adding sugar to skim milk the mixture is pasteurized, then cooled and incubated at 25-30°C (77-86°F) with the above-mentioned mixture of lactic acid bacteria. After fruit is mixed in, the product is cooled and packaged. To obtain improved flavor, consistency, and stability, the inoculum should be made up of the above-mentioned starter culture.

Food Value: Intestinal dysbacteriosis, which was found in 61% of observed children, was caused by conditionally pathogenic microflora, hemolytic form of colon bacillus, streptococcus, staphylococcus, proteus, or candida molds, and was quickly and completely eliminated by using aerin because of the better ability of the microorganisms of aerin to adapt themselves to the intestinal environment. Intensive excretion of vitamin C and of group B vitamins (especially B_1 and B_2) indicates a better supply of vitamins for children taking aerin. Children's growth and development is improved as a result.

References:

Mostovaya, L. A. V. P. Kulchitskaya, N. N. Romanskaya, and S. I. Kochubei. 1982. Biological evaluation of a new cultured milk product. *Proceedings 21st International Dairy Congress* **1**(1):30. Moscow: Mir Publishers.

Romanskaya, N. N., G. S. Dyment, and L. D. Tovkachevskaya. 1983. Method of preparing a starter for cultured products (in Russian). *USSR Patent* 938 894. Cited in *Dairy Science Abstracts*, 1983, **45**(5024):521.

AIRAG (En, Fr, De).

Short Description: Mongolia; traditional beverage of rural populations; mare's milk; efforts are being made to increase mare's milk production and the consumption of airag (Accolas and Aubin 1975).

Food Value: Dietetic and therapeutic applications (e.g., treatment of tuberculosis), of high social, ritual, and religious value. Kumisslike (Mongolian kumiss), with an alcohol content of up to 2.5-3.0%, refreshing. Current per capita consumption estimated at 50 l/year.

Related Product: Kumiss.

References:

Accolas, J. P. and F. Aubin. 1975. The dairy products. In *Rural life in the Democratic Republic of Mongolia* (in French), ed. J. P. Accolas, J. P. Deffontaines, and F. Aubin, pp. 68-69. Paris: Etudes Mongoles, cahier 6.

Chomakov, Kh. 1966. Study of the composition of lactic acid bacteria in koumiss (in Russian). *Molochnaya Promy̆shlennost'* **27**:44-46.

AIRAN (En, Fr, De); *ayran; eiran;* various spellings of the name depending on language or region.

Short Description: Central Asia (peoples of the Altai language region; traditional beverage, key product; milks of cow, goat, and ewe; liquid.

History: Airan is produced by all known Altaic (Turkic) peoples who are linguistically related. Their ancestors were nomads who in the sixth century A.D. founded an empire stretching from Mongolia to the Black Sea. In the northern regions of this empire the word yoghurt was never used (Kiesban 1969). Airan is first mentioned by Rubruk, 1253-1255 (Doerfer 1965). It is also made by Asian peoples (Bashkirs, Kazan, Tatars, Kirghiz, Kazaks)—inhabitants of the areas near the present borders of China and Mongolia (Ränk 1969; Räsanen 1969).

Microbiology: The airan of central Asia is a product of both lactic fermentation and a more or less alcoholic fermentation (yeast development). The bacterial composition of the starter culture used is not the same everywhere.

Manufacture: The parameters of the manufacture and the starter culture used are not the same everywhere, which explains why different statements can be made about the preparation.

Related Products: Airan, Bulgarian; airan, Russian; airan, Turkish.

References:

Doerfer, G. 1965. *Turkic and Mongolian Elements in the Neo-Persian, with Special Reference to Older Neo-Persian Historic Sources such as the Mongolian Period* (in German), Vol. 2, p. 639. Wiesbaden: O. Harrassowitz.

Kiesban, E. In: Földes, L. 1969, p. 517.

Ränk, G. 1969, pp. 6-7.

Räsanen, M. 1969. *Attempt at an Etymological Dictionary of Altaic-Turkic Languages* (in German). Lexica societatis fenno-ugricae, 18:1. Helsinki: Suomalaisen Kirjallisuuden Kirjapaino Oy.

AIRAN, BULGARIAN (En); Airan bulgare (Fr); Bulgarischer Airan (De).

Short Description: Bulgaria; traditional yoghurt beverage; cow's milk.

Microbiology: Yoghurt starter culture.

Manufacture: Made from milk with 1.0% fat. After incubation of pasteurized milk at 40-42°C (104-107.6°F) with 1-2% of yoghurt culture, the coagulum is stirred using a high-speed stirrer, cooled to 5°C (41°F) and bottled. Acidity of the product is not more than 130°Th (1.17% titratable acidity).

Reference: Koroleva, N. S., and M. S. Kondratenko. 1978. *Symbiotic Starters of Thermophilic Bacteria in the Manufacture of Cultured Milk Products* (in Russian). Moscow, Sofia: Pishchevaya Promishlenost.

AIRAN, RUSSIAN (En); Airan russe (Fr); Russischer Airan (De).

Short Description: USSR (the region of Kazakstan, Azerbaijan, Caucasus, Central Asia); traditional product.

History: The original rural preparation in Kazakstan first reported by Radloff (1884) is as follows: fresh cow's milk (brought to boiling), left standing to form cream layer and removal of cream, dilution with ⅓ water, inoculation with previously made airan (called *ujutku*), heating over low-heat fire with constant stirring until coagulation is complete, filling into leather tubing (*sabag*) and incubation for 24 h. To stimulate fermentation, a piece of sheep stomach is added by the Kazaks, Altais, Azerbaijans, Kalmucks of the Volga region, and the peoples of the Caucasus.

Microbiology: Product with a combined lactic acid and alcoholic fermentation (Koroleva 1973).

Manufacture: Little has been reported about the manufacture of Russian airan.

Related Product: Airan.

References:

Koroleva, N. S. 1973. In *Great Soviet Encyclopedia,* a translation of the third edition, p. 34. New York: Macmillan.

Radloff, W. 1884. *Siberien.1.Buch* (in German) (*Siberia.1.Book*) Vol. 1, p. 439. Leipzig.

AIRAN, TURKISH (En); Airan turc (Fr); Türkischer Airan (De).

Short Description: Turkey; traditional beverage made from diluted yoghurt.

Microbiology: Undefined microflora with yoghurt starter bacteria and two species of yeasts (Bey 1925).

Manufacture: Domestic yoghurt preparation with inoculum from previously made airan.

Food Value: Water 94%, fat 1.2%, protein 2%, added salt 0.75%; mineral water 1.1%.

References:

Bey, R. 1925. Bacteriological studies about yoghurt and the bacillus of Turkish yoghurt (in French). *Le Lait* **5**(47):681-690.

Yaygin, H. 1979. Study of the quality of airan cultured milk (in Turkish). Ege Universitesi Ziraat Fakultesi Dergisi. R. Cemil Adam ozel sayisi 27-32.

AKER (En, Fr, De).

Short Description: Tibet; traditional milk brandy; yak milk.

Manufacture: Made by the addition to yak milk of berries from *Nitraria schober* in order to improve the taste and to enhance the sugar and alcohol contents, followed by fermentation and distillation.

Related Product: Araq.

Reference: Amschler, N. 1931. (in German) *Molkerei-Zeitung* **45:**1556-1558. Cited by Demeter, K. 1941, p. 712.

AMASI (En, Fr, De).

Short Description: South Africa; traditional milk-vegetable baby food; cow's milk and maize (corn); beverage for supplementing diet.

Microbiology: Undefined microflora, spontaneous fermentation.

Manufacture: Product is made by allowing fresh cow's milk to acidify spontaneously, followed by adding a filtrate of maize (corn). The corn is first cooked in water, then filtered by pressing through a sieve. There is a risk of pathogens occurring in such a product because unheated raw milk is used as a starting material.

Food Value: Beverage for supplementary feeding for babies.

Reference: Feer, E. 1937. Fermented milks in human nutrition and in modern medicine (in German). In *Volume Jubilaire Louis E. C. Dapples,* p. 192. Vevey: Nestlé and Anglo-Swiss Holding Company, Ltd.

ANIMAL FEEDS AND BACTERIAL CONCENTRATES (En); Affouragement animale et concentrée bactériennes (Fr); Tierfütterung und Bakterienkonzentrate (De).

Short Description: Many European countries, United States, Japan; nontraditional preparations; freeze- or vacuum-dried bacteria concentrates (animal feed supplements).

History: The modern production of livestock places a great stress on the young animal that is taken from its mother after birth and fed an artificial diet often fortified with a high level of antibiotics or other medications. As a result, an imbalance of the gut microflora occurs which in turn may lead to scours, enteritis, or diarrhea; this imbalance is especially pronounced under unsanitary conditions. The inclusion of indigenous lactic acid bacteria in animal feed establishes and maintains a healthful balance of the gut flora.

Microbiology: The different preparations contain the following microorganisms: *Lactobacillus acidophilus,* the species most often used, which is characterized by its antagonism to pathogenic *E. coli,* salmonellae and proteus; *L. delbrueckii* subsp. *bulgaricus; Bifidobacterium thermophilum;* propionic acid bacteria.

Manufacture: Sometimes cultured on milk or whey. Freeze- or vacuum-dried.

Food Value: Lactic acid bacteria are utilized for animal feed supplements and for the treatment of intestinal disorders and diseases in domestic animals. For example:

- Acidophilus preparations are used for prophylaxis and therapy of gastrointestinal disorders in livestock, particularly pigs and calves. Good results were experienced in reducing scours and enteritis in young pigs
- Reportedly, special preparations of *L. bulgaricus* showed good results against enteritis and diarrhea in young pigs. Yoghurt bacteria have also been used as feed supplements for prophylactic purposes
- Bifidobacteria have been suggested as feed supplements, for example, a dried preparation called Korolac B, containing 10^8/g viable *B. thermophilum,* which is used for prophylaxis and therapy in piglets and for prevention of diarrhea in early weaned Holstein bull calves.

The successful use of bacterial preparations requires ingestion of large numbers of viable cells indigenous to the animal being fed; the availability of fermentable carbohydrates; and ingestion of microorganism strains that are antagonistic to pathogenic *E. coli,* salmonellae, proteus, and others.

Antipov and Subbotin (1980) emphasized that such supplements prevent disease by improving the gastrointestinal microflora but have little therapeutic effect.

References:

Antipov, V. A., and V. M. Subbotin. 1980. Use of probiotics—efficacy and prospects (in Russian). *Veterinariya (Moscow)* **12:**55-57. Cited in *Dairy Science Abstracts,* 1982, **44**(8567):914.

Stern, R. M., and A. B. Storrs. 1975. The rationale of Lactobacillus acidophilus in feeding programs for livestock. Paper read at 36th Minnesota Nutrition Conference, 15-16 September.

ARAGAN (En, Fr, De).

Short Description: USSR (Kazakstan and the Central Asian desert); traditional product, sour cream; camel milk.

Reference: Bairamov, D., and V. Gavrichkin. 1983. Camels should be returned to the desert (in Russian). *Konevodstvo i Konnyi Sport No.* **7:**11-12. Cited in *Dairy Science Abstracts,* 1984, **44**(147):17-18.

ARAK (En, Fr, De); Araq; numerous derivations of the name, which means juice or sap, can be found.

Short Description: Turkic, Arab, and Mongolian peoples; traditional milk brandy; cow's, sheep, and goat milk; distilled product of airan or kumiss.

Manufacture: In some instances the product is distilled from fermented molasses or the juice of certain palms. It can be assumed that the art of distilling was still unknown in the thirteenth century. Multiple distillations of fermented sheep milk were denoted as follows in Siberia (Demeter 1942): first distillation, Arak; second distillation, Dang; third distillation, Arsa; fourth distillation, Chorza. The residue after distillation (*poza* or *podza*) is prepared and used like *quarg* (Ränk 1969). There is an alcoholic beverage by the same name (arak) made in East India by fermenting rice inoculated with a type of millet called *ragi.*

Related Product: Milk brandy.

References:

Demeter, K. J. 1942, p. 712.
Johansen, U. 1961. Did the old Turkic peoples drink milk brandy? (in German). *Ural-Altaische Jahrbücher,* Wiesbaden, **23:**227-235.
Ränk, G. 1969, p. 20-21.

ARERA (En, Fr, De).

Short Description: Ethiopia; traditional beverage or heated and whey-drained product; buttermilk from yoghurt.

Microbiology: Yoghurt microorganisms.

Manufacture: After the completed incubation of milk, yoghurt is fermented for another 4 to 5 days at room temperature. It is then beaten or shaken for 30 min to separate the butterfat. The fluid remaining is removed and used as buttermilk; or the buttermilk is heated to produce a cheeselike curd called *ayib.*

Food Value: Yoghurt buttermilk beverage or concentrated yoghurt buttermilk.

Reference: Vogel, S., and A. Gobezie. 1983. In *Symposium on Indigenous Fermented Foods,* Bangkok. Cited by Steinkraus, K. H. 1983, p. 267.

ARLA ACIDOPHILUS MILK (En); Arla lait à l'acidophilus (Fr); Arla Acidophilus-Milch (De); Arla is a company name.

Short Description: Sweden; nontraditional product; cow's milk; set type of fermented milk; dietetic product.

Microbiology: Obtained by fermenting a specially formulated milk with a culture of *Lactobacillus acidophilus.*

Manufacture: This product is made from low-fat milk (0.5% fat), fortified with soybean oil to a total level of 1.0% fat, 5% skim milk protein (containing 70% protein), fat-soluble and water-soluble vitamins, then sterilized at 140°C for 3-4 s. The milk is inoculated with 1% *L. acidophilus* and 0.02% sterile yeast autolysate, followed by incubation at 37°C (98.6°F) for about 24 h, cooling, and cold storage.

Food Value: Arla acidophilus milk is regarded as a dietetic product. It has an acidity of 90-100°Th (0.81-0.90% titratable acidity), a pH of 4.2 and contains about 6×10^9/ml viable acidophilus bacteria.

References:

Alm, L. 1982. The effect of fermentation on nutrients in milk and some properties of fermented liquid milk products. Dissertation, Stockholm University, Stockholm.

Alm, L. 1983. Arla Acidophilus—an updated product with a promising future (in Swedish). *Nordisk Mejeriindustri* **10**(6):390-394, 396. Cited in *Dairy Science Abstracts,* 1986, **48**(4341):512.

ARSA (En, Fr, De).

Short Description: Asia; traditional milk brandy; mixture of fermented mare's milk and skimmed cow's and sheep milks.

Reference: Fleischmann, W., and H. Weigmann. 1932, p. 384.

ASSES' MILK, FERMENTED (En); Lait d'âne fermenté (Fr); Fermentierte Eselsmilch (De).

Short Description: Middle East, various countries; traditional preparations; asses' milk.

History: In earlier times, asses' milk has been used for human consumption (baby food). It has also been used for the preparation of fermented milks, for example, kumiss.

Food Value: The chemical composition of asses' milk is as follows: water 89.6%, protein 2.0%, fat 1.2%, lactose 6.7%, mineral matter 0.5%, total solids 10.4%.

AUTUMN MILK (En); Lait fermenté domestique préparé en automne (Fr); Herbstmilch (De).

Short Description: Germany (Bavaria) and Austria; traditional (eventually whey-drained) product; milk from different farm animals; usually only utilized in soups.

Microbiology: Nondefined, mesophilic lactic-acid producing bacteria.

Manufacture: In Bavaria it has been produced in farm households since the early part of this century and earlier in the following manner: All milk not consumed fresh is allowed to turn sour by placing it in flat dishes. After the cream layer has formed, the cream is removed and used for other purposes. The soured milk portion is poured into a wooden pail of particular conical shape that previously has been treated with boiling water and rubbed on the inside with salt and onions. This milk is allowed to ferment further and additional portions of sour milk are mixed into it periodically. The mixing is done with a wooden spoon or ladle also previously treated with boiling water and rubbed with salt. If so preferred, the whey that forms can be scooped off, or it may be stirred into the curd when removed for making soup.

Food Value: This particular sour or clabbered milk is a traditional ingredient in so-called *Herbstsuppen* (autumn soups). It is used almost entirely in cooking instead of for fresh consumption.

Related Products: Baassmilch (Austria, Salzkammergut), clabber milk, concentrated fermented milks, Setzmilch.

References:
Fleischmann, W., and H. Weigmann. 1932, p. 370.
Martiny, B. 1907, p. 53.

AZI (En); Aigrette (Fr); Sauer oder Azi (De); derived from Latin *acidus, aceticum.*

Short Description: Switzerland, France, Germany; traditional acidified whey mainly used for whey protein denaturation; cow's, goat, sheep milk cheese whey; whey vinegar; agreeable, young winelike taste; clear or straw-yellow in color.

Microbiology: Lactic acid fermentation by *Lactobacillus helveticus* or *L. lactis.* In some cases acetic acid bacteria also develop and cause acetic acid fermentation.

Manufacture: The product is prepared from hard cheese whey that is heated to 90-95°C (194-203°F). Addition of azi (6-8 l/100 l with an acidity of 112.5-125.0°Th or titratable acidity 1.01-1.12%). The heat-precipitated whey proteins are removed and the clear whey is then fermented by the lactobacilli.

Food Value: The vinegar is used for salads and in salad dressings. The precipitated whey proteins are used as high-quality protein to enrich the food called *Zieger.*

Reference: Frehner, O. 1919. *The Swiss-German Language. Terminology in the German Part of Switzerland. The Dairy* (in German), p. 83. Frauenfeld, Switzerland: Verlag Huber u. Co.

B

B*A® (En, Fr, De); B denotes bifidus and A denotes active.

Short Description: France, since 1985; nontraditional fermented milk; cow's milk; set type; plain, natural product or with mandarin orange or mint flavors, with addition of vitamin C; packages of 125 g.

Microbiology: *Bifidobacterium longum* (called bifidus active®), human strains and yoghurt starter culture.

Food Value: The final product contains protein 3.4%, fat 3.0%, lactose 4.0%, calcium 128 mg/100 g, 238 kJ (57 kcal). Flavored products are enriched with vitamin C.

Reference: GIE B*A. Consumer information. GIE B*A, centre 341, F-94596 Rungis, France.

BAASS-MILK (En); Lait baass (Fr); Baassmilch (De).

Short Description: Austria (Salzkammergut region); traditional product; cow's and other types of milk; long-life concentrated sour milk (ingredient for soups).

Reference: Fleischmann, W., and H. Weigmann. 1932, p. 370.

BABY FOODS, FERMENTED (En); Aliments pour bébés (Fr); Säuglingsnahrungsmittel (De).

Short Description: The Netherlands, Switzerland; nontraditional products; cow's milk; fermented, dried baby foods with humanized milk formula.

History: Fermented milk was originally used in the form of buttermilk in The Netherlands in the eighteenth century for feeding babies suffering from gastrointestinal disorders. Subsequently, the Dutch physician Teixeira de Mattos in 1903 enriched buttermilk with small quantities of rice flour or other flour and sucrose and recommended this so-called Dutch buttermilk in the diet therapy of babies.

Microbiology: The starter culture consists of *Streptococcus lactis* subsp. *lactis.*

Half-Skimmed Milk Formula

Manufacture: The lack of uniform composition of natural buttermilk has led to the development of a half-skimmed milk formula, called Eledon (produced in Switzerland). It is made by fermenting the heat-treated concentrated half-skimmed milk (1.2-1.4% fat) with the above-mentioned culture to the desired acidity; this is then homogenized and spray dried.

Food Value: The final product contains viable *S. lactis* subsp. *lactis* organisms and L (+)-lactic acid and may contain added carbohydrates. It is intended for feeding babies suffering from gastrointestinal disorders. Nutritional advantages of a half-skimmed milk formula are improved digestibility due to the fine protein flocculation and partial hydrolysis of the milk protein; the reduced fat content may be beneficial against dyspepsia; reduced content of lactose; the presence of L (+)-lactic acid and of viable *S. lactis* subsp. *lactis* cells.

Clinical experiences in many countries, especially in Brazil, have shown favorable results with fermented half-skimmed milk formula in feeding babies suffering from dyspepsia, diarrhea, dysentery, enterocolitis, toxicosis, and other disorders.

Whole Milk Formula

Manufacture: The best known product is called Pelargon (produced in Switzerland). It is made by fermenting the heat-treated concentrated whole milk with the above-mentioned culture to the desired acidity and then cooled. The formula is mixed with heat-treated carbohydrates, such as malto-dextrin, corn (maize) starch or rice starch, and sucrose and usually vegetable oil; it is then homogenized and spray dried; about 20% of the *S. lactis* subsp. *lactis* cells survive the drying.

Food Value: The product is intended for feeding healthy babies after two months of age.

Related Products (both formulas): Acidophilus baby foods, bifidus baby foods.

References:

Ballarin, O. 1971. The scientific rationale for the use of acidified and fermented milk in feeding infants and young children, *FAO/WHO/UNICEF PAG Doc. 1.14/19*, pp. 1-16. Rome: Food and Agriculture Organization of the United Nations.

Grieder, H. R. 1960. Experiences with a new, biologically acidified baby food (in German). *Praxis* **58:**1236-1238.

BABY KEFIR (En); Kéfir pour bébés (Fr); Säuglingskefir (De).

Short Description: USSR; nontraditional product; cow's milk; liquid (dietetic beverage for babies).

Microbiology: Kefir culture (from which kefir grains have been removed).

Manufacture: Baby kefir is made from a high-quality milk that is standardized to 3.2% fat, homogenized and sterilized at 135-140°C (275-284°F) for 2-5 sec or pasteurized at 90-95°C (194-203°F) for 20 min, then inoculated with 2-3% kefir culture. The milk is fermented at 20-22°C (68.0-71.6°F) for 10-12 h, then ripened at 14-16°C (57.2-60.8°F) for 12 h; this is followed by cooling to refrigeration temperature, bottling (0.2-liter containers) and storage at 0-6°C (32.0-42.8°F).

Food Value: The final product contains 3.2% fat, 8.1% nonfat solids and has an acidity of 80-100°Th (0.72-0.90% titratable acidity). It is intended for artificial and mixed feeding of babies from 6 months of age.

Related Products: Kefir.

Reference: Ivanova, L. N., A. N. Bulatskaya, and A. E. Silaev. 1980. Trial to make baby kefir industrially (in Russian). *Molochnaya Promȳshlennost'* **3:**15-17.

BALDYRGAN (En, Fr, De). *See* Acidophilus baby foods.

BALKAN AND FERMENTED MILK PRODUCTS (En); Les Balkans et les produits laitiers fermentés (Fr); Balkan und Sauermilchprodukte (De).

Short Description: Balkan area (various countries); traditional products; different milk types: ewe's, cow's, goat's, and mixed milks.

History: The Balkan area is comprised of Yugoslavia, Greece, Bulgaria, Rumania, Albania, and the European part of Turkey. It is a significant source of fermented fresh milk products and is historically important with regard to regional varieties and experimentation with milk fermentations. Therefore, Balkan is also the trade name for different fermented products (yoghurt, fresh cheese, etc.).

BASA (En, Fr, De).

Short Description: Yugoslavia (Croatia); traditional product; cow's, ewe's or mixed milks; whey-drained, concentrated plain yoghurt or soured milk; pasty consistency; acidic taste and pleasant flavor.

Microbiology: Yoghurt or mesophilic sour milk microflora.

Manufacture: The product is homemade, usually in the mountain villages. Yoghurt or soured milk is stirred and placed in a hanging cloth bag so that the whey is drained off until the coagulum acquires a desirable consistency.

Food Value: The finished product may be salted or unsalted. Direct consumption with bread, sometimes used in baked foods.

Related Products: Labneh, kurut.

Reference: Rašić, J. Lj. 1982. Special products. *Proceedings 21st International Dairy Congress* **1**(2):151. Moscow: Mir Publishers.

BÄSSRIK (En, Fr, De).

Short Description: USSR (Kalmuck people, Asian part); traditional fermented milk; mixture of mare's and cow's milk.

Reference: Pallas, P. S. 1776. Collection of historical reports about Mongolian people (in German). Vol. 1, 132 pp. St. Petersburg.

BELGUSS (En, Fr, De); trade name.

Short Description: USSR; nontraditional beverage; cow's skim milk and sweet buttermilk.

Microbiology: Inoculation with leaven for fermentation.

Manufacture: A mixture of 1 part skim milk and 1 part buttermilk is filtered, pasteurized at 85°C (185°F), cooled to 28°C (82.4°F), inoculated with 5-10% of leaven; then pepsin powder is added followed by incubation for approximately 12 h until the desired viscosity is obtained. At that time 10-15% sugar and 1 g of vanilla per 20 kg is added.

Food Value: Skim milk, buttermilk, 10-15% sugar, and 0.005% vanilla.

Reference: Nowosselov, N. 1948. Preparation of belguss (in Russian). *Milk-Industry* **9**(4):39-40. Cited in *Chemisches Zentralblatt Berlin,* 1949, **1:**1430.

BEVERAGES (En); Boissons (Fr); Getränke (De).

Short Description: Various countries; traditional and nontraditional products; fermentations of milk (different milk types, mainly cow's milk), whey, natural buttermilk, or mixtures thereof.

Microbiology: Fermented fresh beverages utilize heat-treated whole milk, low-fat milk, or skim milk and are cultured with any one of the following:

- Mesophilic lactic streptococci combined with leuconostoc (e.g., cultured buttermilk, buttermilk, *lättfil,* and related products)
- Thermophilic lactic acid bacteria, such as *Streptococcus thermophilus* and *Lactobacillus delbrueckii* subsp. *bulgaricus* (e.g., yoghurt drinks)
- *L. acidophilus* and optionally other lactic acid bacteria (e.g., acidophilus drinks, acidophilus buttermilk, etc.)
- Bifidobacteria combined with *L. acidophilus* or with other lactic acid bacteria (e.g., cultura drink, mil-mil, biogarde drink)
- Combined mesophilic and thermophilic lactic streptococci (e.g., product *ljubitelkskiy*)
- Kefir grains (single-use kefir starter) or kumiss microorganisms (e.g., kefir and kumiss)
- *L. casei* (e.g., *yakult*)

Fermented fresh beverages from heat-treated natural sweet buttermilk obtained from unripened cream are cultured with mesophilic lactic streptococci incorporating aroma producers or leuconostocs (e.g., cultured natural buttermilk) or *L. acidophilus* as such or together with mesophilic lactic streptococci containing aroma producers and leuconostocs (e.g., acidophilus drinks, acidophilus natural buttermilk).

Fermented fresh beverages from whey (whole or deproteinized) are made by using lactic acid bacteria (e.g., yoghurt-flavored drink containing *L. acidophilus* and *L. casei* in addition to yoghurt organisms); combined with yeasts (e.g., acidophilus yeast beverage, whey kwas, milone); or pure yeasts (e.g., whey champagne, whevit, milk kwas from whey, etc.). Most of the beverages from whey are made by using pure cultures of yeasts, including baker's yeast and brewer's yeasts.

Manufacture: The manufacture of these products involves stirring the coagulum after fermentation to obtain drinkable properties, usually diluting with fruit juice, fruit syrup, or water, and sometimes using supplementary polysaccharide-producing lactic strains to enhance the viscosity of the products.

Traditionally, such products are made for fresh consumption and contain large numbers of viable bacteria; they may also be made as long-life products with a shelf life of up to several months. The latter are made by heat treating the fermented products, which results in the destruction of most of the microorganisms; they usually contain sugar, flavorings, and stabilizers. If the heat treatment was a sterilization process, it is followed by aseptic packaging of the product.

Related Products: See Appendix F for listings of numerous beverages.

References:

Holsinger, V., L. Posati, and E. Devilbiss. 1974. Whey beverages: Review. *Journal of Dairy Science* **57**:849-859.

Rašić, J. Lj. 1987. Other products. In *Milk—The Vital Force,* ed. Organizing Committee of the 22nd International Dairy Congress, pp. 673-682. Dordrecht: D. Reidel.

Rašić, J. Lj., and J. A. Kurmann. 1978, p. 34.

BIFIDER® (En, Fr, De); trade name.

Short Description: Japan; pharmaceutical preparation; cow's milk; used as food in special dietary regimens and in the treatment of gastrointestinal disorders; freeze-dried preparations containing viable cells of *Bifidobacterium bifidum.*

Reference: Rašić, J. Lj., and J. A. Kurmann. 1983, p. 141.

BIFIDOGÈNE® (En, Fr, De); trade name.

Short Description: France; pharmaceutical preparation; cow's milk; used to treat gastrointestinal disorders resulting from antibiotic therapy; freeze-dried preparation containing viable bifidobacteria.

Reference: Bajard, A. 1977. Microbiological control of preparations containing lactic acid bacteria and related organisms (in French). *Revue de l'Institut Pasteur, Lyon* **10**(4):313-332.

BIFIDUS BABY FOODS (En); Aliments au bifidus pour bébés (Fr); Bifidus-Säuglingsnahrungsmittel (De).

Short Description: France, Germany; nontraditional products; cow's milk; fermented, dried, or liquid baby foods with humanized milk formula.

History: Bifidobacteria were used originally in France in 1906 by Tissier who administered a culture of *Bifidobacterium bifidum* to infants suffering from diarrhea (Rašić and Kurmann 1983). The name bifidobacteria originates from possible bifurcated cell forms. Subsequently, bifidobacteria were used in 1948 by Mayer in making baby foods; bifidus milk containing growth-promoting substances was used to feed infants suffering from nutritional disturbances. Bifidobacteria are the predominant intestinal organisms of breast-fed babies. The potentially beneficial roles of these bacteria in the gut include competitive antagonism against pathogens, production of acetic and L(+)-lactic acids, inhibition of nitrate reduction to nitrite, improved nitrogen retention, and weight gain in infants.

Microbiology: The starter culture consists of *Bifidobacterium bifidum* alone or in combination with *Lactobacillus acidophilus* and *Pediococcus acidilactici.*

Manufacture: Commercial production of bifidus baby foods began in 1964. Products incorporating human intestinal strains of bifidobacteria either alone or together with their growth-promoting substances include the following preparations:

- The dried formula product called Lactana-B, containing lactulose and viable *B. bifidum,* produced from modified milk (developed in Germany in 1964).

- The liquid formula product called Bifiline, containing viable bifidobacteria (developed in the USSR in 1982). It is made by using a milk formula called Malutka (*see* Acidophilus baby foods) and selected strains of bifidobacteria. The product is made by fermenting the heat-treated, homogenized milk formula with 5% starter culture containing 0.5% corn (maize) extract at 37°C (98.6°F) for 8-10 h until coagulation and then cooling. The final product is reported to have about 70°Th (about 0.60% titratable acidity) and 10^7-10^9/g viable bifidobacteria.
- The dried formula product called Femilact, containing viable bifidobacteria (developed in Czechoslovakia in 1984) and made by fermenting heat-treated cream (12% fat) with 2-5% mixed culture consisting of *B. bifidum, L. acidophilus* and *P. acidilactici* (1:0.1:1 ratio) at 30°C (86°F) to the desired acidity followed by cooling; heat-treated vegetable oil, lactose, whey protein, and vitamins are added and the mixture is homogenized and spray-dried. The final product when reconstituted has 28°Th (0.25% titratable acidity) and contains 10^8-10^9/ml viable culture bacteria; but the number of viable cells declines by a factor of 10 during storage for 2 months. See Appendix F for listing of different single products.

Food Value: *B. bifidum* is the species most often used. Either alone or together with its in vivo growth-promoting substances is said to modify the gut microflora of bottle-fed babies and infants; protect against enteric infections or side effects of antibiotic therapy; and act as an aid in the therapy of intestinal disorders and enteric infections.

Related Products: Freeze-dried products containing human strains of intestinal bifidobacteria, such as *B. bifidum, B. infantis, B. breve,* and *B. longum.*

References:

Dedicova, L., and J. Drbohlav. 1984. Humanized fermented milk, spray-dried for feeding babies "Femilact" (in French). *IDF Bulletin Doc. 179* (Fermented Milks), posters, p. VI. Brussels: International Dairy Federation.

Koroleva, N. S. 1982. Special products. *21st International Dairy Congress* **2:**146-151. Moscow: Mir Publishers.

Mayer, J. B. 1948. Development of a new baby food containing Thermobacterium bifidum (in German). *Zeitschrift für Kinderheilkunde* **65:**319, 345.

Rašić, J. Lj., and J. A. Kurmann. 1983, pp. 86, 134.

Semenikhina, V. E. 1984. Other fermented milks. *IDF Bulletin Document 179* (Fermented Milks), pp. 120-122. Brussels: International Dairy Federation.

BIFIDUS MILK (En); Lait au bifidus (Fr); Bifidus-Milch (De).

Short Description: Germany; nontraditional fermented milk, key product; cow's milk; set type or stirred; mild acid taste and distinct flavor, slightly spicy and different from the flavor of other fermented milks, its taste may be modified by incorporating fruit, and defects in product consistency and viscosity may be avoided by increasing the milk total solids content to 15-20%.

History: Bifidobacteria were first used in 1948 by Mayer in the manufacture of baby foods. Subsequently, Schuler-Malyoth and co-workers (1968) suggested the first large-scale commercial process for making fermented milks containing bifidobacteria. The product is named according to the bacteria (former name *Lactobacillus bifidus*) used in the fermentation. Bifidobacteria are the predominant intestinal bacteria of breast-fed infants and a major component of the large intestine's flora in human adults. The naturally occurring large numbers of these bacteria in the gastrointestinal tract may be indicative of their importance. Bifidus milk is produced in small quantities in some European countries. Its consumption is linked to alleged dietetic and therapeutic values rather than to its sensory properties.

Microbiology: The product, based on skimmed milk, partially skimmed milk, or whole milk, is made with viable strains of *Bifidobacterium bifidum* or *B. longum* originating from the gut (feces) of healthy humans. Bifidobacteria grow slowly in milk and produce L(+)-lactic acid and acetic acid in an approximate molar ratio of 2:3; small amounts of formic acid, ethanol, and succinic acid are also formed. Their growth and acid production may be improved by using selected strains that are more acid-tolerant and by using a larger inoculum during culturing and/or by adding to the milk such growth-stimulating nutrients as yeast extract or autolysate, pepsin-hydrolyzed milk, corn (maize) extract, or whey protein.

The mother culture is transferred every 2-3 days to ensure maximum activity. Sterilized skim milk, with or without added growth-stimulating nutrients, is inoculated with approximately 10% of a culture and incubated at 37-42°C (98.6-107.6°F) until coagulation and then cooled. Concentrated cultures, with or without growth-stimulating nutrients, may be used as a bulk starter inoculum to avoid subculturing in the laboratory.

Manufacture: The manufacturing process of bifidus milk is the following: standardization of milk, homogenization, heat treatment (80-120°C/176-248°F, 5-10 min.), tempering, inoculation with ca. 10% of *B. bifidum* culture, incubation (37°C/98.6°F for human strains and 42°C/107.6°F for animal strains) until coagulation, cooling, packaging, and cold storage. The final product has a pH of 4.3-4.7 and contains 10^8-10^9/ml viable bifidobacteria whose numbers decline by 2 logs during refrigerated storage for 1-2 weeks. Bifidus milk can be produced as a stirred product as well.

Food Value: Bifidus milk is a food that is claimed to be more easily digested than the milk from which it is made. It has been used as an aid in therapy of gastrointestinal disorders and as a protective means against imbalances in the gut microflora, as well as in the treatment of liver diseases and chronic constipation.

Related Products: The relatively slow acid production in milk by bifidobacteria and the unusual flavor of the final product have led to the development of fermented milk products incorporating other lactic acid bacteria in addition to bifidobacteria (bifidus yoghurt, acidophilus bifidus yoghurt).

Due to the limited shelf life of fresh bifidus milk, freeze-dried pharmaceutical preparations containing large numbers of viable bifidobacteria with or without other lactic acid bacteria are now also available (*see* Freeze-dried preparations).

References:

Mayer, J. B. 1948. Development of a new baby food containing *Thermobacterium bifidum* (in German). *Zeitschrift für Kinderheilkunde* **65:**319-345.

Schuler-Malyoth, R., A. Ruppert, and F. Müller, 1968. A survey of the theoretical and practical principles of using bifidus cultures in the dairy industry. II. Technology of the bifidus culture in the milk processing factory (in German). *Milchwissenschaft* **23:**554-558.

Rašić, J. Lj. and J. A. Kurmann. 1983, pp. 12, 87, 102.

BIFIDUS MILK WITH YOGHURT FLAVOR (En); Lait au bifidus avec saveur au yaourt (Fr); Bifidus-Milch mit Joghurt-Geschmack (De).

Short Description: United Kingdom; nontraditional product; cow's milk; set type; pleasant yoghurtlike flavor that may be modified by adding fruit.

History: The technology of this product was developed in the United Kingdom in 1982 (Marshall et al. 1982) with the purpose of obtaining a dietetic product from specially formulated milk.

Microbiology: *Bifidobacterium bifidum* or *B. longum* culture.

Manufacture: The product is made by mixing equal volumes of the retentate of ultrafiltered sweet whey (conc. × 8) and ultrafiltered skim milk (conc. × 2), then pasteurizing the mixture at 80°C (176°F) for 30 min and cooling to 37°C (98.6°F), followed by adding threonine (0.1%) and starter (2%). The incubation is at 37°C (98.6°F) for 24 h, followed by cooling to 4°C (39.2°F) and cold storage.

Food Value: The final product has about 15% total solids, including 7.3% protein and 1.3% fat, a pH of about 4.7, and an acetaldehyde concentration between 29 and 39 ppm. It contains 10^9/ml viable bifidobacteria whose numbers decline during storage for 21 days at 4°C (39.2°F) to 10^7-10^6/ml viable cells.

References:

Gurr, M. J., V. M. Marshall, and F. Fuller. 1984. Fermented milks, intestinal microflora and nutrition. In *Fermented Milks, IDF Bulletin Doc. 179,* pp. 54-59. Brussels: International Dairy Federation.

Marshall, V. M., W. M. Cole, and L. A. Mabbitt. 1982. Fermentation of specially formulated fermented milk with single strains of bifidobacteria. *Journal of the Society of Dairy Technology* **35:**143-144.

BIFIDUS YOGHURT (En); Yaourt au bifidus (Fr); Bifidus-Joghurt (De).

Short Description: Germany, USA, Japan and several other countries; nontraditional fermented milk; cow's milk; set type (firm body); yoghurt with bifidobacteria; characteristic mild acid flavor, which may be masked or modified in fruit-flavored varieties.

Microbiology: A fermented product obtained by fermenting milk with selected cultures of *Bifidobacterium bifidum* or *B. longum, Streptococcus thermophilus* and *Lactobacillus delbrueckii* subsp. *bulgaricus.*

Manufacture: The product is made either by a simultaneous fermentation or by mixing into cultured yoghurt a separately cultured bifidus milk at a desirable ratio. The manufacturing process of bifidus yoghurt with a mixed culture for simultaneous fermentation is the following: standardization of milk, homogenization, heat treatment, tempering, inoculation with 5-10% of a starter containing *B.bifidum* and yoghurt culture, incubation (40-42°C/ 104.0-107.6°F until coagulation for animal strains and 37°C/98.6°F for human strain cultures), cooling, stirring, packaging, and cold storage.

Food Value: *See* Yoghurt; Bifidus milk.

References:

Rašić, J. Lj. 1986. Other products. In *Milk—The Vital Force,* ed. Organizing Committee of the 22nd International Dairy Congress, pp. 673-682. Dordrecht: D. Reidel.

Rašić, J. Lj., and J. A. Kurmann. 1983, pp. 91, 132.

BIFIGHURT® (En, Fr, De); trade name, denotes presence of bifidobacteria.

Short Description: Germany; nontraditional product; cow's milk; stirred type, smooth creamy consistency; mild taste, bifidus-typical.

Microbiology: Bifighurt is made either exclusively with *Bifidobacterium longum* C KL 1969 (DSM 2054, a slime-forming variant) or *B. longum* and *Streptococcus thermophilus.*

Manufacture: Similar to preparation of bifidus milk or bifidus yoghurt.

Food Value: The final product contains more than 95% L(+)-lactic acid and about 10^7 bifidobacteria/ml.

Reference: Klupsch, H-J. 1986. German Patent 3 120 505, April 24.

BIFILAKT (En, De); Bifilact (Fr); trade name.

Short Description: USSR; nontraditional preparation; used predominantly for children.

Microbiology: *Bifidobacterium* and *Lactobacillus* starter.

Manufacture: Whole milk is brought to boiling and 30 g lactose and 30 g corn (maize) starch dispersed in a small amount of water are added per liter. Then the mixture is sterilized. Bifilakt is prepared from this by the addition (per liter) of 0.05 g lysozyme in physiological saline solution and 50 ml each of the bacterial starter cultures at the appropriate temperature. The mixture is stirred and filled into sterile containers to be incubated at 37°C (98.6°F) for 18-20 h. The finished product is stable for 1 week at 4-10°C (39.2-50°F). It has an acidity of 27-32°Th (0.2475-0.2925% titratable acidity expressed as lactic acid) and a pH of 5.8-6.0.

Food Value: Bifilakt contains 10^8-10^9 viable culture bacteria/ml, as well as protein 4.23% and fat 3.7%. The product has antacid properties and has been demonstrated to be effective in the treatment of gastrointestinal diseases during childhood.

Reference: Dorofeichuk, V. G., A. I. Volkov, N. N. Kulik, G. N. Karaseva, and V. S. Zimina. 1983. Antacid bifilakt preparation and its effectiveness in treatment of chronic-duodenitis and peptic ulcer in children (in Russian). *Voprosy Pitaniya,* No. 6:30-33. Cited in *Dairy Science Abstracts,* 1985, **47**(2123):243.

BIFILINE (En, Fr, De). *See* Bifidus baby foods.

BIOBEST (En, Fr, De); trade name.

Short Description: Germany, since 1980; nontraditional fermented milk; cow's milk; low-fat yoghurt containing bifidobacteria, with the addition of "biogerm" grains (e.g., wheat flakes, whole grain flour) and fruits (e.g., strawberry and red currant); a related product is bifidus yoghurt.

Reference: Rašić, J. Lj., and J. A. Kurmann. 1983, p. 132.

BIODYNAMIC MILK YOGHURT (En); Yaourt au lait biodynamique (Fr); Joghurt aus biodynamischer Milch (De).

Short Description: Europe (e.g., Switzerland, Germany); nontraditional preparations; cow's milk produced according to organic methods.

History: Yoghurt made from milk that has been produced according to organic methods, also known as "biodynamic agricultural methods." There is no legal definition as to what these methods are or must be. Particular requirements address soil preparation, use of fertilizers, and protection of crops from insects, fungi, and other pests.

Manufacture: *See* Demeter (fermented fresh milk) products.

BIOFLORINE® (En, Fr, De); trade name.

Short Description: United States, Switzerland and several other countries; pharmaceutical preparation with *Streptococcus faecium* SF 68, group D; used in the treatment of diarrhea; *see* Paraghurt®.

Reference: Giuliani S. A. Prospectus. Giuliani S. A., Lugano-Castagnole, Switzerland.

BIOGARDE®-FERMENTED MILK (En); Laits fermentés Biogarde® (Fr); *Biogarde®-Sauermilch* (De); trade name.

Short Description: Germany; nontraditional fermented milk; cow's milk; set type (firm body); mildly acidic taste and characteristic flavor, which may be changed by incorporating fruits or other desirable flavorings.

History: The manufacturing procedure for this product was developed by the Bioghurt® Company in Germany based on the studies of Schuler-Malyoth et al. (1968).

Microbiology: The Biogarde® culture contains *Lactobacillus acidophilus, Bifidobacterium bifidum* and *Streptococcus thermophilus.* The first two cultures are human intestinal strains; *S. thermophilus* is incorporated to help acidification.

Manufacture: The process of creating the bulk starter is: dissolve special nutrient medium for Biogarde culture (1.5%) in water; add this solution to the bulk starter milk; heat to 90°C (194°F) for 10 min; cool to 42°C (107.6°F); inoculate with liquid, freeze-dried or deep-frozen Biogarde culture; incubate for 4.5-6.5 h at 41-42°C (105.8-107.6°F) depending on the quantity; cool to and store at approximately 8°C (46.7°F). The manufacture of Biogarde® involves standardization of milk, homogenization, heat treatment at 90°C (194°F) for 10 min or 95°C (203°F) for 5 min, tempering and inoculation with 6% of a bulk starter, mixing well, packaging and incubation at 42°C (107.6°F) for about 3.5 h until coagulation, then cooling.

Food Value: The final product reportedly contains 85-90% L(+)-lactic acid, about 10^7-10^8 cells of *L. acidophilus* per ml and 10^6-10^7 *B. bifidum* per ml in addition to large numbers of *S. thermophilus.*

Related Products: Biogarde® starter cultures are also used in the preparation of various other products, including buttermilk, sauces, muesli (breakfast cereal), beverages, fresh cheeses, ice cream, and desserts.

References:

Klupsch, H. J. 1984. *Fermented Milk Products, Beverages and Desserts* (in German). Gelsenkirchen, Germany: Verlag Th.Mann.

Sanofi Bio-Industries (formerly Biogarde/Bioghurt Company). Consumer information. Sanofi Bio-Industries GmbH, Düsseldorf, Germany.

Schuler-Malyoth, R., A. Ruppert, and F. Müller. The technology of bifidus fermented milks in the dairy industry (in German). *Milchwissenschaft* **23:**554-558.

BIOGARDE® ICE CREAM (En); Crème glacées biogarde® (Fr); Biogarde® Eiskrem (De); trade name.

Short Description: Germany; nontraditional product; cow's milk.

Microbiology: *Lactobacillus acidophilus* and *Bifidobacterium bifidum* cultures (human intestinal strains) and *Streptococcus thermophilus.*

Food Value: Biogarde® ice cream is reported to contain up to 10^8 viable cells of *L. acidophilus* per gram and up to 10^7 bifidobacteria organisms per gram as well as *S. thermophilus.*

Reference: Anonymous. 1986. Ice cream production through 50 years. *North European Dairy Journal,* No. 7.

BIOGHURT® (En, Fr, De); trade name.

Short Description: Germany, Japan, South America, Canada, The Netherlands, Saudi Arabia, Belgium, Austria and several other countries; nontraditional fermented milk; cow's milk; set type (firm body); mildly acidic flavor.

Microbiology: An acidophilus product is obtained by fermenting milk with the culture consisting of *Lactobacillus acidophilus* (human intestinal strains) and *Streptococcus thermophilus.* Formerly the product was made by using a mixed culture of *L. acidophilus* and *S. lactis* var. *taette.*

Manufacture: The procedure for manufacturing the bulk starter is as follows: Dissolve special nutrient medium for Biogarde® (1.5%) in water; add this solution to bulk starter milk; heat to 90°C (194°F) for 10 min.; cool to 42°C (107.6°F); inoculate with liquid, freeze-dried, or deep-frozen Biogarde culture; incubate for 4.5-6.5 h at 41-42°C (105.8-107.6°F) depending on the quantity; cool to and store at approximately 8°C (46.4°F). The manufacture of Bioghurt® involves standardization of milk, homogenization, heat treatment at 90°C (194°F) for 10 min or 95°C (203°F) for 5 min, tempering and inoculation with 6% of a bulk starter, mixing well, packing and incubating at 42°C (107.6°F) for about 3.5 h until coagulation, then cooling.

Food Value: The final product has a pH of about 4.50-4.70 and is said to contain 3×10^8 cells of *L. acidophilus* per ml in addition to large numbers of *S. thermophilus.*

Related Product: Biogarde® products.

Reference: Sanofi Bio-Industries (formerly Biogarde/Bioghurt Company). Consumer information. Sanofi Bio-Industries GmbH, Düsseldorf, Germany.

BIOKYS (En, Fr, De); trade name.

Short Description: Czechoslovakia; nontraditional product; cow's milk; sour-cream-like viscosity (beverage); clean and mild acid taste.

History: The technology of biokys as a health product was developed in Czechoslovakia where this product is commercially produced.

Microbiology: A cultured product obtained by fermenting milk with a mixed culture of bifidobacteria, acidophilus bacteria, and *Pediococcus acidilactici.* The first two cultures are human intestinal strains, whereas *P. acidilactici* is incorporated to help with the acidification.

Manufacture: It is made by using standardized milk (15% total solids including 3.5% fat, which is homogenized and heat-treated, then fermented with 2-5% of a starter consisting of *B. bifidum, L. acidophilus,* and *P. acidilactici* (1:0.1:1 ratio) at 30-31°C (86.0-87.8°F) to the desired acidity, followed by stirring and cooling.

Reference: Hylmar, B. 1978. Cultured milk beverages (in Czech). *Prumysl Potravin* **29**:99-100. Cited in *Dairy Science Abstracts,* 1978, **40**(7122):752.

BIOLACTIN (En, Fr, De); trade name.

Short Description: Bulgaria; nontraditional product; cow's, ewe's or mixed milks; a paste prepared by using a culture of *Lactobacillus delbrueckii* subsp. *bulgaricus;* used to treat some purulent inflammations, lesions, and burns (pharmaceutical preparation).

Reference: Rašić, J. Lj., J. A. Kurmann. 1978, pp. 135, 349.

BIOLACTIS® (En, Fr, De); trade name.

Short Description: Japan; capsule containing *Lactobacillus casei;* pharmaceutical preparation.

Reference: K. K. Yakult Honsha. Consumer information. K. K. Yakult Honsha, 1-1-19, Higshi, Shinbishi, Tokyo, 105 Japan.

BIOLAKT (En, Fr, De); trade name.

Short Description: USSR; nontraditional product; cow's milk; stirred fermented milk; special cultures; clean acid taste, slightly sweet; for feeding young children.

Microbiology: Biolakt is an acidophilus product obtained by fermenting milk with a culture of *Lactobacillus acidophilus.* The process uses a culture of selected intestinal strains of *L. acidophilus* obtained from breast-fed infants. The organisms are characterized by strong antibiotic and proteolytic properties.

Manufacture: The product is made by fermenting standardized, homogenized, pasteurized (90-92°C, i.e., 194.0-197.6°F for 15 min) milk containing 4% sucrose with 2% *L. acidophilus* starter culture at 36-38°C (96.8-100.4°F). Incubation is to an acidity of 70°Th (0.63% titratable acidity) and takes 4-5 h. It is followed by simultaneous stirring and cooling, then packaging and cold storage.

Food Value: The final product has a clean acid taste, is slightly sweet and has an acidity of 80-105°Th (0.72-0.95% titratable acidity). It contains 3.2% fat and 4% sucrose and is intended for feeding young children.

References:

Chokoeva, D. N., V. V. Babich, V. M. Yastreb, and A. A. Knazeva. 1980. The manufacture of "Biolakt" (in Russian). *Molochnaya Promȳshlennost'* **3**:12-15. Cited in *Dairy Science Abstracts,* 1981, **43**(107):21.

Lukovnikova, L. A., and N. P. Pyatkova. 1974. Introduction of new bacterial preparations (in Russian). *Molochnaya Promȳshlennost'* **11**:7-8. Cited in *Dairy Science Abstracts,* 1975, **37**(3957):377-378.

BIOLAKTON (En, Fr, De); trade name.

Short Description: Bulgaria; nontraditional infant food; cow's milk; dried.

Microbiology: Yoghurt starter (*Streptococcus thermophilus* and *Lactobacillus delbrueckii* subsp. *bulgaricus*) together with *S. lactis* starter.

Food Value: Contains 14-15% fat, 31-32% total protein, 39-40% lactose, 6.5-7.0% minerals, and 3.7-4.0% lactic acid.

Reference: Ivanov, I. G., S. Velev, S. Obretenova, and I. Tosheva. 1971. Biologically acidified dry milk for infants (in Bulgarian). Nauchnoizsledovatelski Institut Mlechna Promishlenost, *Vidin* **5**:89-96. Cited in *Dairy Science Abstracts,* 1973, **35**(3336):336.

BIOMILD (En, Fr, De); trade name.

Short Description: Germany (Südmilch, Stuttgart); nontraditional fermented milk; cow's milk; "natural" product without stabilizers; with *Lactobacillus acidophilus* and *Bifidobacterium* cultures. Low in fat, sold in 500-g packages.

BIO-PRODUCTS (En); Produits “bio” (Fr); Bioprodukte (De).

Short Description: Various countries (mainly in Europe); nontraditional products; mainly cow’s milk.

History: A group of commercially made fermented milk products with the prefix *bio* (from bios = life). The meanings of the prefix *bio* and the word *natural* cannot be clearly defined. Some consumers consider use of these terms, when associated with food or agricultural products, misrepresentations.

Food Value: In the strictest physiological sense, all foods are beneficial and provide nutritional components for life-giving body processes. With analytical examination, all foods can be shown to contain some components that may be adverse to life, especially if intake exceeds a certain danger threshold. Because of the impossibility of demonstrating total absence of, for example, environmental pollutants, contaminants, or naturally occurring toxicants, it is impossible for the terms *bio* and *natural* to have any scientific meaning beyond their emotional impact. It is anticipated that in the near future government agencies of some countries will address this issue and, possibly, restrict the use of such meaningless terms as *bio* and *natural* in food marketing.

Related Product: See Appendix F (listing of single products).

BIOTHERAPY, of infantile diarrhea with lactobacillus milk. *See* Lactobacillus milk for biotherapy of infantile diarrhea.

BIO-YOGHURT (En); Bio-yaourt (Fr); Bio-Joghurt (De).

Short Description: Switzerland; nontraditional product; since 1988; made from cow’s milk “organically” produced (i.e., alternative farm production method).

Microbiology: Mild acidifying yoghurt starter culture.

Manufacture: A yoghurt prepared with “biomilk” and with “organically” grown fruits (e.g., strawberries, cherries), usually with twice the amount of fruit and with sugar reduced by ⅓ of regular yoghurt and no additional flavors.

Related Products: Bio-products.

Reference: Miba (Dairy). Consumer information. Miba, Basel, Switzerland.

BIRCHER MUESLI (En, De); Bircher (Fr).

Short Description: Switzerland; nontraditional product; cow's milk or yoghurt, cereal mixtures, nuts, and fruits; pleasant flavor.

History: The term is derived from the German *Mus,* meaning pap or porridge, according to Dr. M. O.Bircher-Benner, physician and nutritional physiologist who in 1897 founded a private clinic, named Bircher-Benner, on a hill near Zürich. Following a number of successful cures with the raw diet, he recognized the importance of "live" food and raw diet.

The Bircher-Muesli, an important homemade, dietetic yoghurt preparation, is the precursor of the now successful commercially made fruit yoghurt preparations.

Manufacture: The original formula for preparing one portion of the breakfast cereal was

Apple (fresh)	150 g
Rolled oats (flakes)	10 g
Sweetened condensed milk	25 g
Lemon juice (½ lemon)	10 g
Nuts	10 g
Water	40 g
Total	245 g

Method of preparation: One tablespoon of rolled oats is softened for 12 hours in 3 tablespoons of water. Alternatively, precooked or instant oat flakes may also be used. One tablespoon condensed milk and the juice of half a lemon are added after soaking. The grated apple is then stirred into it and chopped nuts or almonds are put on top. The dish is consumed as breakfast or as an hors d'oeuvre. Individual ingredients can be exchanged as follows: Condensed milk can be replaced with yoghurt (2-3 tablespoons) or cream (1 tablespoon) or milk (4 tablespoons) and honey (1 tablespoon). Apple can be replaced with strawberries, raspberries, oranges, plums, apricots, or other fruits. Oat flakes can be replaced with wheat, barley, rye, millet, or other grain flakes. Many other such preparations are described in Bircher-Benner diet books. It is stressed that the ingredients be fresh.

Today, Bircher-Muesli preparations are dry breakfast cereal mixtures, always containing nuts and fruits. They are commercially made and sold for household use with the recommendation to add yoghurt prior to eating. A number of stirred yoghurt preparations are commercially available that contain all these ingredients. They are marketed under a variety of trade names, including fiber-enriched diet, yoghurt (these product names are English translations).

Food Value: Bircher-Muesli is the preferred breakfast food for many people because it is not only based on milk but also on raw, uncooked, or relatively unprocessed fruits, nuts, and cereals. It is considered fermented milk-cereal-fruit mixtures, with the nutrients of the various ingredients complementing each other.

Reference: Bircher, H. *75 Savory Yoghurt Recipes* (in German). Freiburg and Salzburg: Humata Publisher.

BRANO MILK (En); Lait brano (Fr); Brano-Milch (De); indicates thickened, fermented milk collected over several days.

Short Description: Bulgaria; traditional product; sheep milk; whey-drained, long-life homemade fermented milk; viscous consistency; pronounced sour taste; alcohol flavor.

Microbiology: The starter culture consists mainly of *Lactobacillus delbrueckii* subsp. *bulgaricus* and *Streptococcus thermophilus.* Supplementary microorganisms in brano milk include lactose-fermenting yeasts.

Manufacture: There are three methods for producing Brano milk:

1. Sheep milk is boiled and incubated with the above culture at 30–35°C (86–95°F). After the milk has curdled, the product is stored overnight in a cold place and then it is transferred into a wooden pail, keg, or barrel to which a new portion of boiled, fresh milk is added every day.
2. The preceding procedure is followed except for the last step. Then a new portion of soured milk instead of fresh milk is added.
3. The sour, clotted milk is strained through a cloth bag and the milk solids collected are put into the wooden barrel. The resulting product is often called "drained milk" and accumulated over several days as in (1).

Related Products: Kurut, labneh.

References:

Gruev, P. 1966. Studies of the microbiology of "brano-milk"—lactoacid product, prepared in certain mountain regions of Bulgaria (in French). *Proceedings 17th International Dairy Congress* E/F:681-688. Brussels: International Dairy Federation.

Koroleva, N. S., and M. S. Kondratenko. 1978. *Symbiotic Starters of Thermophilic Bacteria in the Production of Cultured Milk Products* (in Russian). Moscow and Sofia: Pishchevaya Promishlenost and Technika.

Oberman, H. 1985. Fermented milks. In *Microbiology of Fermented Foods,* ed. Brian J. B. Wood, pp. 167-195. London and New York: Elsevier.

BUFFALO MILK FERMENTED FRESH PRODUCTS (En); Lait buffle fermenté (Fr); Fermentierte Büffel-Milchprodukte (De).

Short Description: Various countries; traditional and nontraditional products; buffalo milk.

History: Buffalo were probably domesticated between 3000 and 2000 B.C.

Manufacture: Several fermented fresh milk products are specifically made from the milk of buffalo or from mixtures of buffalo and other milks. For manufacturing procedure see individual products mentioned in Appendix C, products by milk types.

Food Value: The average composition of milk from water buffalo *(Bubalus bubalis)* is 7.4% fat, 3.6% protein, 0.6% whey protein, 4.8% lactose, 0.8% ash, and 17.2% total solids.

Products: For individual products see Appendix C, products by milk types.

Reference: Ganguli, N. C. 1979. Buffalo milk technology. *World Animal Review* **30**:2-10. Cited in *Dairy Science Abstracts,* 1980, **42**(2155):257.

BULGARIAN ROD CONCENTRATE (En); Concentré de batonnêts bulgares (Fr); Bulgarisches Stäbchen-Konzentrat (De); signifies *Lactobacillus delbrueckii* subsp. *bulgaricus.*

Short Description: Bulgaria; nontraditional product; cow's milk; pharmaceutical preparation.

Microbiology: *Lactobacillus delbrueckii* subsp. *bulgaricus* is an antagonist of potentially pathogenic microorganisms and capable of surviving in the intestinal tract.

Manufacture: *L. delbrueckii* subsp. *bulgaricus* is cultivated in a specific nutrient medium (hydrolyzed milk broth with two aliquots of added water, 3% [w/v] yeast extract, 10% skim milk, and 3% sodium acetate, kept at 45°C [113°F], pH 5.0, for 4-5 h and then neutralized with ammonia). The accumulated biomass is centrifuged and then suspended in sterile milk. Then 5-g portions of the suspension are aseptically dispensed and to each portion is added humanized milk and another special preparation.

Food Value: Babies suffering from severe intestinal microflora problems have been treated with this preparation.

Reference: Kondratenko, M. S., S. Spasov, Y. Stefanova, V. Kochkiva, and S. Z. Kondareva. 1982. Production of Lactobacillus delbrueckii subsp. bulgaricus (Bulgarian rod) concentrate and study of its effects on babies with intestinal disbioses. *Proceedings 21st International Dairy Congress* **1**:298-299. Moscow: Mir Publishers.

BULGARICUM, L.B. (En, Fr, De).

Short Description: United States (California); freeze dried; pharmaceutical preparation.

Microbiology: *Lactobacillus delbrueckii* subsp. *bulgaricus* LB-51.

Therapeutic Value: *L. delbrueckii* subsp. *bulgaricus,* strain LB-51 has been shown to possess antitumorigenic glycopeptides in the cell wall (Bogdanov and Dalev 1975) and has been claimed to stimulate immunological activity in the host. It favorably affects the balance of the gut microflora and the peristaltic function of the gastrointestinal tract as well.

Reference: Bogdanov, I., and P. Dalev. 1975. Antitumor glycopeptides from Lactobacillus bulgaricus cell wall. *FEBS Letters* **57**:3.

BULGARICUM TABLETS (En); Tablettes bulgaricum (Fr); Bulgaricum Tabletten (De).

Short Description: Bulgaria; nontraditional product; cow's milk; pharmaceutical preparation; brown tablets (2.5 cm diameter, 0.4 cm thick) with pleasant sweet-acidulated flavor.

Microbiology: A special starter consisting of "Bulgarian rod," *Lactobacillus delbrueckii* subsp. *bulgaricus* strains.

Manufacture: The production involves extreme heating, 95°C (203°F) for 60 min, of whole, partially skimmed, or skim milk containing 16% total solids and 7% sucrose, inoculation with 5% of *L. delbrueckii* subsp. *bulgaricus* starter, and incubation to coagulation and acidity production of 90°Th (0.81% titratable acidity). The product is then freeze dried, formed into tablets, and packaged under nitrogen.

Food Value: The tablets contain up to 2.5×10^9 viable bacteria/g and have been successfully used in the treatment of various gastrointestinal disorders in children and adults.

Reference: Kondratenko, M. C., A. L. Christova, C. Z. Kondareva, K. Todorov, M. Kozareva, and S. Spasov. 1982. New lactic acid product with increased nutritive, dietetic and curative properties. *Proceedings 21st International Dairy Congress* **1**(1):297-298. Moscow: Mir Publishers. Cited in *Dairy Science Abstracts,* 1983, **45**(3353):378.

BULGARICUS CULTURED BUTTERMILK (En); babeurre acidifié au lactobacille bulgare (Fr); Geschlagene bulgaricus Buttermilch (De).

Short Description: United States; nontraditional product; cow's skim milk.

Microbiology: A cultured buttermilk containing *Lactobacillus delbrueckii* subsp. *bulgaricus* in addition to a cream culture.

Related Product: Bulgaricus milk.

Reference: Hargrove, R. E. 1970. Fermentation products from skim milk. In *Byproducts from Milk,* ed. B. H. Webb and E. A. Whittier, pp. 33-34. Westport, Conn.: AVI

BULGARICUS MILK, BULGARICUS CULTURED BUTTERMILK, BULGARIAN MILK, BULGARIAN BUTTERMILK (En); Babeure acidifié au lactobacille bulgare (Fr); Geschlagene bulgaricus Buttermilch (De); denotes the presence of *Lactobacillus delbrueckii* subsp. *bulgaricus.*

Short Description: Bulgaria, United States, prior to 1940; nontraditional fermented milk; cow's, sheeps, and goat's milk; mainly set type; generally sour but with a more or less characteristic taste; for children especially, a milder tasting variant exists for which also beneficial health claims are made.

Microbiology: The starter culture consists of *L. delbrueckii* subsp. *bulgaricus* with strong acid-producing strains, with mild acid-producing strains, or with hydrocolloid-producing strains in order to prevent whey separation in the high-acid coagulum.

Manufacture: Various manufacturing techniques exist, one of which is as follows: Milk is pasteurized at 85-95°C (185-203°F) for 30-60 min, cooled to 37-42°C (98.6-107.6°F) and then inoculated with 2-5% culture. Incubation is for up to 5 h at 42°C (107.6°F) to 85-155°Th (0.76-1.40% titratable acidity); then the product is cooled to about 6°C (42.8°F). Bulgaricus milk is sometimes diluted 1:1 with cultured buttermilk to reduce its sour taste. Dried bulgaricus milk product is made as follows: Whole, half-skimmed milk containing 16% total solids (which includes 7% sucrose) is pasteurized by holding for 60 min. The milk is then cooled to 45°C (113.0°F) and inoculated with 5% pure culture. After coagulation and attaining an acidity of 90°Th (0.81% titratable acidity), the milk is cooled and freeze-dried. The dried product is formed into tablets and packed under nitrogen.

Food Value: Both the liquid and dry products have been recommended as wholesome foods for babies, children in general, and adults, based on nutritional and curative evaluations conducted by the Academy of Medicine, Sofia, Bulgaria. One gram of the freeze-dried product contains up to 2.5×10^9 viable cells. It has a lactic-acid-like taste and a free amino acid content of up to 100 mg per 100 g. After feeding the product to babies, their feces contain about 1×10^9 viable cells/g. Bulgaricus milk has been employed with success in the treatment of severe bacterial population imbalances. When used for supplementary treatment of leucocytosis and chronic lympholeucocytosis it brought about favorable alterations in the mouth cavity, normalized leucocyte count and also stimulated hematopoiesis. In cases of hyperlipoproteinemia and overweight, treatment with bulgaricus milk brought about a decrease in blood serum lipids, specifically, serum cholesterol dropped by 22%.

Related products: Bulgaricum tablets.

References:

Burke, A. D. 1938. *Practical Manufacture of Cultured Milks and Kindred Products.* Milwaukee, Wis.: The Olsen Publishing Company.

Kondratenko, M. C., A. L. Christova, C. Z. Kondareva, K. Todorov, M. Kozareva, and S. Spasov. 1982. New lactic acid product with increased nutritive, dietetic and

curative properties. *Proceedings 21st International Dairy Congress* **1**(1):297-298. Moscow: Mir Publishers.
Oberman, H. 1985. Fermented milks. In *Microbiology of Fermented Foods,* ed. Brian J. B. Wood, pp. 167-195. London and New York: Elsevier.
Orla-Jensen, S. 1942. *The Lactic Acid Bacteria,* 2nd ed. Copenhagen: Ejnar Munksgaard Publisher.

BURGHUL WITH YOGHURT (En); Burghul avec yaourt (Fr); Burghul mit Joghurt (De); various spellings according to region, for example, burghol, bulgur, burgul.

Short Description: Near East; dried homemade yoghurt burghul (parboiled wheat) mixture; traditional product.

History: Burghul (without yoghurt addition) is one of the oldest processed foods known. It is usually prepared from varieties of hard wheat. Whole wheat grain is boiled in open kettles for 2-4 h until tender. The boiled grain is then spread in the sun to dry. The dry grains are sprinkled with water and scrubbed to help remove the bran. The dried grains are ground and sifted to separate the particles into different sizes. The bran is removed by winnowing with forced air.

Manufacture: Milk is processed into yoghurt in a traditional manner and then mixed with wheat flour, semolina, or parboiled wheat mixture (burghul), shaped into small rolls, and placed in the sun to dry.

Related Products: Fermented milk-vegetable mixture.

References:

FAO. 1982. Food composition tables for the Near East. *Food and Nutrition Paper* **25:**226. Rome: Food and Agriculture Organization of the United Nations.
Tamime, A. Y., and R. K. Robinson. 1985. *Yoghurt,* p. 3. Oxford: Pergamon Press.

BUSA (En, Fr, De).

Short Description: USSR (Turkestan); traditional product; milk (lactic acid fermentation) and rice (alcoholic fermentation) mixture.

Microbiology: Chekan (1922) stated that the alcoholic fermentation of rice was induced by *Saccharomyces busae asiaticae* and a bacterium named

Bacterium busae asiaticae (a lactic-acid-producing bacillus). These microorganisms are today no longer considered as species.

Food Value: Busa contains up to 0.78-1.1% lactic acid and up to 7.1% alcohol by weight.

Related Products: Milk-vegetable mixtures, fermented.

Reference: Chekan, N. 1922. *Zentralblatt für Bakteriologie* (part 2) **2:**74-93.

BUTTERMILK CHAMPAGNE (En); Champagne au babeurre (Fr); Buttermilch-Champagner (De); trivial name.

Short Description: Germany; nontraditional product; cow's buttermilk; champagnelike effervescent beverage.

Manufacture: Preparation from fermented buttermilk serum with addition of white wine after fermentation. When bottled, the beverage becomes better tasting during storage.

Related product: Champagne (*see also* Appendix F).

Reference: Demeter, K. J. 1941, p. 714.

BUTTERMILK, NATURAL OR TRADITIONAL (En); Babeurre (Fr); Buttermilch (De).

Short Description: Regions with milk and butter production; traditional beverage, sometimes dried or concentrated; prepared from different types of milk.

History: Buttermilk is an old traditional product that had its earliest mention in India 800-300 B.C. according to Prakash (1961). The first evidence of butter and buttermilk manufacture is a Sumerian relief (2550 B.C.) on the El-Obed (Tell Ubaid) Temple, depicting the sacred cattle farm of the goddess Kin-Khursag. Priests wearing feather skirts are seen cleaning milk by pouring it through a funnel-shaped sieve. Butter is made in large containers which are rocked back and forth, thus churning cream (presumably soured) into butter and buttermilk.

Manufacture and Microbiology: There are two methods for preparing butter or obtaining buttermilk: (1) in earlier times through deliberate acidification of milk, but also from naturally soured milk (Takamiya 1978), and (2) from cream as is common today. Natural or true buttermilk refers to the serum that remains after cream is churned into butter. In some countries, for example, Finland, the USSR and Switzerland, natural buttermilk is either used as a beverage for humans, sometimes flavored, or is used for animal feed. In the United States and other countries, buttermilk is usually concentrated or dried and is used as an ingredient in the manufacture of process cheese, ice cream, and bakery and confectionery products.

Natural buttermilk may be sweet (nonfermented) or acidic (fermented). Sweet buttermilk is obtained during butter manufacture from pasteurized nonfermented (sweet) cream. This product may be cultured for the purpose of producing a refreshing and dietetic drink. It is made from sweet-cream buttermilk by fermenting with cultures of mesophilic lactic streptococci incorporating aroma producers or leuconostocs (*see* Cultured natural buttermilk) or combined with *Lactobacillus acidophilus* (*see* Acidophilus natural buttermilk) or with pure cultures of *L. acidophilus* (*see* Acidophilus drinks). Acid buttermilk, also called true buttermilk, is obtained during butter manufacture from sour or cultured cream; the pasteurized cream is fermented with 1-2% of a mixed culture consisting of *Streptococcus lactis* subsp. *lactis, S. lactis* subsp. *cremoris, Leuconostoc mesenteroides* subsp. *cremoris* and/or *S. lactis* subsp. *diacetilactis* at 21-23°C (70-73.4°F) for 14-16 h, followed by cooling. Acid buttermilk contains around 0.7% lactic acid and traces of diacetyl in addition to protein, minerals, and about 80% of the nonfermented lactose. Acid buttermilk may be cultured if its acidity is not more than 50°Th (0.45% titratable acidity); the amount of starter culture used is less than with the culturing of sweet buttermilk.

Food Value: Natural buttermilk contains approximately 91.3-91.5% water, 0.5-0.7% fat and 8.0% nonfat solids, and also traces of lecithin and cholesterol due to the presence of fat globule membrane material. Variations in the composition may occur depending on the nature of the cream and processing. It is claimed to have easy digestibility and has other interesting characteristics. For example, it is the key ingredient of the so-called red buttermilk paint traditionally used by Pennsylvania Dutch farmers to paint their red barns. Presumably, the lecithin acts as an emulsifier for the red mineral pigment.

References:

Kosikowski, F. V. 1978. Whey utilization and whey products. *20th International Dairy Congress* Conferences 50 ST. Brussels; International Dairy Federation.

Kosikowski, F. V. 1984. Buttermilk and related fermented milks. *IDF Bulletin Doc.179 (Fermented Milks),* pp. 116-119. Brussels: International Dairy Federation.

Prakash, O. M. 1961. *Food and Drinks in Ancient India.* Delhi-6: Munshi Ram Manohar Lal, Oriental Bookseller and Publishers.
Takamiya, T. 1978. Contribution to the history of food and food preparation by the shepherd people of Middle and Interior Asia (in German). *Inaugural Dissertation* 110p. Munich: Ludwig-Maximilian Universität.

BYO (En, Fr, De); trade name.

Short Description: France, since 1987; nontraditional product; cow's milk; set type fermented milk, packages of 125 g.

Microbiology: *Bifidobacterium longum* (called bifidus activ) and lactic acid bacteria.

Food Value: Protein 3.6%, fat 3.5%, carbohydrate 5.2%, and calcium 132.8 mg/100 g; 279.2 kJ (67.2 kcal) per 100 g.

Reference: Gervais—Danone. Consumer information. F-93202 Levallois—Perret, France.

C

CABARET (En, Fr, De); trade name.

Short Description: United Kingdom; nontraditional product; cow's milk; fruit-flavored yoghurt fortified with a viscous, creamy, alcohol-containing preserve; the product is available in strawberry-champagne, orange-cointreau, black currant-rum and red cherry-brandy flavors.

Reference: Anonymous. 1985. *Milk Industry* **85:**11.

CAMEL'S MILK FERMENTED PRODUCTS (En); Laits fermentés à partir de lait de chameau (Fr); Sauermilchprodukte aus Kamelmilch (De).

Food Value: The gross composition of camel's milk is as follows: 87.67% water, 3.45% protein, 3.02% fat, 5.15% lactose and 0.71% ash (Lampert 1965). Data on the chemical composition of camel's milk have been reported by several workers and there is wide variation (Abu-Lehia 1987). Similar physiological properties with some of the same therapeutic values (Khann 1988) as mare's milk have been attributed to camel's milk.

Products: Fermented camel's milk products are listed in Appendix C.

References:

Abu-Lehia, I. H. 1987. Composition of camel milk. *Milchwissenschaft* **42:**368-371.
Khann, N. D. 1988. Camel as a milk animal. *Indian Farming* **36:**39-40. Cited in *Dairy Science Abstracts,* 1988, **50**(1804):199.
Lampert, L. M. 1965. *Modern Dairy Products,* p. 13. New York: Chemical Publishing Company, Inc.

CAUDIAUX (En, Fr, De); denotes good-tasting acid.

Short Description: France (Normandy region); traditional, rural, long-life fermented milk; cow's milk.

Microbiology: Skimmed, boiled milk is added periodically to spontaneously fermenting milk in a barrel and salted.

Food Value: Long-life product used for making soups that also contain vegetables and flour.

Related Products: Concentrated fermented milks, cellarmilk.

Reference: Gidon, F. 1983. About souring of foods. Caudiaux, a soured milk from the lower part of Normandie (in French). *Bulletin de la Société des Antiquaires de Normandie* **45:**266-281.

CELLARMILK (En); Laits de cave (Fr); Kellermilch (De); Kjälder Mjölk, Kjaeldermelk, Kaeldermelk.

Short Description: Scandinavia, Germany, Eastern Europe, Switzerland; traditional product; different types of milk; long-keeping fermented milk.

Microbiology: Norway cellar milk: lactic acid bacteria and *Torula* yeasts.

Manufacture: Different manufacturing procedures are possible. Mainly skim milk (the creamy layer taken away after cooking) or milk at the end of the lactation period is cooked, then cooled and inoculated with a portion of fermented milk from the preceding manufacture. After the milk has curdled, the product is stored overnight in a cold place and then transferred into barrels of different volumes up to 1000 liters to which a new portion of either cooked milk or soured milk is added every day. Temperature for storage in the cellar is 10-15°C (50-59°F). The product keeps several months or more.

Food Value: Norway cellar milk: contains about 0.5% ethanol and 2% or more lactic acid.

Related Products: Brano milk, zimne.

References:

FIL-IDF Dictionary. 1983, p. 129.

Kurmann, J. A. 1986. Yoghurt made from ewe's and goat's milk. *IDF Bulletin 202* (Production and Utilization of Ewe's and Goat's Milk), pp. 153-166. Brussels: International Dairy Federation.

CHAKKA, INDIAN (En); Chakka indien (Fr); Indischer Chakka (De).

Short Description: India (western part); traditional product; buffalo milk, cow's milk, mixed or reconstituted dried milk; whey-drained product; acidic flavor.

Microbiology: The starter culture consisting of *Streptococcus lactis* subsp. *lactis* is added and mixed in at a rate of 1%.

Manufacture: A product obtained by partial separation of whey from the traditional Indian fermented milk called dahi. Chakka is produced as a basic ingredient for the preparation of a fermented product called shrikand. It is made mainly from fresh buffalo milk, standardized to 6% fat, pasteurized at 71°C (159.8°F) for 10 min, and cooled to 28-30°C (82.4-86°F). The incubation with the above-mentioned culture is at 28-30°C (82.4-86°F) for 15-16 h until an acidity of 80-90°Th (0.7-0.8% titratable acidity) is produced. The coagulum is then broken and put in a cloth bag to drain the whey for 8-10 h; during this time the curd may be gently squeezed to facilitate whey separation.

Food Value: The final product contains about 60% moisture, 22% fat, 10.3% protein, 4.4% lactose, 1.0% and 2.3% lactic acid. The composition of chakka, however, may vary considerably depending on the kind of milk used for manufacture, the acidity of the curd, and the extent of whey removal.

References:

Indian Standard Institution. 1980. Specification for chakka and shrikand. *Indian Standard IS: 9532-1980,* p. 9. New Delhi: Indian Standard Institution.

Kulkarni, M. B., I. G. Chavan, and N. D. Belhe. 1987. Chemical composition of chakka whey. *Indian Journal of Dairy Science* **40:**65-69.

Sukumar, D. 1980. *Outlines of Dairy Technology.* Delhi: Oxford University Press.

CHAKKA, RUSSIAN (En); Chakka russe (Fr); Russischer Chakka (De).

Short Description: USSR (Tadzhikistan); nontraditional product; cow's milk; whey-drained product; pasty texture; acidic flavor.

Microbiology: The starter culture consisting of mixed thermophilic and mesophilic lactic acid streptococci (1:1 ratio) is added and mixed in at a rate of 5%.

Manufacture: A product obtained by fermenting milk with lactic acid streptococci with partial separation of whey. It is made from milk standardized to 2% fat or from skim milk, pasteurized at 85-87°C (185-188.6°F) for 20-30 s, and cooled to 36-38°C (96.8-100.4°F). The incubation with the above-mentioned culture is until 70-90°Th (0.63-0.81% titratable acidity) is produced, and then the coagulum is cut and left for 15-20 min for partial separation of whey. After whey drainage the curd is put in a cloth bag and pressed until the moisture content reaches 70% or 75% depending on the product and then packed.

Food Value: The final product has a maximum acidity of 220°Th (1.98% titratable acidity) if made from whole milk and 230°Th (2.07% titratable acidity) for the skimmed product.

Related Products: Syuzma and concentrated milks.

Reference: Bogdanova, G. J., and E. A. Bogdanova. 1974. *New Whole Milk Products of Improved Quality* (in Russian). Moscow: Pishchevaya Promishlenost.

CHAL (En, Fr, De); câl, different spellings.

Short Description: USSR (Central Asia/Kazakhstan, Turkmenistan); traditional product; camel's milk; effervescent beverage; liquid; specific taste.

Microbiology: The starter culture consists of thermophilic lactobacilli and streptococci.

Manufacture: Traditional manufacture of chal involves the use of raw camel's milk as such or diluted with warm water (1:1 ratio), inoculation with 10-40% of the previously fermented product and incubation at 25-30°C (77-86°F). The milk is coagulated in 3-4 h, and after ripening for 8 h a characteristic taste is obtained. It is suspected that lactose-fermenting yeasts participate in the process of fermentation and ripening of chal in addition to lactic acid bacteria.

Modern manufacture of chal involves the use of pasteurized camel's milk, as well as starter culture of thermophilic lactobacilli and streptococci (no data exist on the use of lactose-fermenting yeasts).

Related Product: Shubat.

References:

Bogdanov, V. M. 1957. *Microbiology of Milk and Milk Products* (in Russian). Moscow: Pishchevaya Promishlenost.
Martinenko, N. I., S. G. Yagodinskaya, A. A. Akhundov, K. C. Charyev, and O. Khumedov. 1977. Content of trace elements, copper, manganese, molybdenum in culture of chal and camels' milk and their clinical significance (in Russian). *Zdravookhranenie Turkmenistana* **3**:20-22. Cited in *Dairy Science Abstracts,* 1978, **40**(7802):824.
Maurizio, A. 1933, p. 81.

CHAMPAGNE (En, Fr, De); common name, from the French word *champagne.*

Short Description: Several European countries; nontraditional beverage; cow's milk; despite the alcohol content these products should not be considered wines. "Champagne" is a sparkling wine and the name is reserved for French products originating from a particular region. The word is not allowed to be used for fermented milk products. No such named product is currently in existence.

Microbiology: Lactic acid- and alcohol-producing microorganisms.

Manufacture: See Kefir; Kumiss. Production of carbon dioxide.

Related Products: Depending on the fermented raw material we could distinguish between buttermilk-, milk- and whey-champagne.

CHANKLICH (En, Fr, De).

Short Description: Middle East; traditional product; milk of cows, buffalo, ewes or combinations; dried product; characteristic flavor.

Microbiology: Yoghurt microflora.

Manufacture: A fermented sun-dried product obtained from a concentrated yoghurt, chanklich is made by mixing the curd of the concentrated product with herbs and spices, then shaping into balls and partially drying in the sun. The dry balls are usually packed in glass containers and covered with olive oil.

Food Value: Normally consumed with bread and olives as an appetizer.

Related Products: Labneh, milk-plant mixtures.

Reference: Tamime, A. Y., and R. K. Robinson. 1978. Some aspects of the production of a concentrated yoghurt (labneh) popular in the Middle East. *Milchwissenschaft* **33:**209-212, 568.

CHEKIZE (En, Fr, De).

Short Description: USSR (Turkmenistan); traditional product; cow's milk; whey-drained, pasty product.

CHHACH (En, Fr, De).

Short Description: India; the liquid product obtained when dahi is churned, heavily agitated, or processed with a separator; it may be sweetened, spiced, or consumed directly.

CHHANA (En, Fr, De).

Short Description: India; nontraditional product; cow's milk, buffalo's milk or mixed milk; (artificially) acidified and whey-drained milk product.

Microbiology: No starter culture microorganisms, but possible presence of contaminants.

Manufacture: An acid-coagulated milk curd product practically devoid of whey, made from whole milk or partially skimmed milk, coagulated by the addition of an acid solution (lactic or citric), including lemon juice, while the milk is boiling. The whey is drained off and discarded or fed to animals.

Related Product: Chhana is used for the preparation of rasgolla.

Reference: Vaghela, M. N. 1984. Evaluation of the effects of superheated condensed milk and additives on the quality of chhana and rasgolla from buffalo milk. Master's thesis. Anand, India: Gujarat Agricultural University.

CHORSA (En, Fr, De).

Short Description: USSR (Kirgizian people); traditional brandy; *see* Milk brandy and Arak.

Reference: Martiny, B. 1907, p. 23.

CHURPI (En, Fr, De).

Short Description: Nepal; traditional product; a dried buttermilk.

CLABBER, clabbered milk, bonny clabber (En); Lait acidifié spontané (Fr); Dickmilch (De).

Short Description: Ireland, England, United States; traditional fermented milk.

History: Clabber has been consumed by the Irish since the earliest recorded history (*see* Spontaneously Soured Milks). Clabber is the American term for sour, curdled milk that was popular before the advent of milk pasteurization. This term is also the English/American equivalent of the German *Dickmilch* or *dicke Milch.* The name comes from the Irish *bainne clabair* (thick milk) and the product became known as "bonny clabber" throughout England.

Manufacture: It is milk that has soured naturally, usually overnight into a thick gellike consistency.

Food Value: In the southern United States it was often eaten with sugar, or even with black pepper and cream.

Related Products: Spontaneously soured milks.

References:

Fitzgibbon, T. 1976. *The Food of the Western World.* New York: Quadrangle/The New York Times Book Co.

Mariani, J. F. 1983. *The Dictionary of American Food and Drink.* New York: Ticknor and Fields.

Martiny, B. 1907, p. 65.

CLOTTER MILK (En); Lait caillé (Fr); Klotz-Milch (De); denotes clumpy, lumpy milk.

Short Description: Germany; traditional product; homemade, skimmed soured milk.

Reference: Martiny, B. 1907, p. 65.

COCKTAIL YOGHURT (En); Yaourt cocktail (Fr); Joghurt-Cocktail (De).

Short Description: Germany; nontraditional yoghurt preparation; cow's milk; yoghurt with added fruit, possibly cream, liqueur, rum, or ethyl alcohol; or yoghurt with added ice cream and strawberries. See yoghurt preparations and dessert products.

Reference: Rašić, J. Lj., and J. A. Kurmann. 1978, p. 343.

COLOSTRUM MILK, FERMENTED (En); Lait colostrum fermentés (Fr); Fermentierte Kolostrummilch (De).

Short Description: Canada, Brazil, Mexico; nontraditional veterinary preparation for animal feeds.

Microbiology: The starter culture consists of *Streptococcus lactis* subsp. *lactis.*

Manufacture: Colostrum is the milklike fluid secreted by a female mammal immediately after birth of the young. In the cow, it takes about 72 hours for the composition of colostrum to change into that of milk. It has been reported that colostrum milk fermented with the above-mentioned culture when fed to neonatal calves produced better weight gains than naturally fermented colostrum (*see* Animal feed and bacterial concentrates).

Reference: Drevjany, L. A. 1983. The feeding of fermented colostrum to neonatal calves. *McGill University (Montreal) Dissertation Abstracts International B (Sciences and Engineering)* **44**(5):1286.

COOKED YOGHURT (En); Yaourt cuit (Fr); Gekochter Joghurt (De).

Short Description: Turkey; traditional product; goat milk; heat-treated after culturing.

Microbiology: Yoghurt starter bacteria.

Manufacture: Yoghurt is heated, then salt is added and the product is packaged in jars and covered with a layer of oil or another fat, such as lard.

Related Products: Kis yoghurt, peskütan.

Reference: Yöney, Z. 1965. Technological investigations on preserved yoghurts (in Turkish). *Ankara Univ. Zir. Fak. Yill* **15:**65-84. Cited in *Dairy Science Abstracts,* 1970, **32**(1479):218.

COSTORPHINE CREAM (En); Lait acidifié spontanée, concentré (Fr); Saure Dickmilch (De).

Short Description: United Kingdom (Scotland); traditional product; cow's milk; long keeping quality, obtained by fermenting milk with a nondefined microflora; concentrated by whey separation (through bottom drainage).

Reference: Martiny, B. 1907, p. 25.

COW'S MILK FERMENTED PRODUCTS (En); Laits fermentés à partir de lait de vache (Fr); Sauermilchprodukte aus Kuhmilch (De).

Short Description: Numerous countries; traditional and nontraditional products; cow's milk.

History: World cow's milk production was about 450 million metric tons (200 lbs/year for each of the 5,000 million persons inhabiting the earth) in 1989. Milk production figures of other species are not available, but dairy cattle produce by far more milk than the other animals domesticated for milk production.

International Dairy Federation statistics show that about 70% of all fermented milks in the world for which data exist are made and consumed in India (16,790 metric tons in 1985). Accurate percentages are not known, but close to 10% of this milk is contributed by cows (*Bos taurus*), 80-85% by water buffalo (*Bubalus bubalis*), and the remainder by zebus (*Bos indicus*), yaks (*Bos grunniens*), and especially goats (*Capra hircus*).

Food Value: The gross composition of cow's milk (*Bos taurus*) is 87.3% water, 3.9% fat, 3.2% protein (2.6% casein and 0.6% whey proteins), 4.6% lactose and 0.7% ash. The energy value is 270.6 kJ (66 kcal) per 100 g.

CREAM CULTURE FERMENTED PRODUCTS (En); Produits fermentés préparés avec la culture de beurrerie (Fr); Mit Butterkultur hergestellte Sauermilchprodukte (De).

Short Description: Numerous countries; nontraditional fermented milk; cow's milk; made by using a cream culture.

Microbiology: The cream culture consists of *Streptococcus lactis* and/or *S. lactis* subsp. *cremoris* and often with low acetaldehyde-producing strains of *S. lactis* subsp. *diacetylactis.*

Related Products: Cultured buttermilk, filmjölk, lättfil, sour cream. *See also* Appendix F.

CREAM TURO (En); Fromage frais avec crème acidifiée (Fr); Sauerrahm-Quarg-Mischprodukt (De).

Short Description: Hungary; traditional product; sour cream and fresh whey-drained fermented milk mixture.

Microbiology: A special heat-resistant, slime-producing strain of *Streptococcus thermophilus* is utilized; it greatly improves rheological properties, especially in the manufacture of cream turo.

Manufacture: Cream turo is a product that can be made at home by mixing turo (a kind of fresh cheese) and sour cream. In industrial manufacture the cream is coagulated within 5-6 h at 37°C (98.7°F) and then packaged at high temperature, 60°C (140°F). Final acidity is about 100-150°Th (0.9-1.35% titratable acidity).

Food Value: Fat content 10.5%, total solids content 26.5% and salt content 0.6%.

Reference: Obert, G. 1984. Manufacturing technology of cream turo with extended shelf life, and isolation of a special strain of lactic acid bacteria which improves product consistency (in Hungarian). *Teijipar* **33**:47-48. Cited in *Dairy Science Abstracts,* 1985, **47**(5402):601.

CRÈME FRAÎCHE, FERMENTED (En); Crème fraîche (Fr, De); means cream fresh and cold.

Short Description: France (orginally), Italy; traditional product; the cream's high fat content makes it highly desirable for many culinary preparations.

Microbiology: Cream culture (*see* Cream culture fermented products).

Manufacture: Crème fraîche is a heavy cream with a slightly developed lactic acidity, 17.5-19.0°Th (0.15-0.17% titratable acidity), as the result of added cultures. Its pH is 6.2-6.3 and it displays a fine diacetyl and lactic acid flavor, as well as a viscous, smooth texture.

Food Value: The fat content is usually around 50% (minimum 39%), protein 1.6%, lactose 2.1%, ash 0.3% and salt 0.15%. A major use is for premium culinary preparations, such as topping for fresh strawberries, raspberries and other fruits, in sauces and gravies, and as a dessert item.

Related Product: Sour cream (high-fat).

References:

Kosikowski, F. V. 1977. *Cheese and Fermented Milk Foods,* p. 57. Ann Arbor, Mich.: Edwards Brothers.

Schulz, M. E. 1981. *Lexicon of Milk-Based Foods* (in German), p. 128. Munich: Volkswirtschaftlicher Verlag.

CROWDIES (En); Préparation lait-céréal à base de caillé lactique (Fr); Sauermilch-Getreide-Zubereitung (De).

Short Description: United Kingdom (Scotland); traditional preparation; cow's buttermilk; mixed with oatmeal, pasty texture.

Microbiology: *See* Cream culture fermented products.

Manufacture: Product 1: A product or dish obtained when buttermilk is poured into finely ground oatmeal until the consistency, after stirring, is similar to that of pancake batter. At one time, this was universal breakfast food in Scotland.

Product 2: The name was also applied to other foods of the porridge type. Crowdies is also the name for a soft cheese from soured milk curd or buttermilk. When made in Aberdeen, it is specifically called Aberdeen crowdies. Often salt and cream and sometimes caraway seeds are mixed in.

Related Product: Milk-cereal preparations (*see* Appendix F).

Reference: Fitzgibbon, T. 1976. *The Food of the Western World.* New York: Quadrangle/The New York Times Book Co.

CULTURA® (En, Fr, De); denotes cultured; trade name.

Short Description: Denmark; nontraditional product with selected strains of bacteria that are native to the human gastrointestinal tract; cow's milk; set type; firm body; specific characteristic flavor and mild acid taste.

History: The potentially beneficial roles of *Lactobacillus acidophilus* and bifidobacteria in the human gut, as pointed out in many reports including the publication about bifidobacteria by Rašić and Kurmann (1983), have led to the development of new products including that called Cultura®. The manufacturing procedure of Cultura® was developed in Denmark in 1985. It is commercially produced by a Danish dairy processing company.

Microbiology: The starter culture consists of selected human intestinal strains of *Bifidobacterium bifidum* and *L. acidophilus.* The strains are separately cultivated and mixed at a desirable ratio when inoculating into milk. Using frozen concentrated cultures would do away with subculturing in the laboratory and preparation of bulk starter.

Manufacture: Cultura® is made by fermenting homogenized heat-treated, protein-enriched whole milk with the above-mentioned culture at 37°C (98.6°F) for about 16 h until the desired acidity is obtained, followed by cooling. It is made by the set method. The product has a shelf life of at least 20 days after production and is sold in plastic containers of 150 ml and 500 ml.

Food Value: The final product contains more than 10^8 cells of *Lactobacillus acidophilus* and more than 10^8 cells of *Bifidobacterium bifidum* per ml.

Related Products: AB-fermented milks.

References:

Hansen, R. 1985. Bifidobacteria have come to stay. *North European Dairy Journal* **3**:79-83.

Rašić, J. Lj. 1983. The role of dairy products containing bifidobacteria and acidophilus bacteria in nutrition and health. *North European Dairy Journal* **4**:80-89.

Rašić, J. Lj., and J. A. Kurmann, 1983.

CULTURA® DRINK, (En, Fr, De); trade name.

Short Description: Denmark; nontraditional product; cow's milk; fermented milk beverage; mild acid taste.

History: The procedure for making this product was developed in 1985 in Denmark where it is now commercially made and distributed by Danmaelk, the dairy processing company.

Microbiology: The starter culture consists of *Bifidobacterium bifidum* and *Lactobacillus acidophilus.*

Manufacture: Cultura® drink is made by fermenting homogenized heat-treated, partially skimmed milk with a blend of highly concentrated and frozen cultures of *B. bifidum* and *L. acidophilus* at 37°C (98.6°F) for approximately 16 h to the desired acidity, followed by stirring, cooling, and packaging. The final product has a characteristic flavor and a mild acid taste; it is available in 0.5-liter cartons and has a shelf life of at least 20 days after the production date.

Related Product: Cultura®.

Food Value: *See* Acidophilus milk; Bifidus milk.

Reference: Hansen, R. 1985. Bifidobacteria have come to stay. *North European Dairy Journal* **3:**79-83.

CULTURED BUTTERMILK, buttermilk, commercial buttermilk, cultured skim milk, cultured milk (En); Babeurre artificiel, babeurre battue, lait écremé acidifié (Fr); Geschlagene Buttermilch (De).

Short Description: United States (origin), mainly produced in the United States, the Netherlands, Denmark, Czechoslovakia, Germany; nontraditional product; cow's milk (skim milk, lowfat milk, milk); beverage; liquid; pleasant mildly acidic aroma and taste, may contain added butter granules (*see* Flake buttermilk), fruits, and flavors.

History: In the mid-1950s, whenever the demand for "true buttermilk" exceeded its supply, a similar product was made by culturing milk with suitable bacteria.

Cultured buttermilk is not a by-product of churning. The term is somewhat misleading; *cultured milk* or *cultured skim milk* would be more appropriate designations. Much of the confusion and public misunderstanding may be removed in the future as a result of the U.S. Food and Drug Administration (FDA) rule on January 30, 1981, to do away with the term *cultured buttermilk.* The FDA has now established standards for identifying cultured milk, cultured low-fat milk, cultured skim milk. The product has been produced since the early 1900s but gained commercial significance in the United States only after World War II. Sales have decreased considerably since the mid-1950s. Buttermilk consumption data for other countries have been compiled by the International Dairy Federation since the late 1970s. Unfortunately, the IDF statistics on cultured buttermilk are in some instances mixed in with true buttermilk, as sold in some countries or with other fermented milks. True buttermilk does not seem to be available to consumers in the United States.

Microbiology: The starter culture consisting of *Streptococcus lactis* and/or *S. lactis* subsp. *cremoris* in combination with *Leuconostoc mesenteroides* subsp. *cremoris* is mixed in at a rate of 1%. Low acetaldehyde-producing strains of *S. lactis* subsp. *diacetilactis* may be incorporated; some manufacturers have standardized the taste and aroma of their cultured buttermilk by adding a definite amount of a separate culture of *L. mesenteroides* subsp. *cremoris* at the time of inoculation. *L. mesenteroides* subsp. *cremoris* is a citrate fermenter and a definite diacetyl producer. At concentrations of 2-4 ppm, diacetyl imparts the characteristic "buttery" note of buttermilk and, with the other taste and aroma components, creates the totality of flavor in a high-quality cultured buttermilk.

Manufacture: Cultured buttermilk is made from skim milk or lowfat milk (approximately 1.7% fat and 9-12% nonfat solids, homogenized and pasteurized to 85°C (185°F) for 30 min or 95°C (203°F) for 3-5 min. The incubation with the above-mentioned culture is at 22°C (71.6°F) for 14-16 h until a pH of 4.6 is obtained, followed by agitating curd and packaging at 5°C (41°F); distribution is within 24 h (see also Fig. 3).

Related Products: Filmjölk, lättfil.

References:

Burke, A. D. 1926. Industrial buttermilk. *Oklahoma Agricultural Experiment Station Bulletin 156.* Cited in Le Lait, 1929, **9**:553.

Emmons, D. B., and S. L. Tuckey. 1967. *Cottage Cheese and Other Cultured Milk Products,* Pfizer Cheese Monograph, Vol. III. New York: Chas. Pfizer and Co.

Kosikowski, F. V. 1977. *Cheese and Fermented Milk Products,* 2nd ed. Ann Arbor, Mich.: Edwards Brothers, Inc.

Kosikowski, F. V. 1984. Buttermilk and related fermented milks. *IDF Bulletin Document 179,* pp. 116-119.

Office of the Federal Register. 1987. *Code of Federal Regulations,* Title 21 (Food and Drugs). Washington D.C.: U.S. Government Printing Office.

Vedamuthu, E. R. 1982. Fermented milk. In *Fermented Foods,* ed. A. H. Rose, pp. 199-224. London: Academic Press.

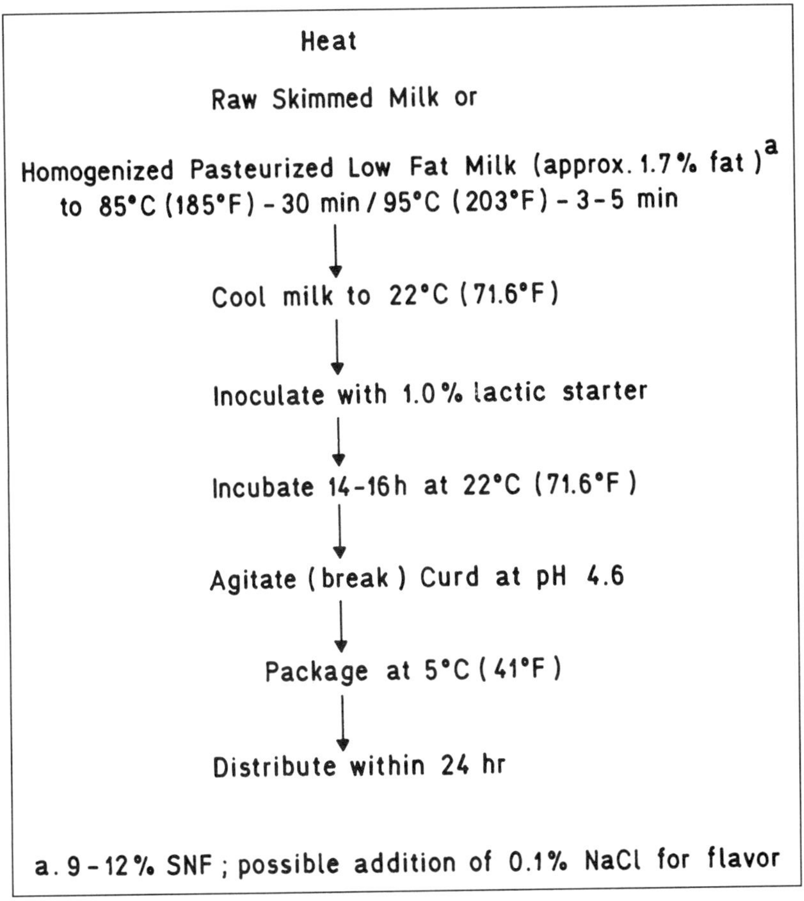

FIGURE 3. Flow diagram of cultured buttermilk. (From *F. V. Kosikowski, 1984,* IDF Bull. 179, *p. 117* [*Brussels: International Dairy Federations*].)

CULTURED CREAM, SOUR CREAM (En); Crème acidifiée (Fr); Sauerrahm (De).

Short Description: Various countries; traditional product; usually cow's milk; used as a topping or dressing on salads, fruit, and baked potatoes, in flavored dips, and in cooked and baked foods; set or stirred; firm consistency, viscous; pleasant aroma, acidic taste.

Microbiology: A cream culture consisting of *Streptococcus lactis* and/or *S. lactis* subsp. *cremoris* in combination with *Leuconostoc mesenteroides* subsp. *cremoris* is added at a rate of 1-2%. Low acetaldehyde-producing strains of *S. lactis* subsp. *diacetilactis* (diacetyl: acetaldehyde ratio greater than 3:1 and less than 4.5:1) often are incorporated in the starter. A proper balance between the acid and aroma producers is necessary.

In certain cases, cultured cream is made by using starters containing polysaccharide-producing strains intended to improve viscosity of the product or by using *Lactobacillus acidophilus* alone or in combination with a cream culture (*See* Acidophilus cream, cultured) or by using a combined cream culture and yoghurt culture.

Manufacture: Cultured cream is made from cream obtained by separating milk or from whole milk fortified with cream to the desired fat content. Culturing conditions and starter bacteria (see Microbiology above) are similar to those of cultured buttermilk. The cream is homogenized, pasteurized at 80°C (176°F) for 30 min or at 90°C (194°F) for 2-5 min and cooled to the inoculation temperature. The incubation is at 21-24°C (69.8-75.2°F) for 12-16 h until 70°Th (0.6% titratable acidity) is produced, followed by slow agitation, cooling, and packaging. Some manufacturers incubate the cream in retail packs to avoid deterioration in viscosity during cooling and packaging of the product (Kosikowski 1977; Vedamuthu 1982).

Food Value: Sour cream may be sold plain or flavored. It has a milk fat content between 12 and 30%, but some types have less than 12% fat (smetanka, cultured cream) or more than 30% fat and some types contain milk fat in combination with vegetable oil (acidophilus cream). In the United States, the milk fat content should be not less than 18% and in flavored types it should be not less than 14.4%.

Related Products: Smetana; Smetanka, cultured cream; Acidophilus cream, cultured.

References:

Kosikowski, F. V. 1977. *Cheese and Fermented Milk Foods,* 2nd ed. Ann Arbor, Mich.: Edwards Bros. Inc.

Obermann, H. 1985. Fermented milks. In *Microbiology of Fermented Foods,* ed. B. J. Wood, pp. 167-195. London: Elsevier.

Vedamuthu, E. R. 1982. Fermented milks. In *Fermented Foods,* ed. A. H. Rose, pp. 199-224. London: Academic Press.

CULTURED MOLD MILK (authors' designation) (En); Laits fermentés aux moisissures (Fr); Mit Schimmelpilzen fermentierte Milch (De).

Short Description: United States (Hawaii, Honolulu); nontraditional product; cow's milk or other milk, vegetable-base extract (e.g., soy extract);

agreeable flavor and aroma all of its own with a creamy texture and a faintly sweet but essentially bland taste.

Microbiology: *Saccharomycopsis* spp. and *Rhizopus* spp. (both are fungi). Preparation of the culture called rice wine: Rice is heated in a closed container with steam until it is completely cooked, usually about 40–50 min. The cooked rice is then cooled to about 43°C (109.4°F). Next, a starter culture of *Saccharomycopsis* spp. and *Rhizopus* spp. is added to warm water at 43°C (109.4°F) and thoroughly mixed. The resulting aqueous culture suspension is then incorporated throughout the cooked rice. The ratio of starter culture to rice is in the range of 0.0002 to 0.0015 part culture to 1 part rice, calculated on a dry rice weight basis. Warm water is then added to make a final ratio of 5 parts inoculated rice to 1 part water, based on weight. This mixture is filled and sealed into a suitable airtight container with sufficient headspace to accommodate the gases from the fermenting mass. Incubation is at about 43°C (109.4°F) for a period of 24-48 h or until a wine is produced that has an acceptable taste. At the completion of fermentation, all solids are removed from the fermented mass leaving the rice wine, which is bottled and stored under refrigerated conditions at about 5°C (41°F).

Manufacture: Milk is pasteurized and homogenized. Rice wine prepared as described above is added to the milk. Incubation is at about 43°C (109.4°F) for a period of 3 to 4 hours. However, different temperatures and incubation times may be used depending on the operator and consumer taste preferences. Completed incubation is evidenced by coagulation of the milk. The finished product is stored under refrigeration, usually at about 5°C (41°F) and has a shelf life of about 3 to 4 weeks if heat-treated after incubation.

Food Value: Related to bacteria-fermented milk products.

Related Product: Gua-nai.

Reference: Kao, A. H. 1987. Preparing naturally sweet yogurt with *Saccharomycopsis* spp. and *Rhizopus* spp. United States Patent 4,714,626.

CULTURED MOLD-CONTAINING FERMENTED MILKS (En); Laits fermentés aux moisissures (Fr); Fermentierte Milch mit Schimmelpilzen (De).

Short Description: United States (tentative details of production); nontraditional product; cow's milk.

History: There is no evidence worldwide for the existence of any traditional fermented milks prepared solely with molds. However, recently several new products have been developed that are exclusively made with mold starter cultures (*see* Cultured mold milk; Gua-nai).

Microbiology: The genera and species utilized for these products are as follows: genus Aspergillus (*A. oryzea*), genus Amylomyces (*A. rouxii*), genus Geotrichum (*G. candidum*), genus Pleurotus (*P. ostreatus*), genus Rhizopus (*R. oryzea, R. javanicus*).

Some fermented milks contain a mixed microflora of different species: mainly *Geotrichum candidum* (*see* Viili), or milk-vegetable mixtures produced with mold-fermented vegetables (the molds are usually destroyed by subsequent heating).

Manufacture: Certain general characteristics are or should be recognized in the production of fermented preparations involving molds:

Coagulating the milk takes several days, perhaps more than a week.
The mold(s) used must not be producers of mycotoxins.
Some governments have regulations requiring approval prior to use or special status for the mold, such as "generally recognized as safe" (GRAS).
The taste is usually that of moldy food.
The color of the resulting product may vary and can be affected by manufacturing procedures.
Surface color of the product may differ from internal color, especially after storage.
The possibility exists that certain species or strains of mold may be hazardous to human health. This potential should be recognized, but there should be no blanket discrimination against molds in future developments with fermented milk products.

Food Value: Some strains could have some probiotic properties, such as antitumor effect or anticholesterolemic effect.

Related Products: Cultured mold milk; gua-nai.

References:

Demeter, K. J., and H. Elbertzhagen. 1968. *Fundamentals of Dairy Microbiology* (in German). Hildesheim: Th. Mann GmbH.

Kurmann, J. A. 1989. Microorganisms of fermented milks, general aspects, bacteria, yeasts, moulds, less-known strains and microorganisms. In *Fermented Milks: Current Research.* International Congress, Palais des Congrès, Paris, France, December 16-18, 1989, pp. 3-10. London: Eurotext. (Also in French, published by John Libbey, Paris.)

Riviere, J. 1975. *Industrial Applications in Microbiology* (in French). Paris: Masson et Cie.

Yoneya, T., and Y. Sato. 1982. Basic research on the production of alcoholic cultured milk using *Rhizopus javanicus,* an organic acid- and alcohol-producing fungus (in Japanese). *Japanese Journal of Dairy and Food Science* **31:** A167-A173. Cited in *Dairy Science Abstracts,* 1983, **45:**679-680.

CULTURED NATURAL BUTTERMILK (En); Babeurre fermenté (Fr); Gesäuerte Buttermilch (De).

Short Description: USSR; nontraditional product; sweet buttermilk based on cow's milk; liquid, beverage; acidic taste, agreeable flavor.

Microbiology: Culture of mesophilic lactic acid streptococci incorporating aroma producers.

Manufacture: A beverage obtained by fermenting natural buttermilk (the by-product of sweet cream butter manufacture) with cultures of lactic streptococci. It is made from sweet buttermilk, pasteurized at 85-87°C (185-188.6°F) for 5-10 min or at 90-93°C (194-199.4°F) for 2-3 min, cooled to 22-26°C (71.6-78.8°F) and inoculated with 1-5% of mixed culture. The incubation takes 12-18 h, until the acidity reaches 85-90°Th (0.76-0.81% titratable acidity), and is followed by stirring, cooling, and packaging.

Food Value: The final product has a maximum acidity of 90-110°Th (0.81-1.00% titratable acidity); it contains not more than 0.5% fat and not less than 8.0% nonfat solids.

Related Products: Sometimes acid buttermilk (the by-product of ripened cream butter manufacture) is used for culturing provided that its initial acidity does not exceed 50°Th (0.45% titratable acidity); the amount of starter culture is less than that used in the culturing of sweet buttermilk.

Reference: Bogdanova, G. J., and E. A. Bogdanova. 1974. *New Whole Milk Products of Improved Quality* (in Russian). Moscow: Pishchevoĭ Promȳshlennosti.

D

DAHI (En, Fr, De); dadhi, dadih, numerous derivatives of the name can be found.

Short Description: India; traditional product; buffalo, cow's and mixed milk; set, stirred, or liquid as beverage; a yoghurtlike fermented milk; numerous preparations; pleasant flavor, acidic taste or acid sweet taste.

History: One of the oldest traditional fermented milks, mentioned 300 B.C. to 75 A.D., Maurya and Sunga period (Prakash 1961).

Microbiology: The starter culture consists of mesophilic lactic streptococci-containing aroma producers and with or without leuconostocs or with added *Lactobacillus delbrueckii* subsp. *bulgaricus* or *Streptococcus thermophilus* or both.

Manufacture: The product is made for direct consumption or for churning (usually the top creamy portion) into indigenous butter called *makkhan.* When made for direct consumption, dahi is produced from whole or skim milk and is consumed as a part of daily meals or as a beverage; in the latter case the product is stirred and usually mixed with water and consumed with or without sugar. Commercially available products include mild acid, acid, sweetened, and carbonated dahi.

A *mild acid dahi,* is made from cow's, buffalo, or mixed milk, which is standardized to 2.5-3.0% fat and 10% nonfat solids. The milk is homogenized, pasteurized at 80-90°C (176-194°F) for 15-30 min, cooled to 22-25%°C (71.6-77°F) and inoculated with 1-3% of a mixed culture consisting of

Streptococcus lactis and/or *S. lactis* subsp. *cremoris* and *S. lactis* subsp. *diacetilactis* in combination with or without *Leuconostoc* species. It is then mixed well, filled into containers and incubated at 22-25°C (71.6-77°F) for 16-18 h until 70-80°Th (0.60-0.70% titratable acidity) is produced, followed by cooling and cold storage. The final product has an acidity of 85-95°Th (0.75-0.85% titratable acidity).

Acid dahi is made in the same manner as a mild acid dahi, except that the starter culture contains *L. delbrueckii* subsp. *bulgaricus* or *S. thermophilus,* or both, in addition to mesophilic lactic streptococci and aroma producers. The final product has an acidity of 110°Th or more (1.0% or more titratable acidity). A good-quality dahi is yellowish creamy-white when based on cow's milk and creamy white for buffalo milk. It is smooth and firm and has a shelf life of one week under cold storage at 5-10°C (41-50°F).

Sweetened dahi is produced and sold in the eastern part of India under the name *misti dahi, lal dahi,* or *payodhi.* It has a characteristic brown color, a cooked and caramelized flavor, and a firm body. It is made from cow's milk or mixed cow and buffalo milk containing 6.25% added sucrose. This milk is boiled and then cooled. Intense heating causes water loss and concentration of the milk solids as well as browning. Artificial color and caramelized sugar are added during manufacture. The milk is fermented with a culture incorporating mesophilic lactic streptococci and aroma producers at room temperature for 15-16 h, followed by cooling and cold storage.

A *carbonated dahi* with a shelf life of 15-30 days without refrigeration has been developed. It is made from good-quality milk, which is heat-treated at 100°C (212°F) for 3-5 min, cooled to 30-35°C (86-95°F), then inoculated with 1.0% of a mixed culture consisting of *Streptococcus lactis* and/or *S. lactis* subsp. *cremoris* together with *S. lactis* subsp. *diacetilactis.* After mixing, the milk is bottled (up to the neck of the bottle) and then carbon dioxide is bubbled through the milk at 1 psi for 1 min, followed by capping and incubating at 25-30°C (77-86°F) for 16-18 h until a firm coagulum has been produced.

Food Value: Similar to yoghurt.

Related Product: Yoghurt.

References:

Prakash, O. M. 1961. *Food and Drinks in Ancient India.* Delhi-6: Munshi Ram Manohar Lal. Oriental Booksellers and Publishers.

Rangappa, K. S., and K. T. Achaya. 1975. *Indian Dairy Products,* 2nd ed. Bombay: Asia Publishing House.

Sukumar, D. 1980. *Outlines of Dairy Technology,* pp. 404-410. Delhi: Oxford University Press.

DA-RA (En, Fr, De).

Short Description: Tibet; traditional domestic product; buttermilk.

Reference: Ränk, G. 1969, p. 10.

DEMETER (FERMENTED FRESH MILK) PRODUCTS (En); Produits de la marque Demeter (Fr); Demeter Produkte (De); in Greek mythology Demeter is the goddess of agriculture; trade name.

Short Description: Switzerland; nontraditional product; alternative fermented fresh milk products; cow's milk; production at the farm; set or stirred; characteristic fresh flavor, pleasant tasting products.

History: Milk for these products is obtained according to Dr. Rudolf Steiner's biodynamic agricultural production method. All Demeter products (fermented milks, quarg, etc.) are made according to strict requirements, such as raw milk, special culture selection.

Microbiology: The starter culture consists of *Streptococcus lactis* or yoghurt bacteria (*S. thermophilus* and *Lactobacillus delbrueckii* subsp. *bulgaricus*). The first is used for the manufacture of mesophilic fermented milks and the second for the manufacture of yoghurt. The first is prepared as follows: Raw cow's milk is incubated in a glass container at 25°C (77°F) until spontaneous coagulation occurs, followed by laboratory isolation of *S. lactis* bacteria. Several transfers of the cultures are made into sterile, filtered skim milk. At the farm, cultures are put in a glass container, then fresh raw milk is added; incubation is at 25°C (77°F) until coagulation occurs (max. 10–12 h). Mother culture is prepared thus: After coagulation, ⅓ is transferred into another glass container, fresh raw milk is added and it is incubated at 25°C (10–12 h) until coagulation occurs, followed by cooling in a refrigerator. The culture is transferred every 2-4 days to renew the culture.

The other ⅔ of the coagulum is used in production. It is distributed over several glass containers in sufficient amounts to allow for effective inoculation.

Yoghurt culture is prepared in the conventional manner (*see* Yoghurt).

Manufacture: Mesophilic fermented milk is made by inoculating raw milk in the milk cans with 1% of a culture, filling the inoculated milk into glass jars or any other vessels, incubating at 25°C (77°F) for 10–12 h until coagulation occurs, then stirring if a stirred product is desired, and cooling to 4-6°C (39.2-42.8°F). The product has a shelf life of 1-2 weeks. Yoghurt is made from pasteurized milk that is inoculated with 2-4% of a yoghurt culture, followed by incubation at 42°C (107.6°F) until coagulation occurs (about 3 h), stirring if a stirred product is desired, adding homemade fruit concentrates, and cooling to 4-6°C (39.2-42.8°F).

Demeter mesophilic fermented products are made from raw milk, and despite great care, a potential health risk caused by contamination with pathogens exists. Therefore, tests for the absence of pathogenic bacteria in raw milk and in the finished products must be performed on a regular basis. Milk to be fermented spontaneously (for producing cultures) always contains enough acid-producing bacteria to bring about souring. However, a certain and variable number of yeast cells and other bacteria is always present and characteristic for these types of product.

Food Value: Similar to yoghurt.

Reference: Hauert, W. 1981. *Quality Aspects of Demeter Milk and Demeter Quarg* (in German). Dornach, Switzerland: Verlag Produzentenverein für biologisch-dynamische Landwirtschaftsmethode.

This entry was contributed by Dr. W. Hauert, Ittigen, Bern.

DESSERT PRODUCTS (En); Produits desserts (Fr); Dessertprodukte (De).

Short Description: Various countries; nontraditional products; mainly cow's milk; products with special pleasant organoleptic properties such as viscosity, consistency, flavor, and appearance.

History: There is a long tradition of preparing desserts from milk and cream but not from fermented milk. Commercially produced desserts based on fermented milk have become popular in many parts of the world since the early 1980s. They are generally considered the last course of a meal and, in addition to cheese and fruit, include sweets such as puddings, pastries, frozen confections, and others.

Microbiology: The starter culture consists of yoghurt bacteria: *Streptococcus thermophilus* and *Lactobacillus delbrueckii* subsp. *bulgaricus.*

Manufacture: Several problems can arise in the manufacture of dessert products:

Excessive lactic acid development during manufacture or afterward during storage (after-souring) can negatively affect the organoleptic properties (flavor, color, texture)

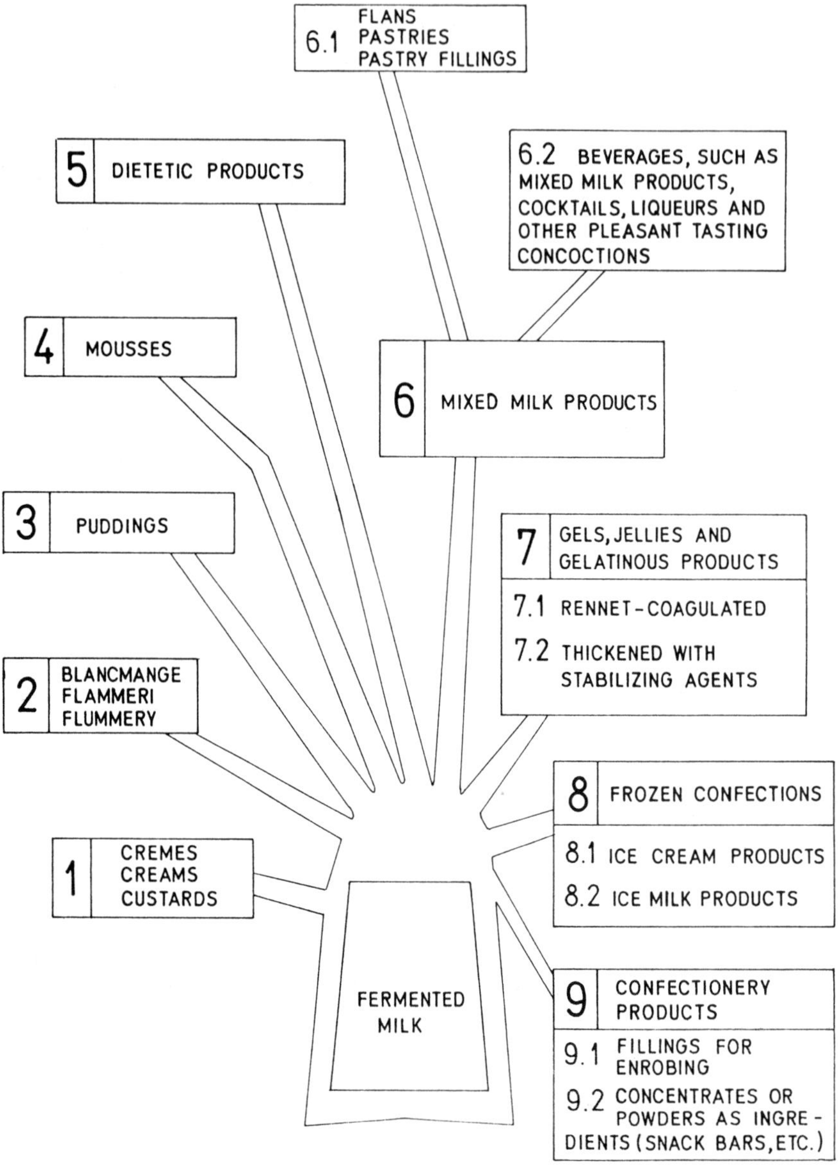

FIGURE 4. Family tree and types of fermented milk product desserts. (From *J. A. Kurmann, 1985, A review of dairy desserts,* Deutsche Molkerei-Zeitung, *106, p. 646.*)

Specific manufacturing steps or techniques may have to be adapted, such as use of stabilization systems
Hot packing or hot filling is not possible, since it would destroy the culture organisms. (Hot-packing is recommended when longer shelf life is desired for conventional products).

Dessert products can be classified according to traditional culinary terminology, which differentiates mainly by consistency and taste (Fig. 4). It is reasonable to assume that a pleasant-tasting fruit yoghurt or flavored yoghurt can be considered a dessert product (see point 6 in Fig. 4). Currently, most fermented milk-based desserts are derivatives of yoghurt (e.g., frozen yoghurt, yoghurt flans, yoghurt cremes), probably because of the relative rapidity of yoghurt incubation and the general absence of contaminating microorganisms.

Food Value: The most important advantages of fermented milk-based dessert products are

Better storage ability (shelf life) because of the lactic acid content
Presence of an active culture with selected, live bacteria
Better digestibility because some milk components have been partially degraded by the bacterial fermentation
New taste sensations can be created because milk fermentation results in the development of new flavorful substances.

Related Products: Cocktail yoghurt, frozen yoghurt, jellied yoghurt, soft-serve yoghurt, yoghurt chocolate, yoghurt ice, yoghurt pudding, yoghurt sherbet.

References:

Kurmann, J. A. 1985. A review of dairy desserts (in German). *Deutsche Molkerei-Zeitung* **106**(21):646-653.
Montagné, P. 1961. *Larousse Gastronomique, The Encyclopedia of Food, Wine and Cookery.* New York: Crown Publishers.
Rašić, J. Lj., and J. A. Kurmann. 1978, pp. 325-362.

DETSKAYA (En, Fr, De).

Short Description: USSR; nontraditional product; cow's milk; an acidophilus product intended for children older than 1 year (*see* Acidophilus cream, cultured).

DIABETIC YOGHURT (En); Yaourt pour diabetique (Fr); Joghurt für Diabetiker (De).

Short Description: Western countries; nontraditional product; cow's milk; plain or flavored yoghurt.

Microbiology: The starter culture consists of yoghurt bacteria, such as *Streptococcus thermophilus* and *Lactobacillus delbrueckii* subsp. *bulgaricus.*

Manufacture: Diabetic yoghurt is prepared from skim milk or partially skimmed milk. When flavored, the product is sweetened with fructose instead of sucrose or glucose. Fructose is 150% as sweet as sucrose and is allowed for diabetics up to 60 g/day. Sometimes saccharin (350 times sweeter than sucrose) is used in combination with fructose. Nonsugared flavorings such as fruit flavors, vanilla, or coffee are typically used rather than fruit syrups or fruit concentrates, which are high in sugar content.

Food Value: Low-calorie product for diabetics. Sucrose or glucose not used for sweetening the product.

Reference: Rašić, J. Lj., and J. A. Kurmann. 1978, pp. 103, 350-351.

DIETETIC FERMENTED MILKS (En); Laits fermentés diétetiques (Fr); Diätetische Sauermilchprodukte (De).

Short Description: Various countries and continents; nontraditional products; mainly cow's milk; for special dietetic and nutritional purposes.

Microbiology: The starter culture consists of yoghurt bacteria such as *Streptococcus thermophilus* and *Lactobacillus delbrueckii* subsp. *bulgaricus,* or other lactic acid bacteria or the appropriate selected intestinal bacteria depending on the product to be made.

Manufacture: Many of the available dietetic products could be prepared in a fermented version or based on fermented products to make them more digestible, more wholesome, and better tasting. If the fermented products were prepared with the appropriate selected intestinal bacteria starter culture, they would enhance the dietetic value.

Dietetic fermented milk products can be grouped according to their chemical composition and their physiological application in the diet. Included in the grouping by chemical composition are fermented milk products with a reduced content of one or more milk components (fat, calories, carbohydrate, body electrolyte) and fermented milk products fortified with one or more dietary components (fiber, vitamins, phospholipids, minerals, other bioactive substances). The physiological application in the diet includes baby food, food for special diet, fortifier and dietetic breakfast or beverages, and special preparation for athletes.

The possibilities regarding dietetic preparations of fermented milk products have not been systematically or sufficiently investigated. Only a small number of dietetic products have been developed for adults (weight reduction, athletic activity, etc.) and specifically for elderly people; and only a few of the presently manufactured dietetic preparations are based on fermented milk. It is anticipated that this will be different in the future as dietetic, health, and food industry professionals increasingly recognize the importance of fermented milks as a growth area in geriatrics and medicine.

Food Value: Dietetic foods are formulated and prepared to meet the particular nutritional needs of persons whose normal processes of assimilation or metabolism are modified or for whom a particular effect is to be obtained by the controlled intake of food, food components, or certain nutrients. They may be formulated for patients suffering from physiological disorders or for healthy persons with additional needs (Bender 1975). Dietetic foods are mainly used as components of a well-balanced and selected diet.

We feel that a dietetic food should meet the following requirements:

- Address a specific nutritional disorder
- Feature the presence or absence of at least one component whose nutritional significance in health and disease is known
- Conform to the respective national legislative and regulatory requirements for dietetic products
- Be manufactured with care, in order to avoid heat damage or physical damage to ensure purity of raw material and additions, to guarantee acceptable quality of the final product (bacteriological, chemical, nutritional, and organoleptic), and to prevent interaction with packaging materials
- Display appropriate labels (including designation of dietetic product for use in a special diet, as well as medical contraindication)

References:

Bender, A. E. 1975. *Dictionary of Nutrition and Food Technology.* London: Butterworths.
Kurmann, J. A., and J. Lj. Rašić. 1988. Technology of fermented special products. *IDF Bulletin Doc. 227* (Technology of Fermented Milks), pp. 101-109. Brussels: International Dairy Federation.
Rašić, J. Lj., and J. A. Kurmann. 1978, pp. 350-356.
Rašić, J. Lj., and J. A. Kurmann. 1983, pp. 81-101.

DIPHILUS MILK (En); Lait diphilus (Fr); Diphilus Milch (De); denotes milk with both *Lactobacillus acidophilus* and *Bifidobacterium* spp.

Short Description: France; nontraditional product; cow's milk; specific taste and aroma.

Microbiology: The starter culture consists of intestinal strains of *Lactobacillus acidophilus,* optionally supplemented with *Bifidobacterium bifidum.*

Manufacture: *See* Acidophilus milk; Bifidus milk.

Food Value: Used in the therapy of intestinal disorders.

Related Products: Acidophilus milk products, bifidus milk products.

Reference: Prevot, A. R. 1971. Cultured acidophilus milk for dietetic use in hospitals (in French). *Information Dietetique* **8**(2):19-21.

DNEPRYANSKY, DNEPROVKSII (En, Fr, De); refers to the river Dnieper.

Short Description: USSR (Ukraine); nontraditional product; cow's milk; symbiotic starters of lactic acid streptococci and leuconostocs; for manufacturing low-fat fermented milk products; viscous consistency and agreeable flavor.

Microbiology: The starter culture consists of selected strains of *Streptococcus lactis, S. lactis* subsp. *cremoris,* and *Leuconostoc* spp. isolated from original fermented milks and is characterized by acid and aroma production as well as the ability to produce viscous consistency (presence of polysaccharide-producing strains). Starters containing strains in symbiotic relationships are called *Dnepryansky* (*see* Smetanka and Aerin). Strains of *Lactobacillus casei, L. acidophilus,* and *Acetobacter aceti* were introduced in some types of these starters.

Manufacture: Used for the production of low-fat fermented milk products. The manufacture is similar to that of products fermented by mesophilic microorganisms.

Reference: Romanskaya, N. N., R. S. Bashkirova, G. S. Dyment, L. D. Tovkachevkaya, and S. J. Kochubey. 1982. Improvement of low-fat cultured milk products. *Proceedings 21st International Dairy Congress* **1**(1):306. Moscow: Mir Publishers.

DOUGH (En, Fr, De).

Short Description: Iran, Afghanistan; traditional product; cow's milk and probably other milks; beverage comprising yoghurt and water (sometimes whey is added) plus spices and/or salt.

References:

Kosikowski, F. V. 1978. Cultured milk foods of the future. *Cultured Dairy Products Journal* **13**(3):5-7.

Lamé, H., and M. Hekmati. 1978. Contribution to the study of changes in Iranian dough during storage: changes in nitrogen. *Proceedings 20th International Dairy Congress* **E:**785-786. Brussels: International Dairy Federation.

DRIED YOGHURT (En); Yaourt seché (Fr); Joghurt getrocknet (De).

Short Description: Middle East; traditional product in some sense; cow's milk or other kinds of milk; direct consumption following reconstitution with water or use in manufacturing other foods and preparations; dried product; increased storage life.

Microbiology: The starter culture consists of *Streptococcus thermophilus* and *Lactobacillus delbrueckii* subsp. *bulgaricus.*

Manufacture: Drying yoghurt is a very old method of its preservation. Presently yoghurt is dried using modern technology and equipment. The powder is reconstituted by mixing and stirring with the original amount of water. Generally, two methods of drying can be used: freeze drying or spray drying.

Freeze drying involves removing water from the frozen product under a high vacuum (*see* Freeze-dried Preparations). The final product is a fine powder containing about 2% moisture. It is packaged in suitable containers, preferably in an atmosphere of inert gas (e.g., nitrogen). The reconstituted product has a weaker consistency than the original yoghurt and the intensity of flavor in plain yoghurt is reduced when compared to the initial yoghurt. Spray drying involves removing water from concentrated yoghurt using a high-temperature chamber. The temperature of spray drying yoghurt, however, is lower than in the case of drying milk. The survival rate of culture bacteria can be expected to be about 20%. The final powder contains approximately 4% moisture and has a pH value of 4.2-4.5. The weak consistency of the reconstituted product may be improved by using stabilizers.

Food Value: Similar to yoghurt.

Related Products: Labneh anbaris and kashk.

Reference: Rašić, J. Lj., and J. A. Kurmann. 1978, pp. 293-295.

E

EGG-MILK PRODUCT, CULTURED (En); Produit fermenté aux oeufs-lait (Fr); Fermentiertes Eier-Milchprodukt (De).

Short Description: Australia; nontraditional product; egg-milk; according to the patentholder the objective was to make available an alternative yoghurt product.

Microbiology: The starter culture consisting of *Lactobacillus delbrueckii* subsp. *bulgaricus* LB_1 and *Streptococcus thermophilus* TS_2 is mixed in at a rate of 3.33% and 1.66%, respectively.

Manufacture: After incorporating the ingredients (except egg) in the milk at 50°C (122°F), the mixture is homogenized at about 2500 lb/in^2, heated to about 85°C (185.0°F) for 30 min, and then cooled to 45°C (113.0°F) with the egg being stirred in, preferably at less than 70°C (158.0°F). The incubation of mix with added culture is at 45°C (113.0°F) to pH 4.9-5.0. After the product is cooled to 4°C (39.2°F), flavors may be incorporated.

Food Value: This patented egg-milk product contains (by wt.) 15% liquid whole egg (or equivalent powdered egg), 6% dried skim milk, 4% sugar, 0.2% stabilizer (e.g., hydrocolloid gums), and 74.8% milk with preferably 3% fat.

Reference: Mackenzie, K. A. 1983. Cultured egg-milk product. *UK Patent* 2,116,819. cited in *Dairy Science Abstracts,* 1984, **44:**46.

EGYPTIAN FERMENTED MILK PRODUCTS (En); Produits laitier fermentés de l'Egypt (Fr); Ägyptische fermentierte Milchprodukte (De).

Short Description: Egypt; traditional and nontraditional products; mainly buffalo milk.

History: Certain soured milk products are frequently mentioned in medical prescriptions used by the ancient Egyptians (Darby et al. 1977).

Manufacture: In Egypt the following fermented fresh milk products are made: laban khad or laban kerbah, sour buttermilk; laban rayeb or laban matrad, sour milk, partially skimmed; laban zeer, drained (concentrated) buttermilk made from laban khad; labneh or lebneh; laban zabady; kishk made from laban zeer; kishk seiamy, vegetable product; zabady, the major Egyptian fermented milk, a kind of yoghurt made in Egypt and Sudan.

References:

Abou-Donia, S. A. 1984. Egyptian fresh fermented milk products. *New Zealand Journal of Dairy Science and Technology* **19:**7-18.

Darby, W. J., P. Ghalioungui, and L. Grivetti. 1977. *Food: The Gift of Osiris.* Vol. 2, p. 775. London, New York: Academic Press.

El-Gendy, Sh. M. 1983. Fermented foods of Egypt and the Middle East. *Journal of Food Protection* **41:**358-367.

ELEDON (En, Fr, De). *See* Baby foods, fermented.

ELVIT (En, Fr, De), trade name.

Short Description: Czechoslovakia; nontraditional beverage; cow's milk; for preschool children; liquid.

Microbiology: The starter culture consisting of *Streptococcus lactis, S. lactis* subsp. *cremoris, Leuconostoc mesenteroides* subsp. *cremoris* and/or *S. lactis* subsp. *diacetilactis* contains an activated culture of *Propionibacterium freudenreichii* subsp. *shermanii* and additionally a kefir or acidophilus culture.

Manufacture: Elvit is a fermented milk beverage biologically enriched with vitamin B_{12} and folic acid. It is made by fermenting milk with the abovementioned cultures.

Food Value: Researchers have reported that vitamin enrichment was most pronounced when a kefir culture was used with the beverage, resulting in thirty times more vitamin B_{12} and seven times more folic acid than is found in sweet unfermented milk. When the beverage was made with an acidophilus culture instead, the vitamin enrichment was less.

In clinical tests extending over three years and involving preschool children, an enhanced resistance against some respiratory diseases was observed.

Reference: Černa, J. 1984. Elvit—fermented milk beverage (in Czech). *Prumysl Potravin* **35**:192-195. Cited in *Dairy Science Abstracts,* 1985, **47**:14.

ENPAC (En, Fr, De), trade name.

Short Description: United Kingdom; freeze-dried *Lactobacillus acidophilus;* pharmaceutical preparation.

Microbiology: The starter culture consists of intestinal strains of *L. acidophilus* claimed to be resistant to ten different antibiotics.

Manufacture: A pharmaceutical preparation obtained by freeze drying a culture of *L. acidophilus.* The product contains large numbers of viable *L. acidophilus* cells.

Therapeutic Value: The product is used both during and after antibiotic therapy as a protection against deleterious side effects of antibiotic therapy in infants. Reportedly beneficial clinical results have been obtained against staphylococcal enterocolitis and moniliasis (Gordon et al. 1957; Hawley et al. 1959). The preparation is also used in the treatment of diarrhea and other gastrointestinal disorders.

References:

Gordon, D., J. Macrae, and D. M. Wheater. 1957. A lactobacillus preparation for use with antibiotics. *Lancet* **273**:899-901.

Hawley, H. B., P. A. Shepherd, and D. M. Wheater. 1959. Factors affecting the implantation of lactobacilli in the intestine. *Journal of Applied Bacteriology* **22**:360-367.

Hoffmann, H. M. 1956. Clinical experiences with infant dyspepsia and a preparation containing antibiotic-resistant *Lactobacillus acidophilus* organisms, with particular emphasis on antibiotic therapy (in German). *Die Medizinische Woche* **7**:296.

ENTEROCOCCI-CONTAINING FERMENTED FRESH MILK PRODUCTS (En); Produits contenant des entérocoques (Fr); Enterokokken enthaltende Produkte (De).

Short Description: Various countries; traditional (with fecal bacterial infection) and nontraditional products; various milk types.

Microbiology: *Enterococcus,* a proposed genus of gram-positive organisms, facultative anaerobic lactic acid streptococci-like bacteria. Actually they belong to the genus *Streptococcus* and are also classified as Lancefield group D streptococci. Members of the genus (e.g., *Streptococcus feacalis* resp. *E. faecalis* and *S. faecium* resp. *E. faecium*) occur in the lower human and animal intestine. They can usually grow at 10-45°C (50.0-113.0°F), in 6.5% NaCl and at pH 9.6. The metabolism, with exceptions, is typically fermentative.

Food Value: The enterococci (*S. faecalis* and *S. faecium*) may act as opportunistic pathogens (implicated in some cases of bacteremia, and gallbladder, wound, and urinary tract infections, as well as subacute bacterial endocarditis in man) and in some cases as food-borne bacterial pathogens. Only seriously selected strains may be utilized for food uses. Some researchers critically consider the use of enterococci for human consumption (Bryant 1979). Enterococci are frequently encountered as contaminants of industrially prepared fermented fresh milk products. They may also be found as components of homemade fermented fresh milk products without a defined microflora.

Products: Enterococci-containing products are dairy products (*see* Labneh) and pharmaceutical preparations (*see* Biofforine and Paraghurt).

References:

Bryant, F. L. 1979. Other bacteria. In *Foodborne Infections and Intoxications,* ed. H. Riemann, and F. L. Bryant, pp. 242-249. New York: Academic Press.

El-Samragy, Y. A., E. O. Fayed, A. A. Aly, and A. E. A. Hagrass. 1988. Properties of labneh-like products manufactured using *Enterococcus* starter culture as novel dairy fermentation bacteria. *Journal of Food Protection* **51:**386-390.

Kurmann, J. A. 1989. Microorganisms of fermented milks, general aspects, bacteria, yeasts, mould, less-known strains and microorganisms. In *Fermented Milks: Current Research.* International Congress, Palais des Congrès, Paris, France, December 16-18, 1989, pp. 3-10. London: Eurotext. (Also in French, published by John Libbey, Paris.)

Saraswat, D. S., G. W. Reinhold, and W. S. Clark. 1965. Relationship between enterococcus, coliform and yeast and mold counts in butter. *Journal Milk Food Technology* **28:**245-249.

ERGO (En, Fr, De).

Short Description: Ethiopia; traditional, yoghurtlike product; microbial state is not well defined; fresh milk is incubated at room temperature for 2 days until soured in a kettle previously treated by inverting over a piece of smoldering olive wood.

Reference: Vogel, S., and A. Gobezie. 1977. Ethiopian fermented milks. *Symposium on Indigenous Fermented Foods.* Bangkok, Thailand. Cited in Steinkraus, K. H. 1983. *Handbook of Indigenous Fermented Foods,* p. 267. New York and Basel: Marcel Dekker.

EUGALAN TÖPFER FORTE (En, Fr, De); trade name.

Short Description: Germany; nontraditional product; cow's milk; freeze-dried pharmaceutical preparation.

Microbiology: Contains viable bifidobacteria.

Food Value: The product is fat free and gluten free. The composition per 100 g is 9.40 g milk protein, 0.70 g plant protein, 62.60 g lactose, 5.90 g lactulose, and 3.00 g mineral salts.

Therapeutic Value: Used in the management of liver cirrhosis and to treat chronic constipation, as well as disturbed balance of the intestinal flora after antibiotic and irradiation therapies.

Reference: Rašić, J. Lj., and J. A. Kurmann. 1983, p. 139.

EUGA-LEIN TÖPFER (En, Fr, De); trade name.

Short Description: Germany; nontraditional product; cow's milk; freeze-dried pharmaceutical preparation (produced since 1973).

Microbiology: Contains viable bifidobacteria.

Food Value: The product is gluten free. The composition per 100 g is 1.80 g fat, 15.10 g protein, 71.10 g carbohydrate, 3.10 g mineral salts, 4.80 g dietary fiber, 0.20 g vitamin C.

Therapeutic Value: Gluten free. Used to treat chronic constipation (instead of with purgatives) in children.

Reference: Rašić, J. Lj., and J. A. Kurmann. 1983, p. 139.

EWE'S MILK FERMENTED PRODUCTS (En); Produits fermentés à base de lait de brebis (Fr); Fermentierte Schafmilchprodukte (De).

Short Description: Numerous countries; traditional and nontraditional products; ewe's milk.

History: Domestication of sheep occurred approximately 8500 B.C. in Zawa Tschemi Schanidar (now Iraq). Exact date of first milking of ewes is unknown. The first sour milk from ewe's milk was probably prepared before 2000 B.C. Today, sheep are found all over the world, but they do not tolerate a very hot climate. The length of lactation is 250-280 days with very variable milk production among the breeds and different animals, for example, 80-100 kg/ewe/year.

Microbiology: The starter culture consists of *Streptococcus thermophilus* and *Lactobacillus delbrueckii* subsp. *bulgaricus.*

Manufacture: *See* Yoghurt.

Food Value: The milk's chemical composition varies: 13.3-19.3% total solids, 5.0-7.4% fat, 4.7-6.5% protein, 4.5-4.6% casein, 4.2-4.8% lactose, and 0.8-1.0% mineral matter.

Products: The most widely known products are mentioned in Appendix F.

References:

Kurmann, J. A. 1986. Yoghurt made from ewe's and goat's milk. *IDF Bulletin 202* (Production and Utilization of Ewe's and Goat's Milk), pp. 153-166. Brussels: International Dairy Federation.

Leonard, J. N. 1982. *The First Farmer* (in German). Amsterdam: Life Books.

Rašić, J. Lj., and J. A. Kurmann. 1978, pp. 71-73, 96, 107.

EWE'S MILK YOGHURT WITH A CREAMY SKIN LAYER ON TOP (En); Yaourt au lait de brebis avec une peau de lait cremeuse (épaisse) à la surface (Fr); Schafsmilch-Joghurt mit einer rahmangereicherten, dicken Milchhaut an der Oberfläche (De).

Short Description: Greece, Turkey, Yugoslavia; traditional product; ewe's milk; firm consistency and pleasant taste.

Microbiology: The starter culture consisting of *Streptococcus thermophilus* and *Lactobacillus delbrueckii* subsp. *bulgaricus* is mixed in at a rate of 2-3%.

Manufacture: The product is made from whole, filtered ewe's milk, which is pasteurized at 95°C (203°F) for 5-10 min in an open vat, followed by warm filling into plastic containers or earthenware bowls, cooling to 40-50°C (104-122°F) and inoculation with a yoghurt culture. The incubation is at 42-45°C (107.6-113.0°F) for about 3 h until 90-105°Th (0.81-0.95% titratable acidity) is produced, followed by cooling to 4-5°C (39.2-41.0°F) in a refrigerated chamber, and preparation for distribution (lids placed on plastic containers).

Reference: Kurmann, J. A. 1986. Yoghurt made from ewe's and goat's milk. *IDF Bulletin 202* (Production and Utilization of Ewe's and Goat's Milk), pp. 153-166. Brussels: International Dairy Federation.

F

FECAL STREPTOCOCCI. *See* Enterococci-containing fermented fresh milk products.

FELISOWKA (En, Fr, De).

Short Description: Poland; nontraditional product; sparkling beverage; fresh buttermilk, derived from cow's milk.

Microbiology: The starter culture consists of a cream culture (*Streptococcus lactis* and/or *S. lactis* subsp. *cremoris* in combination with *Leuconostoc mesenteroides* subsp. *cremoris* and/or *S. lactis* subsp. *diacetilactis*) and a baker's yeast.

Manufacture: Natural buttermilk is postacidified to 112.5-125°Th (1.013-1.125% titratable acidity), then 8% sucrose and bakers' yeast are added and the mixture is fermented at 18-20°C (64.4-68°F) to 137.5-150°Th (1.237-1.350% titratable acidity) and 0.4-1.2% alcohol content, followed by cooling to 8°C (46.4°F). There are no indications about how the buttermilk is postacidified.

References:

Lang, F., and A. Lang. 1971. Fermented dairy beverages and specialties. *Milk Industry* **69**(2):13-15.

Urbanski, Z. 1970. *Manufacture of Milk Beverages* (in Polish). Produkcja napajow mlecznych. 89 pp. Warsaw: Zaktad Wydawictw CRS. Cited in *Dairy Science Abstracts,* 1971, **33**(3940):594.

FEMILACT (En, Fr, De). *See* Bifidus baby foods.

FERMENTED MILK-VEGETABLE MIXTURES (En); Mélanges lait-végétaux fermentés (Fr); Fermentierte Mischung von Milch mit Pflanzengewächsen (De).

Short Description: Various countries; traditional and some nontraditional products; various milk types (cow's, goat's, ewe's milk).

Manufacture: The manufacturing technology of fermented milk-vegetable mixtures is still in the early stages of development. Skim milk, natural buttermilk, or whey may be used as the milk component, and soy protein or a specific soy isolate may be the vegetable component.

Food Value: These products constitute excellent sources of low-cost protein and are also suitable for dietetic purposes. For example, they could be claimed to contain dietary fiber, phospholipids, or could be recommended for the prophylaxis and treatment of certain allergic reactions and for diabetics.

Products: Traditional and nontraditional products with varying proportions of different vegetable mixtures and fermented milks include fermented milk and addition of nonfermented vegetables (for example, amasi, Bircher-Muesli, burghul, fura, peskütan, rasogolla, and tamarroggt) and milk and vegetable fermentations (for example, kishk, trahana, Turkish trahana, busa).

References:

Cadena, M. A., and R. K. Robinson. 1978. Factors affecting the quality of fermented milk-wheat mixtures. *Proceedings 20th International Dairy Congress,* **E:**994-995.

Kurmann, J. A., and J. Lj. Rašić. 1988. Technology of fermented special products. *IDF Bulletin 227* (Fermented Milks), pp. 101-109. Brussels: International Dairy Federation.

FILLED MILK FERMENTED FRESH MILK PRODUCTS (En); Lait maigre enrichit avec de la graisse végétale (Fr); Mit Pflanzenfett angereicherte Magermilch (De).

Short Description: Various countries; nontraditional products; filled milk.

Manufacture: Filled milk is a recombined milk into which fat other than milkfat has been incorporated (usually reconstituted skim milk powder with vegetable fat).

Products: Fermented fresh milk products made from filled milk include yoghurt made in Chad.

Reference: *FIL-IDF Dictionary.* 1983, p. 92.

FILMJÖLK (En, Fr, De), signifies sour and set milk.

Short Description: Sweden; nontraditional product; cow's milk, partially skimmed milk; utilized as a beverage; stirred, viscous; clean acid taste.

History: Filmjölk is a typical Nordic fermented milk product manufactured in Sweden since the 1930s and represents the leading fermented milk product on the Swedish market.

Microbiology: The starter culture consisting of *Streptococcus lactis* (a), *S. lactis* subsp. *cremoris* (b), *S. lactis* subsp. *diacetilactis* (c) and *Leuconostoc mesenteroides* subsp. *cremoris* (d) at a ratio of 85:15 (a + b + c = 85; d = 15) is mixed in at a rate of 1-2%.

Manufacture: The product, made from milk standardized to 3.0% fat, is homogenized and pasteurized at 90-91°C (194.0-195.8°F) for 3 min. The incubation is at 20-21°C (68-69.8°F) for 20-24 h, followed by stirring, cooling, and packaging.

Food Value: A high-quality filmjölk has a smooth, viscous body without signs of whey separation, an acidity of 90-95°Th (0.81-0.85% titratable acidity), a pH of 4.5 and a count per ml of about 9×10^8 colony-forming units (viable starter bacteria).

Related Products: A-fil milk, cultured buttermilk, lactofil, ropy filmjölk.

References:

Alm, L. 1982. The effect of fermentation on nutrients in milk and some properties of fermented liquid milk products. Dissertation, p. 12. University of Stockholm, Stockholm.

Anonymous. 1953. Filmjölk a new dairy product (in Swedish). *Mejeritidskrift för Finlands Svenskbygd* **15:**141-143. Cited in *Dairy Science Abstracts,* 1954, **16:**194.

FINNISH FERMENTED FRESH MILK PRODUCTS (En); Produits laitiers fermentés finlandais (Fr); Finnische fermentierte Milchprodukte (De).

Short Description: Finland; traditional and nontraditional products; mainly cow's milk.

History: The most important traditional (homemade) fermented fresh milk products in Finland according to Forsén (1966) are harmaa, jamakka (East Finland, incubated in an oven), kesävelli, kirnupiimä (buttermilk, obtained from buttermaking), kokkeli (East Finland, incubated in an oven chamber), piimä (designation for sour milk in general), pitkapimä (viscous, ropy milk prepared in North and West Finland), taikkuna (barley flour is used with viili or piimä), viili (ripened in special cups; whole milk; short ropy in East Finland, long ropy in West Finland), viilipiimä (cultured milk beverage, whole milk).

Products: The major commercially produced fermented milk products in Finland are mesophilic fermented milks with different fat percentages; acidophilus piimä (A-piimä); acidophilus yoghurt (A-yoghurt); natural buttermilk (Kirnupiimä); slightly viscous (ropy) fermented milks with different fat percentages (eaten with a spoon, with addition of various sweetened fruit or berry preserves, often with special names according to the manufacturer—Valio, Kotisaari, Hj. Ingman, for example), called viili and kevytviili in cups; pitkäpimä (as viscous drink); sour cream (kermaviili); smetana (crème fraiche); yoghurt (with different sweetened fruit or berry preserves); biokefir.

References:

Forsén, R. 1966. The Long Milk, Pitkäpiimä (in German). *Mejeriitieteelinen Aikakauskirja (Finnish Journal of Dairy Science)* **26**(1):10-11.

Martiny, B. 1907, p. 63.

FLAKE BUTTERMILK (En); Babeurre granuleux (Fr); Flockige Buttermilch (De).

Short Description: United States; nontraditional product; cow's milk; beverage; special type of cultured buttermilk (buttermilk with butter flakes or granules).

Manufacture: Popular in several regions of the United States and manufactured by only a few dairies, rendered butteroil is dribbled into the cultured buttermilk. This practice is mainly a symbolic gesture to remind consumers of "true" buttermilk, the by-product of ripened cream butter manufacture (which cultured buttermilk is meant to simulate).

Related Product: Cultured buttermilk.

Reference: Cronshaw, H. B. 1947. *Dairy Information* (a textbook). London: Dairy Industries Ltd.

FREEZE-DRIED PREPARATIONS (En); Produits lyophilisés (Fr); Lyophilisierte Produkte (De).

Short Description: Various countries; nontraditional products; cow's milk; preparations of microorganisms or fermented milks; used for the manufacture of fermented milk products or for direct consumption for prophylactic or therapeutic purposes; dried.

Manufacture: A process referring to freezing and drying of cultures of some microorganisms and of certain types of heat-sensitive materials (e.g., blood plasma) or foodstuffs, including fermented milks. The bacteria culture is suspended in a suitable medium, e.g., milk, serum, or other protein-rich substances that permit noncrystalline solidification to occur during the freezing process. Subsequently, bacterial cells and/or foodstuff materials are dehydrated under vacuum (reduced pressure). During the process water passes from the solid state to the vapor phase. Freeze-dried materials are very hygroscopic and porous and, therefore, reconstitute rapidly with water or other aqueous liquids. They are usually stored under an inert gas (e.g., nitrogen or carbon dioxide) or under reduced pressure to prevent atmospheric oxidation.

References:

Bender, A. E. 1970. *Dictionary of Nutrition and Food Technology.* London: Butterworths.

Singleton, P., and D. Sainsburry. 1981. *Dictionary of Microbiology.* New York: John Wiley.

FROZEN YOGHURT (En); Glace au yaourt (Fr); Joghurt-Eiskrem (De).

Short Description: Various countries; nontraditional product; cow's milk; direct consumption after thawing; frozen product, plain or fruit-flavored; prolonged storage life; taste and flavor similar to that of nonfrozen yoghurt.

History: The term *frozen yoghurt* has taken on specific commodity significance since the 1970s, especially in the United States. For frozen yoghurt there is either a hard-frozen product, similar to ice cream, or a soft-frozen and dispensed product, similar to soft ice milk or custard.

Microbiology: The starter culture consists of *Streptococcus thermophilus* and *Lactobacillus delbrueckii* subsp. *bulgaricus* (*see* Yoghurt).

Manufacture: Frozen yoghurt (both a dipped hard-frozen product and a dispensed soft-frozen product exist) is made by first making regular yoghurt and then mixing it with a pasteurized stabilizer/sweetener blend prior to freezing in ice-cream-making equipment. Fruits and flavorings, if desired, are also added at the time of freezing. Soft-frozen yoghurt is more popular than the hard-frozen kind in North America, usually dispensed into edible cones or plastic cups in fast-food restaurants. Vanilla and chocolate were the preferred flavors in the 1970s and 1980s, but various fruit flavors were also sold.

Freezing yoghurt is only feasible with the stirred types. In set yoghurt ice crystals will damage the gel structure during freezing and thawing and cause whey separation (syneresis), especially if freezing is slow. Plain stirred yoghurt with 13-14% total solids or fruit yoghurt with 20-25% total solids can be successfully frozen at −26° to −30°C. Addition of stabilizers has a similar effect as does a high total solids content. Rapidly frozen yoghurt may be stored at −26°C for 3 to 12 months. Thawing of the product should be carried out in cold storage, e.g., at −5°C to −7°C for 24 to 36 h.

Products: Yoghurt ice, sherbet, soft-serve ice cream from yoghurt.

Reference: Rašić, J. Lj., and J. A. Kurmann. 1978, pp. 293, 344, 345.

FRU-FRU (En, Fr, De); trade name.

Short Description: Austria; nontraditional product; cow's milk.

History: Fru-fru was developed by the dairy processing firm MIAG, in Vienna, before World War II (1939-1945) and described by Becker in 1969.

Microbiology: The starter culture consisting of *Streptococcus lactis* and/or *S. lactis* subsp. *cremoris* in combination with *Leuconostoc mesenteroides* subsp. *cremoris* and/or *S. lactis* subsp. *diacetilactis* is mixed in at a rate of 2%.

Manufacture: Fru-fru is a cultured milk product obtained by incubating milk (containing 5% fat, 1% nonfat milk solids, and 2% sucrose) with the above-mentioned culture. Incubation is in 220-ml retail containers. Thirty g strawberry jam or preserve is added to the cooled product before applying the lid.

Related Products: Mesophilic fermented milks.

Reference: Becker, F. 1969. Milk drinks and fermented milk products—methods and equipment used in Austria (in German). *Oesterreichische Milchwirtschaft* **24** (13):1241-1243. Cited in *Dairy Science Abstracts,* 1969, **31** (3280):500.

FRULATI (En, Fr, De); denotes fruit milk; trade name.

Short Description: Brazil; nontraditional beverage; acid buttermilk; pleasant flavor; heat-treated after culturing.

History: Frulati was developed in Brazil in 1978 (Martins 1978).

Manufacture: Frulati is made from acid buttermilk (natural buttermilk) to which 10% sucrose is added, as well as a stabilizer to keep the protein dispersed. After pasteurization at 90°C (194°F) for 5 min it is cooled to 5°C (41°F) and mixed with frozen concentrated fruit juice. This syrup is then mixed with carbonated water in a ratio of 1:2 (final pressure is 2.5 atm) and bottled and capped. The final product has a pH of about 4.0.

Reference: Martins, A. C. H. 1978. Frulati: a new Brazilian carbonated drink, made from buttermilk. *Proceedings 20th International Dairy Congress* **E:**942.

FURA (En, Fr, De).

Short Description: Africa, East Sudan; traditional product; popular beverage; a mixture of millet and sour milk (*see* Milk-vegetable mixtures, fermented).

Reference: Hintze, K. 1934. *Geography and History of Nutrition* (in German). Leipzig: Thieme Verlag.

G

GALAZYME (En, Fr, De); signifies milk fermented with yeasts; also called milk champagne.

Short Description: France; nontraditional product; more or less skimmed cow's milk; sparkling beverage.

History: One of the first preparations of a nontraditional fermented milk product (Dujardin 1887).

Microbiology: Baker's or champagne yeast.

Manufacture: The homemade preparation is as follows: Twenty g sugar and 6 g of baker's or champagne yeast is added to one liter of milk that is tempered to 25-30°C (77.0-86.0°F). This mixture is filled into thick-walled glass bottles (2/3 full), which are then closed like champagne bottles and stored horizontally at 10-12°C (50.0-53.6°F). After three days the beverage becomes frothy and can be consumed.

Related Products: Champagne, fermented milk beverage.

Reference: Dujardin. 1887. Mentioned in Martiny, B. 1907, p. 41.

GANDŽLIK (En, Fr, De).

Short Description: USSR (Azerbaijan); traditional, national product; cow's milk and cream; stirred product, fruit-flavored; direct consumption; smooth, creamy consistency; acidic taste, pleasant flavor.

Microbiology: Culture of thermophilic lactic acid bacteria (no composition given).

Manufacture: The product is made from a mixture of whole milk and cream, standardized to 9.0% fat, which is homogenized, pasteurized at 85-87°C (185.0-188.6°F) for 5-10 min, then cooled to 38-40°C (100.4-104.0°F). It is inoculated with 5% of above-mentioned culture. Incubation to an acidity of 75-80°Th (0.67-0.72% titratable acidity) takes 6-7 h and is followed by stirring, adding 10% fruit syrup and mixing, then cooling to 8-10°C (46.4-50.0°F) and bottling.

Food Value: The final product has 8% fat and an acidity of 80-100°Th (0.72-0.90% titratable acidity).

Related Product: Cream-enriched fermented milk products.

Reference: Bogdanova, G. J., and E. A. Bogdanova. 1974. *New Whole Milk Products of Improved Quality* (in Russian). Moscow: Pishchevoĭ Promȳshlennosti.

GEFILUS® (En, Fr, De); trade name.

Short Description: Finland; nontraditional fermented fresh milk group with probiotic properties; cow's milk.

History: In March 1990, Valio Dairy Cooperative, Finland, launched a new series of probiotic fermented milk types with the brand name Gefilus®.

Microbiology: Gefilus® products are based on the use of *Lactobacillus acidophilus* GG strain (named after Sherwood Gorbach and Barry Goldin, Tufts University School of Medicine, Boston, Massachusetts, which fulfills the requirements generally expected for probiotic lactic acid bacteria strains. Of human origin, it has the ability to survive passage through the stomach and upper bowel and to colonize and establish itself in the intestine (adhesive strain).

Products: *Gefilus® low-lactose fermented milk* is an unflavored, spoonable milk product that has a firm consistency and a refreshing, slightly sour taste. Fructose is used as a sweetener and the fat content is 1.5%. Some of its lactose is hydrolyzed. It is a typical breakfast or snack product that can be eaten with fruit or cereals. Its shelf life is three weeks.

Gefilus® whey drink is an apricot-peach product that is fruity and rich-tasting. The whey supplements the taste of the fruit and improves the nutritional value of the beverage. It has a shelf life of three weeks.

Food Value: Gefilus® products already have considerable clinical research documentation to back up the marketing program. Gefilus® products have been tested in different intestinal disorders, such as travelers' diarrhea, antibiotic diarrhea, different children's intestinal disorders, and to ease the symptoms of patients suffering from constipation.

References:

Goldin, B. R., and S. L. Gorbach. 1984. The effect of milk and lactobacillus feeding on human intestinal bacterial enzyme activity. *American Journal of Clinical Nutrition* **39:**756-761.

Meriläinen, V., and J. Setälä. 1990. Lactic acid bacteria research leads to new applications in Finland. *Scandinavian Dairy Information* **2:**28-29.

GEROLAKT (En, Fr, De).

Short Description: USSR; nontraditional product.

Microbiology: Streptosan, a new starter intended for the manufacture of Gerolakt, was developed by selecting high-activity strains of lactic acid bacteria (cocci), in liquid and freeze-dried form.

Manufacture: Coagulation time for pasteurized or sterilized milk at 37°C (98.6°F) is 8-16 h when an acidity of 80-90°Th (0.72-0.81% titratable acidity) is reached. The number of viable lactic acid bacteria is 10^8/ml.

Nutritive Value: The product is recommended for the diet of older persons.

Reference: Gritsenko, T. T., A. K. Raradii, and N. K. Kovalenko. 1987. New bacterial starter (in Russian). *Molochnaya Promȳshlennost* **11:**37-38. Cited in *Dairy Science Abstracts,* 1988 **50**(4605):518.

GIBNEH-LABANEH (En, Fr, De).

Short Description: Middle East; traditional product; cow's, ewe's, goat's, or mixed milk; directly consumed with bread; dried product; pleasant taste and flavor.

Microbiology: *See* Labneh.

Manufacture: A fermented product prepared by adding salt to concentrated yoghurt (labneh), then shaping into balls and drying in the sun. The dry balls are packaged in glass jars and covered with olive oil. Reportedly the product has outstanding organoleptic properties.

Related Product: Kashk.

Reference: Gordin, S. 1970. Milking animals and fermented milks of the Middle East and their contribution to man's welfare. *Journal of Dairy Science* **63:**1031-1038.

GIL (En, Fr, De).

Short Description: Israel; nontraditional product; cow's milk; coagulated (set) preparation.

Microbiology: Mesophilic lactic acid bacteria starter culture from Chr. Hansen's Laboratorium A/S, Roskilde, Denmark (no composition given).

Manufacture: Prepared from pasteurized milk (3% fat) and 3% added starter mentioned above. After filling in cups (170 ml), incubation is at 25°C (77.0°F) for 6-8 h, followed by cooling to 5-7°C (41.0-44.6°F).

Related Products: Mesophilic fermented milk products, cultured buttermilk.

Reference: Sadovski, A. Y., I. Formeman, and S. Gordin. 1978. Low-temperature growth of starter organisms in a fermented milk product. *Proceedings 20th International Dairy Congress.* **E:**542-544. Brussels: International Dairy Federation. Cited in *Dairy Science Abstracts,* 1978, **40**(4921):524.

GIODDU (En, Fr, De).

Short Description: Italy (Sardinia); traditional product; ewe's, goat's, sometimes cow's milk; also consumed as beverage; more or less firm consistency; acid and eventually piquant taste.

Microbiology: *Home manufactured products:* The microflora is varied and is composed of *Lactobacillus delbrueckii* subsp. *bulgaricus, Streptococcus thermophilus, S. lactis, S. faecalis* and *S. faecium.* Traditional gioddu prepared in the rural areas contains a high proportion of enterococci and mesophilic lactic streptococci (Arizza et al. 1983). Grisconi (1905) reported the presence of a yeast and a lactic bacillus.

Industrial manufactured products: yoghurt starter culture, sometimes baker's yeast is included.

Manufacture, Home: The traditional home manufacture, which is of ancient origin, can be summarized as follows: The milk is kept boiling for shorter or longer periods in order to produce a gioddu of less firm or firmer consistency, respectively. When the milk has cooled to about 30-40°C (86.0-104.0°F), gioddu from the previous day's production is added at a rate of 1-2%. The Sardinian name for this starter is *madrighe.* In the past, a container made from cork was used to keep the inoculated milk at a fairly constant incubation temperature. Today it is customary to use a metal container covered with woolen fabrics and kept in a lukewarm place for 3 to 7 hours. After incubation, the gioddu is cooled and is ready for consumption.

Manufacture, industrial: Industrial manufacture is similar to that of making yoghurt. Pasteurized milk is cooled to 40-45°C (104.0-113.0°F), inoculated with 1-2% yoghurt starter culture, and incubated at 40°C (104.0°F) for at least 3-4 h.

Related Product: Miciuratu, name used in northern Italy for gioddu.

References:

Arizza, S., A. Ledda, P. G. Sarra, and F. Dellaglio. 1983. Identification of lactic acid bacteria in "Gioddu" (in Italian) *Scienze e Tecnica Lattiero-Casearia* **34**(2):87-102. Cited in *Dairy Science Abstracts,* 1984, **46**(1278):147.

Grisconi, G. 1905. A new fermented milk, "Gioddu," easy to prepare in hospitals (in Italian). *Annali di Medicina navale* vol. 2, fasc. 3. Cited in *Le Lait,* 1905-1906, **5**:93-94.

GOAT MILK FIRM YOGHURT (En); Yaourt ferme au lait de chèvre (Fr); Stichfester Ziegenmilch-Joghurt (De).

Short Description: Several countries; traditional product; sometimes with goaty flavor.

Microbiology: Yoghurt starter bacteria.

Manufacture: Goat milk differs from other milks in that its curd is semiliquid. Unless considerable changes are made, goat milk does not lend itself to the manufacture of yoghurt with a degree of firmness similar to that of cow's milk. Therefore, goat milk, unless concentrated, is generally not used to make industrially prepared fermented milk products. Sometimes goat milk is concentrated by ultrafiltration (Abrahamsen 1986).

However, goat milk is perfectly suited for the production of fruit and cereal yoghurts (birchermuesli type) known commercially under various names, such as bircherMuesli, and fruit, sports, all-fruit yoghurt, etc. The consistency of such products is mainly caused by the presence and swelling of the added ingredients (natural dietary fibers, dried fruits, ripe fruits rich in pectin). Usually a recipe that has a minimum of added ingredients but produces considerable thickening is chosen. However, the weight of all the added ingredients must not exceed 25% of the total weight if the product is still to be known as yoghurt.

Related product: Yoghurt.

References:

Abrahamsen, R. K. 1986. Yoghurt from ultrafiltrated goat's milk. *IDF Bulletin 202* (Production und Utilization of Ewe's and Goat's Milk), pp. 171-174. Brussels: International Dairy Federation.

Duitschaever, C. L. 1978. Yoghurt from goats' milk. *Cultured Dairy Products Journal* **13**(4):20-24.

Flanagan, J. F., and V. H. Holsinger. 1985. Procedure for making yoghurt and cheese from goat's milk on the small farm. *Cultured Dairy Products Journal* **20**(2):6-7.

Kurmann, J. A. 1986. Yoghurt made from ewes' and goats' milk. *IDF Bulletin 202* (Production und Utilization of Ewe's and Goat's Milk), p. 156. Brussels: International Dairy Federation.

Rašić, J. Lj., and J. A. Kurmann. 1978, p. 70-74.

GOAT MILK LIQUID YOGHURT (En); Yaourt boisson au lait de chèvre (Fr); Trinkjoghurt aus Ziegenmilch (De).

Short Description: Several countries; traditional beverage; possibly develops a goaty flavor during storage.

Microbiology: Yoghurt starter bacteria.

Manufacture: The manufacturing procedure is the same as for yoghurt. For a number of reasons, goat milk is very suitable for the manufacture of an excellent yoghurt beverage. Goat milk curd is softer than that of other milks, it is more aqueous and less firm. It is more easily stirred and, after this treatment, much smoother and more homogeneous than cow's milk liquid yoghurt.

In Switzerland, small quantities of yoghurt beverages are prepared with the addition of black or red currant syrups (berries of the region). When homogenized, a very stable coagulum is obtained and the product displays an appealing freshness (Kurmann 1986).

Food Value: The beverage is claimed to be light and easily digested, most likely because of the relatively small size of goat milk fat globules.

Related Products: Airan, a yoghurt-based beverage, is also produced from goat milk.

References:

Kurmann, J. A. 1986. Yoghurt made from ewes' and goats' milk. *IDF Bulletin 202* (Production and Utilization of Ewe's and Goat's Milk), p. 156. Brussels: International Dairy Federation.

Kurmann, J. A. 1987. Special goat's milk products prepared in Switzerland. *IDF Goats & Ewes Newsletter,* May, pp. 3-4. Brussels: International Dairy Federation.

Rašić, J. Lj., and J. A. Kurmann. 1978, pp. 70-74.

GOAT MILK, MESOPHILIC FERMENTED MILK (En); Lait fermenté mésophile au lait de chèvre (Fr); Mesophile Sauermilch aus Ziegenmilch (De).

Short Description: United States, Norway; traditional product; possibly goaty flavor.

Microbiology: Mesophilic DL-type starter (*Streptococcus lactis, S. lactis* subsp. *cremoris* and subsp. *diacetylactis, Leuconostoc mesenteroides* subsp. *cremoris*), with mucogenic strains.

Manufacture: Because of the soft consistency of acid-coagulated goat milk curd, attempts have been made to thicken such curd with the aid of mucogenic cultures. The resulting product is more viscous and even stringy and can be achieved with, for example, viili cultures. A production test using goat milk and mesophilic cultures was carried out by Hardland and Hoffmann (1974). Since a milk with a low solids content yields a more mucous and thicker product than a milk with a high solids content, goat milk is well suited to the use of mucogenic cultures.

Related Products: Mesophilic fermented fresh milk products.

References:

Hardland, G., and T. Hoffmann. 1974. Structural modification of goat's milk for production of fermented milks. *Proceedings 19th International Dairy Congress* **E:**740-741. Brussels: International Dairy Federation.

Rysstad, G., and K. Abrahamsen. 1983. Fermentation of goat's milk by two DL-type mixed strain starters. *Journal of Dairy Research* **50:**349-356.

GOAT MILK PRODUCTS (En); Produits laitiers à partir de lait de chèvre (Fr); Ziegenmilchprodukte (De).

Short Description: Various countries; traditional products; goats' milk; set and stirred products.

History: Goats were domesticated approximately 7500 B.C. in Ganj-Dareh (now Iran). Although the date of first milking and milk utilization is unknown, the first deliberately soured goat milk probably was prepared before 2000 B.C. Today goat breeds are found worldwide. Goats tolerate heat better than sheep. The length of goat lactation period is 250-305 days and the approximate milk production is 300-900 kg/year.

Manufacture: A manufacturing problem with goat milk is that consistency of the coagulum is too soft, therefore it is not suitable for making set and stirred yoghurt, but goat milk is very suitable for the preparation of fermented beverages.

Food Value: The chemical composition of goat milk varies with breed and other factors, but goat milk contains approximately 11.0-13.0% total solids, 3.0-5.5% fat, 2.9-4.6% protein, 3.8-5.1% lactose and 0.69-0.81% ash (mineral matter). The nutritional value is similar to cow's milk, but goat's milk, in general, especially goat milk protein and fat, is claimed to be more easily digested.

Products: Goat milk fermented products are those prepared from goat milk only (usually in liquid form) such as airan, goat milk liquid, and firm yoghurt and those prepared from mixtures of goat milk, sheep milk and cow's milk, such as gioddu, kefir, and mazun. It is possible to prepare more or less firm goat fermented milks: goat milk firm yoghurt, goat milk, mesophilic fermented milk, and zabady.

References:

Haenlein, G. F. W. 1984. *Extension Goat Handbook.* Washington, D.C.: U.S. Department of Agriculture.

Kurmann, J. A. 1986. Yoghurt made from ewes' and goats' milk. *IDF Bulletin 202* (Production and Utilization of Ewe's and Goat's Milk), pp. 153-169. Brussels: International Dairy Federation.

Leonard, J. N. 1982. *The First Farmers* (in German) Sixth German Printing. Amsterdam: Time-Life Books.

GOUBASHA (En, Fr, De).

Short Description: Sudan; traditional beverage; buffalo milk; table beverage especially in the winter months. It is a laban rayeb to which some cream is added and then diluted with water.

Reference: El-Gendy, Sh. M. 1983. Fermented foods of Egypt and the Middle East. *Journal of Food Protection* **46:**366.

GRUSHEVINA (En, Fr, De); grushavina, grusevina; denotes coagulated milk.

Short Description: Yugoslavia; traditional product; cow's, ewe's or mixed milk; used for direct consumption or in cooked and baked foods; set type or paste; acidic taste, agreeable flavor.

Microbiology: Mesophilic lactic acid bacteria.

Manufacture: A soured milk product homemade in some mountain villages of Yugoslavia. It is made from fresh milk derived from healthy animals and is soured spontaneously or with the addition of starter (with above-mentioned starter bacteria) taken from the coagulated product made the previous day. Sometimes the whey is partially removed from the coagulated milk by using a cloth bag.

Reference: Rašić, J. Lj. 1982. Special products. *Proceedings 21st International Dairy Congress* **2:**151, discussion. Moscow: Mir Publishers.

GUA-NAI (En, Fr, De); named after the developer, Guan Jiahuai.

Short Description: This particular product was developed in Dr. J. R. Brunner's laboratory at Michigan State University in 1986; nontraditional key product; custardlike consistency; unique fruity flavor that can be enhanced by garnishing with raisins, sliced almonds, or other garnishes.

Microbiology: Three filamentous fungy *Amylomyces rouxii, Rhizopus oryzae,* and *Aspergillus oryzae* and one yeast species, *Endomycopsis burtonii* were isolated from the Chinese wine cake employed as the starter for Gua-nai manufacture.

Manufacture: Gua-nai is a sweet gel produced from either whole or skim milk by the enzymatic activity of "sweet leavening," a fermented mass produced by a mixture of filamentous molds and yeast derived from Chinese wine cake.

The manufacture involves the production of a bulk culture (i.e., sweet leavening) of wine cake flora propagated on steamed, sweet rice. The starch of sweet rice consists principally of amylopectin, a preferred substrate for the endogenous amylases elaborated by the culture fungi. After a six-day semianaerobic fermentation at 30°C (86°F) the liquid phase of the culture is characterized by its light amber color, a pH of approximately 3.8-4.0, an alcohol content of 8-10% and a pleasant fruity flavor. At this stage, yeast accounts for essentially all of the viable organisms present.

Following clarification, the liquid phase is added to cold, pasteurized milk (5-8% v/v) previously fortified with 1-2% skim milk powder, 5-10% sugar or corn syrup solids (according to taste) and 0.2% gelatin (optional). The mixture is filled into retail-size containers and incubated at 40°C (104°F) for 1-3 h to achieve the desired gel characteristics. Gua-nai should be held at refrigerator temperature until consumed. A shelf life of approximately ten days can be expected before the gel structure shows indications of syneresis.

The pH of the product is approximately 6.2 and the alcohol content 0.3-0.4% w/v depending on the amount of liquid leavening added. The milk itself does not ferment to any appreciable extent. Coagulation is brought about by the enzyme(s) of the culture produced on the rice, mainly species of *Rhizopus* and other filamentous fungi.

Related Product: Chinese wine cake (related microflora).

References:

Guan, J., and J. R. Brunner. 1987. "Gua-nai"—an oriental-style dairy food. *Cultured Dairy Products Journal* **22**(2):16.

Onyeneho, S. N., J. A. Partridge, J. R. Brunner, and E. S. Beneke. 1988. The microbiology of gua-nai. *Cultured Dairy Products Journal* **23**(1):10, 12-13.

Onyeneho, S. N., J. A. Partridge, J. R. Brunner, and E. S. Beneke. 1988. Ethanol production, milk clotting and proteolytic activity of the fungi obtained from Chinese wine cake. *Cultured Dairy Products Journal* **23**(2):6-8.

GWEDEN (En, Fr, De); Gros lait; denotes set or clabbered milk (clotted milk).

Short Description: France (Bretagne); traditional, whey-drained product; usually cow's milk; concentrated fermented milk with a creamy consistency and slightly ropy.

Microbiology: Nondefined mesophilic lactic acid bacteria.

Manufacture, Home: Fresh milk is, immediately after milking and at 35-36°C (95.0-96.8°F), inoculated with 5% of gweden from the previous day's production. After 18 h the cream is skimmed off the top and is used to make butter. The coagulum is filled into small cloth pouches for whey drainage. It is stored in a cool room until ready for consumption.

Related Product: Clabber milk.

Reference: Mazé, P., cited by Rolet, A. 1920. *Ancillary Dairy Processing Industries, Utilization of By-Products and Waste Products* (in French) pp. 178-189. Paris: J.-B.Baillière et Fils.

H

HARMAA (En, Fr, De).

Short Description: Finland; traditional beverage; consumed by Karelian farmers; consists of sour milk mixed with whey and water.

Reference: Martiny, B. 1907, p. 51.

HOME-PREPARED FERMENTED MILK PRODUCTS (En); Préparations de laits fermentés à domicile (Fr); Im Haushalt hergestellte Sauermilchprodukte (De).

Short Description: Various countries; traditional and nontraditional products; various milk types; different fermented milk products, mainly yoghurt types.

History: From time immemorial, nomads as well as settled peoples have prepared fermented milk products in relatively small quantities as minor or major adjuncts to their diets. Household manufacture of these products has gradually diminished in so-called developed areas of the world as industrial production has increased.

Microbiology: With home-prepared fermented milk products a distinction can be made between traditional products made with wild, relatively undefined, cultures according to ancient methods based on empirical knowledge and traditional products prepared with defined and controlled starter cultures that are now widely available worldwide.

Products: Traditional products with undefined or wild cultures include airan (original preparation), baassmilch, ergo, and gioddu (traditional manufacture). Demeter sour (fermented fresh milk) products exemplify modern traditional manufacture.

Traditional products made with defined cultures include yoghurt, airan, and the like.

References:

Kurmann, J. A. 1984. The production of fermented milk in the world. *IDF Bulletin 179* (Fermented Milks) pp. 16-19. Brussels: International Dairy Federation.

Rašić, J. Lj., and J. A. Kurmann. 1978, p. 303.

HONEY CLABBERED MILK (En), Lait acidifié au miel (Fr); Dickmilch mit Honig-Zusatz (De).

Short Description: Ancient Babylonia; key preparation, popular food; a milk-cereal mixture of clabbered (sour) milk, flour and honey (or date syrup), or a paste made from clabbered milk, honey, and oil, with various other ingredients added.

Related Products: Milk-cereal preparations, fermented.

Reference: Hintze, K. 1934. *Geography and History of Nutrition* (in German), p. 16. Leipzig: Georg Thieme.

HONEY-YOGHURT (En); Yaourt au miel (Fr); Honig-Joghurt (De).

Short Description: Various countries; traditional fermented milk preparation; sheep's, goat's, and cow's milk; proportion of honey according to taste; yoghurt preparations.

Microbiology: Yoghurt starter culture microorganisms.

Manufacture: Sheep milk yoghurt is very compatible with honey and the resulting product is of premium quality due to the relatively high fat content of sheep milk and its high concentration of total solids. However, care must be exercised in its preparation to avoid too high an acidity in the yoghurt, since that would partially mask the delicate flavor of the honey. If natural honey is used, it should be stirred into the yoghurt after fermentation or be layered over or under the yoghurt when incubated in the retail container (sundae-style). The type of yoghurt in a particular market depends on consumer preferences. In another process, the honey is added to the milk before pasteurization. With such a procedure there may be effects of the honey on the growing, incubating organisms (e.g., increased osmotic pressure) or a weakening of the honey flavor.

References:

Brown, G. D., and F. V. Kosikowski. 1970. How to make yoghurt. *American Dairy Review* **32**(4):60-62.

Kurmann, J. A. 1986. Yoghurt made from ewes' and goats' milk. *IDF Bulletin 202* (Production and Utilization of Ewe's and Goat's Milk) pp. 157. Brussels:International Dairy Federation.

Schei, S. A., and R. K. Abrahamsen. 1985. Honey yoghurt from cows' and goats' milk (in Norwegian). *Meieriposten* **74**(7):189-191. Cited in *Dairy Science Abstracts,* 1987, **49:**241.

HUMAN MILK, FERMENTED FRESH PRODUCTS (En); Lait humain fermenté (Fr); Fermentierte Frauenmilch (De).

Short Description: Various countries; no traditional fermented milk products based on human milk are known; sometimes spontaneously acidified when kept in a container after collection.

Manufacture: Except for human milk bank purposes, the milk is usually not collected. Occasionally, a nursing mother may express breast milk and keep it for later feeding or discard it. Small manual pumping and collecting devices are available for that purpose.

Because of the low casein content of 0.4-0.6%, acidified human milk does not form a coagulum suitable for soured beverages. Instead, only flakes of curd particles appear.

Food Value: The average chemical composition of human milk is 6.0-7.1% lactose, 1.1-1.5% protein, 2.5-4.8% fat, 0.15-0.25% ash (Schultz 1965).

Reference: Schulz, M. E. 1965, **1:**343-345.

HUMANIZED MILK, FERMENTED (En), Lait humanisé fermenté (Fr); Fermentierte humanisierte Milch (De).

Short Description: Several countries; nontraditional products; special cow's milk formula, different types.

History: In order to make cow's milk resemble human milk in its behavior during passage through the digestive tract of infants, various alteration procedures have been suggested. The most important change made is a reduction in the amounts of casein and calcium and/or in the ability of the milk to form a hard clot in the stomach.

Manufacture: There are three types of acidified humanized milk.

1. Modified milk. The content and quality of components of cow's milk is adapted as much as possible to resemble human milk with regard to protein (specifically casein), fat, lactose, vitamins, and minerals. Example: Femilact.
2. Partly modified milk. The carbohydrate of this milk has been changed substantially; lactose is complemented by adding maltodextrin. Example: lactane B.
3. Formula milk for infants 3 months to approximately 12 months old or older. This type of milk is specifically formulated to specific nutrient and growth requirements at a particular age. Example: Pelargon.

In all these milk types the concentrations of vitamins and minerals are adjusted to the requirements of the babies. Particularly important are calcium for the formation of bone and iron for the formation of red blood corpuscles.

HUNGARIAN FERMENTED FRESH MILK PRODUCTS (En), Produits laitiers fermentés (frais) de la Hongrie (Fr); Ungarische gesäuerte (frische) Milchprodukte (De).

Short Description: Hungary; traditional and nontraditional products; milk of cows, goats, and ewes.

Products: In Hungary the most important fermented fresh milk products are

Aludttej (Hungarian), Clabber milk (En), Dickmilch (De). Only a small quantity is produced by the milk processing industry, but even today in the small family farms a large amount of it is prepared by spontaneous souring.

Iró (Hu), natural buttermilk (En). It is only produced and sold by the dairy industry, where it is made in small quantities as a by-product of cultured cream butter manufacture.

Joghurt (Hu), yoghurt (En). Routinely produced and sold by the dairy industry.

Kefir (Hu, En). Only produced and sold by the dairy industry.

Sóstej (Hu), salty milk (En). *See* Sóstej.

Tarhó (Hu, En). A national yoghurt type. *See* Tarhó.

Teijföl (Hu), cultured cream (En). Among all the fermented dairy products, teijföl is made in the largest quantity. The importance of the product as prepared by spontaneous souring in the small family farms is decreasing.

References:

Balatoni, M., and F. Ketting. 1981. *Dairy Handbook* (in Hungarian). Budapest: Mezögazdasági Kiadó.

Gratz, O. 1925. *Milk and Milk Products* (in Hungarian). Budapest: Eggenberger-féle könvyvkereskedés.

Tomka, G. 1943. Producing tarhó yoghurt at home (in Hungarian). *Tejgazsdasag* **3:**(5):110–114.

This entry was contributed by Dr. Ferenc Ketting, Budapest.

HUSLANKA (En, Fr, De); Hooslanka.

Short Description: USSR (eastern Carpathian Mountains), Bukowina (northeast Carpathian Mountains), also prepared by certain Rumanian ethnic groups; traditional product; sheep milk, cow's milk, or mixed, also skim milk.

Microbiology: The microflora, according to Supinska and Pijanowski (1937) is composed of *Lactobacillus delbrueckii* subsp. *bulgaricus,* Streptococcus cells, and lactose-fermenting yeasts.

Manufacture: Huslanka is a homemade product. Milk is evaporated to 25% of its original volume. This high-solids medium at 40–45°C (104–113°F) is poured into an elongated wooden vat (called *berbenitza*) the inner surface of which is covered with a portion of earlier prepared huslanka. After 24 h of fermentation at room temperature the vat is tightly closed. The final product can be stored over several years.

Food Value: The lactic acid content of huslanka is nearly 2%, the alcohol concentration about 0.5% and there is slight saturation with CO_2. It has been used as a food, especially in times of sparse milk production.

Related Product: Concentrated fermented milks (with preconcentrations).

References:

Brodny, N. 1911. *Oesterreichische Molkerei-Zeitung* **18:**321.

Oberman, H. 1985. Fermented milks. In *Microbiology of Fermented Foods,* ed. Brian J. B. Wood, vol. 1, p. 183. London and New York: Elsevier.

Supinska, J., and E. Pijanowski. 1937. The manufacture of huslanka (in Polish). *Polish Agricultural and Forestry Annual* **38:**209-224.

HYDROLYZED-LACTOSE FERMENTED FRESH MILK PRODUCTS (En). Produits laitiers fermenté (frais) avec de lactose hydrolyzé (Fr); Gesäuerte (frische) Milchprodukte mit hydrolysierter Laktose (De).

Short Description: Various countries; nontraditional products; cow's milk; mainly yoghurt types.

Microbiology: Lactose can be hydrolyzed by various means and the production of many products is possible.

- Single-stage (usual) or multiple-stage fermentations (Robert 1978) with β-galactosidase-producing microorganisms (some lactobacilli, *Streptococcus thermophilus,* some bifidobacteria)
- Pretreatment of milk with β-galactosidase-producing yeast, e.g. *Saccharomyces fragilis* (Edelstein et al. 1978)
- Products manufactured with β-galactosidase enzyme-treated milk (Günther 1983).

Food Value: Lactose-intolerant (lactase-deficient) persons usually consume little or no milk. It is with these consumers in mind that lactose-reduced products have been developed, including fermented milk products.

References:

Edelstein, D., E. W. Nielsen, and E. Refstrup. 1978. Fermentation of lactose in milk with *Saccharomyces fragilis* (in German). *Proceedings 20st International Dairy Congress* **D:**1047. Brussels: International Dairy Federation.

Günther, E. 1983. Process for producing yoghurt with a reduced lactose content (in German). *German Patent* 3,146,198.

Robert, J. G. 1978. Process for the manufacture of a food based on milk (in German). *German Patent* 2,725,731.

I

INDIAN FERMENTED FRESH MILK PRODUCTS (En); Produits laitiers fermentés de l'Inde (Fr); Indische fermentierte Milchprodukte (De).

Short Description: India; traditional and nontraditional products; buffalo's, cow's, goat's and ewe's milk.

History: Food and drink in ancient India included fermented fresh milk products, for example, buttermilk, dahi, sour milks and different mixtures, which are described by Prakash (1968).

Manufacture: There are wide variations in the home and factory manufacture of all Indian products. A good overview is presented by Chakraborty (1983).

Products: The most important traditional and nontraditional fermented fresh milk products of India are mentioned in Appendix B.

References:

Bandyopadhyay, A. K., and B. N. Mathur. 1987. Indian milk products: a compendium. In *Dairy India 1987,* ed. P. R. Gupta, pp. 212-218. Delhi: Dairy India, A-25 Priyadarshini Vihar.

Chakraborty, B. K. 1983. Modernizing traditional milk products technology. In *Dairy India 1983,* ed. R. M. Acharya, R. P. Aneja, P. H. Bhatt, and R. K. Patel, pp. 105-108. Delhi: Dairy India, A-25 Priyadarshini Vihar.

Prakash, O. 1968. *Food and Drinks in Ancient India (from earliest time to 1200 A.D.).* Nai Sarak, Delhi-6: Munshi Ram Manohar Lal.

INFLORAN BERNA (En, Fr, De); trade name.

Short Description: Switzerland, Italy; nontraditional product; pharmaceutical, freeze-dried preparation.

Microbiology: The product contains 10^9 cells of *Lactobacillus acidophilus* and 10^9 cells of *Bifidobacterium bifidum.*

Therapeutic Value: Used against digestive disturbances in formula-fed babies and in the treatment of enterocolitis (acute unspecified and chronic), acute enterocolitis after antibiotic therapy, and chronic constipation.

Reference: Branca, G., E. Salvaggio, S. Manzara, and F. M. Paone. 1979. Research on the administration of antibiotic resistant Bifidobacterium bifidum and Lactobacillus acidophilus after antibiotic treatment (in Italian). *Clinica Veterinaria* **90**(5):565-578.

INGREDIENTS: FERMENTED MILK PRODUCTS IN OTHER FOODS (En); Ingrédiants: addition de laits fermentés dans d'autres aliments (Fr); Zusätze: Sauermilchprodukte zu anderen Lebensmitteln (De).

Short Description: Different countries; in most of the cases nontraditional products; mainly cow's milk.

Products: The subject "dairy ingredients in foods" was the title of a special FIL-IDF seminar dealing with many food groups and subgroups. The same overview can be applied to fermented milk products. However, single milk components are rarely prepared from fermented milks.

Fermented milk products (buttermilk, whey, yoghurt, kefir, etc.) can be and are utilized as ingredients in the following product groups: sauces, meat products, spreads, beverages, cereal products, and confectionery and bakery products. The other categories (fish products, simulated foods, food additives, dietetic foods) either are already or will soon be outlets for fermented milk products.

See also Yoghurt as Ingredient in Other Foods.

Buttermilk is frequently used as an important or characterizing ingredient in, for example, bread (buttermilk bread) and other bakery products, creme, flammery (Swiss flammeri), jelly, honey, soups, puddings, quarg, sauces, farinaceous pastes, pancakes, and other preparations.

Food Value: There is growing consumer interest in foods containing milk product components because of their recognized valuable properties (functional, nutritional, physiological, and organoleptic). Their use is only limited by their price, which generally is lower for by-products (e.g., fermented buttermilk or whey) and higher for other dairy ingredients obtained through fermentations.

References:

FIL-IDF. 1982. Seminar proceedings Luxembourg, May 1981. *IDF-Bulletin Document* **147**:1-103. Brussels: International Dairy Federation.

Kurmann, J. A., and J. Lj. Rašić. 1988. Technology of fermented special products. *IDF Bulletin 227* (Fermented Milks), pp. 101-109. Brussels: International Dairy Federation.

Schulz, M. E. 1981. *Lexicon of Milk-Based Foods* (in German), pp. 89-91. Munich: Volkswirtschaftlicher Verlag.

IREK-MAI (En, Fr, De).

Short Description: USSR (central Asian republics); traditional fermented milk products; camel milk; kumisslike product.

Reference: Schulz, M. E. 1965, vol. 1, p. 450.

TABLE 7 The Chemical Composition of Some Israeli Fermented Fresh Milk Products

Product Name	Fat (%)	Protein (%)	Carbo-hydrate (%)	Total Solids (%)	Calories (kcal)	Calories (kJ)
Leben or Zivdah						
Skim	0.1	3.2	3.7	8.6	30	125.6
Low-fat	3.0	3.1	3.5	11.5	57	238.6
Plain	4.5	3.0	3.5	12.8	70	293.1
Yoghurt						
Plain	4.0	3.3	3.5	12.5	65	272.1
Fruit-flavored	3.0	3.2	12.5	20.5	90	367.8
With fruit added	1.5	2.4	19.5	75.0	100	418.9
Revion, Lebenit	1.5	3.2	3.6	10.5	48	201.0
Shamenet (cultured sour cream)	15.0	2.5	3.1	22.0	150	628.0

Source: Dr. Ionel Rosenthal, personal communication

ISRAELI FERMENTED FRESH MILK PRODUCTS (En); Produits laitiers fementés (frais) d'Israël (Fr); Sauermilchprodukte aus Israel (De).

Short Description: Israel; traditional and nontraditional products.

Food Value: The chemical composition of some Israeli fermented fresh milk products is given in Table 7.

Products: In Israel gil, kefir, kosher milk products, leben or zivdah, lebenié, revion, lebenit (cultured buttermilk), shamenet (cultured sour cream), yoghurt, and zabad are manufactured.

Reference: Israel, Standards Institution of Israel. 1987. (Cultured milk products). Israel Standard SI 285. Cited in *Dairy Science Abstracts,* 1988, **50**(3097):344.

This entry was contributed by Dr. Ionel Rosenthal,
Volcani Center, Bet-Dagan, Israel.

J

JALEBI (En, Fr, De).

Short Description: India; traditional fermented milk-cereal preparation; key product.

Microbiology: *Lactobacillus fermentum* (6×10^8/g), *Streptococcus lactis* (6×10^8/g). *L. büchneri* (3.2×10^8/g) and *S. faecalis* (6×10^8/g) have been isolated from fermented jalebi. *Saccharomyces* spp. have been found.

Manufacture: To make jalebi, refined flour (maida), dahi, and water are combined into a cohesive batter and fermented 14 to 16 h. The fermented batter pieces are deep-fat-fried in spiral shapes, immersed in sugar syrup for a minute or two, and are then ready for consumption. During fermentation the pH decreases from about 4.4 to 3.3. Both amino nitrogen and free sugar decrease. There is also a 9% volume increase in the batter.

Food Value: Popular dessert product.

Related Products: Milk-plant mixtures, fermented (mixed milk-cereal fermentation).

References:

FOA. 1982. Food composition tables for the near East. *FAO Food and Nutrition Paper 26,* p. 229. Rome: Food and Agriculture Organization of the United Nations.

Ramakrishnan, C. F. 1979. Studies on Indian fermented foods. The *Barodi Journal of Nutrition (India)* **6:**1-57.

Steinkraus, K. H. 1983. *Handbook of Indigenous Fermented Foods,* pp. 275-276. New York and Basel: Marcel Dekker.

JAMAKKA (En, Fr, De).

Short Description: Finland (South Karelia); traditional product made in rural areas, whey-drained; cow's milk.

Microbiology: Spontaneous mesophilic lactic fermentation.

Manufacture, Domestic: After milk has been soured by incubating on a warm stove, a soft coagulum develops. When cream is skimmed off and whey is also eliminated, the coagulum becomes firm.

Food Value: The coagulum (quarglike product) is consumed with warm milk.

Related Products: Concentrated fermented milks.

References:

Grotenfeld, G. 1906. *Contribution to the Knowledge of Domestic Milk Preparations in Finland* (in Swedish). Kuipo: K. Malmström.

Martiny, B. 1907, p. 57.

JHOPA MILK FERMENTED PRODUCTS (En); Lait jhopa fermenté (Fr); Gesäuerte Jhopa-Milchprodukte (De).

Short Description: Nepal; traditional products; jhopa is a yaklike animal that provides milk for human consumption; jhopa milk is used to make dahi.

Reference: Tokita, F., A. Hosono, T. Ishida, F. Takahashi, and H. Otani. 1980. Variety and manufacturing methods of native milk products in Nepal. *Journal of the Faculty of Agriculture, Shinsu University* **17**(2):119-123.

JOVO-COCKTAIL (En); Cocktail jovo (Fr); Jovo-Cocktail (De); trade name.

Short Description: Czechoslovakia; nontraditional yoghurt-based beverage with fruit pulp addition; cow's milk.

Reference: Hylmar, B. 1978. Cultured milk beverage (in Czech) *Prumysl Potravin* **29**(2):99-100. Cited in *Dairy Science Abstracts,* 1978, **40**(7122):752.

JUB-JUB (En, Fr, De).

Short Description: Lebanon; traditional yoghurtlike food.

Manufacture: Yoghurtlike food that may be concentrated and mixed with herbs, spices, and cereals, or, conversely, diluted with water and mixed with spices and herbs to make a soft-drink-like beverage.

Related Products: Milk-plant mixtures.

Reference: Kosikowski, F. V. 1977. *Cheese and Fermented Milk Foods,* 2nd ed., p. 70. Ann Arbor, Mich.: Edward Brothers, Inc.

JUNKET (En); Lait fermenté emprésuré (Fr); Dickmilch (gesäuert) (De).

Short Description: Denmark; nontraditional product; cow's milk.

Microbiology: Starter culture containing *Streptococcus lactis* subsp. *cremoris* (75%), *S. lactis* subsp. *diacetylactis* (20%) and *Leuconostoc mesenteroides* subsp. *cremoris* (5%).

Manufacture: Milk is coagulated with rennet and with weak acid production by above-mentioned starter culture composition.

Reference: Larsen, J. B. 1982. The content of carbohydrates and lactic acid in cultured liquid milk products. In *Proceedings 21st International Dairy Congress* **1**:299-300. Moscow: Mir Publishers.

K

KARAGHURT (En, Fr, De).

Short Description: Area between western Asia and Turkmenistan; traditional liquid whey product; whey serum obtained during the preparation of keshk (*see* Keshk).

References:
Fleischmann, W., and H. Weigmann. 1932, p. 371.
Martiny, B. 1907, p. 58.

KARAKOSMOS (En, Fr, De).

Short Description: Mongolia; traditional milk brandy; fermented mare's milk; mentioned as early as the thirteenth century.

Reference: Maurizio, A. 1933, p. 89.

KASHK (En, Fr, De); Kaskg.

Short Description: Middle East/Iran; traditional whey-drained product; ewe's milk or mixed ewe's and cow's milk; consumption with bread; dried product, salted; pleasant taste and aroma.

Microbiology: The starter culture consists of *Streptococcus thermophilus* and *Lactobacillus delbrueckii* subsp. *bulgaricus,* a yoghurt culture.

Manufacture: A fermented homemade milk product produced mainly from ewe's milk which is partially skimmed, boiled, and fermented with a yoghurt culture obtained from the previous day's yoghurt. After fermentation, the product is put in a cloth bag to remove much of the whey. The curd is salted and then shaped into balls of various sizes and dried in the sun (*see* Gibneh labneh).

Food Value: The final product contains the following in average percentages with the ranges in parentheses: 4.40% (2.0-8.3) moisture, 7.9% (2.8-14.5) fat, 3.8% (2.4-5.1) ash, 54.4% (39.6-67.8) protein and 8.7% (4.2-14.0) common salt (Taleban and Renner 1972). Prior to consumption, usually with bread, the balls are moistened with water and often flavored with various spices (*see* Keshk).

Related Products: Kishk, gibneh labneh, keshk.

Reference: Taleban, H., and E. Renner. 1972. Studies on kashk, an Iranian milk product (in German). *Milchwissenschaft* **27:**753-756.

KATYK (En, Fr, De); different spellings. The name *katyk* is used for the various traditional and industrial products mentioned below.

Beverage

Short Description: USSR (Ukraine); buffalo milk; alcohol-containing beverage similar to kumiss and made with *Torula* spp.

Reference: *FIL-IDF Dictionary.* 1983, p. 128.

Short Description: USSR (Kazakhstan); a quarg made from airan and consumed as beverage (called *quatig*) after dilution with water.

Reference: Kisban, E. In Földes, L. 1969, p. 524.

Cultured Milk

Short Description: USSR (Kazakhstan); of nutritional importance; the starter culture consists of *Streptococcus thermophilus, S. lactis* subsp. *diacetilactis* and *Lactobacillus helveticus* or *L. delbrueckii* subsp. *bulgaricus* in a ratio of 1:1:1.

Reference: Khakimova, Sh., and A. A. Rasulova. 1983. Synthesis of vitamins during batch and continuous cultivation of lactic acid bacteria in milk for the production of katyk (in Russian). *Izvestiya Vyyhikh Uchebnyk Zavedenii Pishch. Teknol.* **3**:55-57. Cited in *Dairy Science Abstracts,* 1984, **46**(2315):269.

Fermented Milk

Short Description: Central Asia; traditional, very popular fermented milk (*see* Ouzbek).

Long-life Fermented Milk

Short Description: Bulgaria; traditional product; ewe's milk; a barrel is filled with raw milk and boiled milk in a defined ratio, then salt is added and the barrel is closed. Katyk is only made in the early autumn months and consumed during the winter. The product develops a very viscous, quarglike consistency. It is a concentrated fermented milk, similar to brano milk.

Reference: Vakarelski, C. In Földes, L. 1969, p. 548.

Dried Katyk

Short Description: USSR (Tartar people); dried yoghurtlike product.

Reference: Fleischmann, W., and H. Weigmann. 1923, p. 371.

Quarg

Short Description: USSR (Kazakhstan); traditional quarg product.

Reference: Radloff, W. 1884. Cited by Ränk, G. 1969, p. 7.

KAYMAK (En, Fr, De); Kajmak; a word of Turkish origin meaning cream; also called *skorup.*

Short Description: Yugoslavia (originally in central and eastern areas); traditional fermented milk product; cow's, ewe's, and mixed milks; directly consumed with bread, sometimes in cooked and baked foods; pasty, salted; agreeable taste, specific aroma.

Manufacture: Kaymak is homemade in some villages of Serbia, Bosnia, Herzegovina, and Montenegro. It is made from curdled cream taken from boiled milk during the cooling process. The cream is naturally enriched with denatured whey proteins. Kaymak is salted with 2% salt and is put in special wooden containers for ripening at 15-18°C (59-64.4°F). Sometimes kaymak is also made from ewe's milk or mixed cow's and ewe's milks.

Food Value: The product made from cow's milk contains 40-60% fat, 6-9% protein, about 1% lactose, 3-4% minerals, 2% common salt, and about 40% moisture; the acidity ranges from 65°Th (0.58% titratable acidity) in the fresh, unripened product to 100°Th (0.90% titratable acidity) in the fully ripened product. Kaymak is either consumed when it is 2-3 days old, after 20-30 days of ripening, or after several months of storage. Sometimes it is used in such national cooked specialties as *cicvara* and *popara.*

References:

Gutschy, L. 1927. Contribution to studies of the mycology of Kaymak (in French). *Le Lait* **7:**113, 256.

Vitkovic, D., and J. Lj. Rašić. 1963. Let's look at Yugoslavia. *Dairy Industries* **4:**324-328.

KEFIR (En, De); Kéfir (Fr). The name is derived from *kefy* in the Caucasus region or *kef* from the Turkish, denoting pleasant taste.

Short Description: USSR (northern Caucasus); traditional product, key product; cow's milk, goat's milk, or other types of milk; direct consumption; stirred beverage; creamlike consistency; characteristic taste and aroma.

History: Kefir is a very old fermented milk. There is no record of the date or time when it was first made. It originated in the northern Caucasus region and two tribes, the Ossetians and Karbadinians, are specifically mentioned as the first kefir manufacturers. These people made refreshing drinks from cow's milk and goat's milk in their homes using kefir grains. The kefir grains had probably developed accidentally and their function and value was apparently recognized. Fresh milk was fermented in leather bags (made from goat skins), which were hung in the house during the winter and outside during the summer. Whenever some of the fermented drink was taken from the bags a new lot of fresh milk was added. The product was sour and contained variable amounts of alcohol and carbon dioxide (Schulz 1968; Koroleva 1975). In the second part of the nineteenth century kefir making spread to eastern and central Europe and from there to other regions. At present, kefir is produced in large quantities in the USSR and in smaller quantities in Poland, Germany, Sweden, Rumania, and other countries.

Microbiology: Traditional manufacture of kefir involves the use of kefir grains. In the Orient these grains are sometimes called "Millet of the Prophet." They resemble miniature cauliflower florets: They are yellowish in color, 1-2 mm to 3-6 mm or more in diameter, and their surfaces are rough and convoluted. Kefir grains contain a complex microflora in an immobilized system that is composed of a polysaccharide called *kefiran* that is of microbial origin. Some milk fat and denatured milk protein is usually associated with this polysaccharide matrix. The microflora includes lactic acid streptococci, leuconostocs, thermophilic lactobacilli, mesophilic lactobacilli, yeasts (lactose-fermenting and lactose-nonfermenting), and often also acetic acid bacteria.

The microbial composition of kefir grains may vary depending on the origin of the grains and the method of their cultivation. The different microbial species include: *Streptococcus lactis, S. lactis* subsp. *cremoris, Leuconostoc* spp., *Lactobacillus kefir, L. casei,* sometimes *L. acidophilus, Candida kefyr, Kluyveromyces marxianus* subsp. *marxianus,* other kinds of yeasts, and *Acetobacter aceti.* When milk containing kefir grains is incubated, microorganisms are shed from the grains into the milk and begin growth and fermentation. Kefir culture (mother culture) from which the grains have been removed cannot be used for successive propagation because the balance of the microflora is lost. A good-quality kefir culture is prepared by the proper cultivation of the grains. One milliliter (milk without grains) of such a culture contains 10^8-10^9 lactic acid streptococci, 10^7-10^8 leuconostocs, 10^5 thermophilic lactobacilli, 10^2-10^3 mesophilic lactobacilli, 10^5-10^6 yeasts, and 10^5-10^6 acetic acid bacteria (Koroleva 1982).

Manufacture: Kefir is manufactured with a kefir culture or with a production culture.

To prepare using a kefir culture, skim milk is pasteurized at 95°C (203°F) for 10-15 min, then kefir grains are added to the milk (inoculation ratio is 1:30 to 1:50), followed by incubation at 18-22°C (64.4-71.6°F) for about 24 h and stirring two times. Kefir grains are sieved out and added to a new lot of milk; the cultured milk without grains represents a kefir culture.

To prepare kefir with production culture, skim milk is pasteurized at 95°C (203°F) for 10-15 min, then kefir culture is mixed in at a rate of 2-3%. The incubation is at 20-22°C (68-71.6°F) for 10-12 h, followed by ripening and slow cooling at 8°C (46.4°F) for about 12 h. The use of kefir culture in manufacturing kefir results in better product properties than does the use of production culture. However, adding small amounts of kefir culture to the production culture improves the flavor of the final kefir.

Kefir is made mainly from cow's milk, although sometimes other milks are used.

Food Value: The final product has an acidity of 110-115°Th (1.0% titratable acidity), and contains 0.01-0.1% ethanol and small amounts of carbon dioxide. One milliliter of a good-quality kefir contains 10^9 lactic streptococci, 10^7-10^8 leuconostocs, 10^7-10^8 thermophilic lactobacilli, 10^4-10^5 yeasts and 10^4-10^5 acetic acid bacteria (Koroleva 1975). In addition to whole milk kefir, there are also low-fat and skim milk kefirs, all usually containing not less than 11% nonfat solids. There is also kefir with 6% fat or more. Fruit kefir, kefir from buttermilk, freeze-dried kefir (*see* Kefir, freeze-dried) are other products.

Kefir is regarded as a refreshing, nutritious drink. It is suitable for inclusion in special dietetic programs. Wherever available, it has also been used as an aid in the therapy of gastrointestinal disorders.

Related Products: Kefir-products.

References:

Klupsch, H. J. 1984. *Sour Milk Products—Mixed Milk Beverages and Desserts* (in German). Gelsenkirchen-Buer: Th. Mann.

Koroleva, N. S. 1975. *Technical Microbiology of Whole Milk Products* (in Russian). Moscow: Pishchevoy Promÿshlennosti.

Koroleva, N. S. 1982. Special products. *Proceedings 21st International Dairy Congress* **2:**146-151. Moscow: Mir Publishers.

Mann, E. 1984. Kefir and kumys (in French). *Revue Laiterie Francaise* **429:**38-39.

Schulz, M. E. 1968. One hundred years of kefir in Northern Europe and its importance today (in German). *Deutsche MolkereiZeitung* **13.**

KEFIR, FREEZE-DRIED (En); Kefir lyophilisé (Fr); Lyophilisierter Kefir (De).

Short Description: USSR; nontraditional product; cow's milk; consumption after reconstitution; dried product.

Manufacture: The product is made from pasteurized milk containing 36% total solids, which is fermented using kefir grains (*see* Kefir). The kefir is then freeze-dried according to the following method: the rate of freezing is 0.9°C per minute, the final temperature of freezing is between −15°C and −18°C, and the maximum temperature on the surface is between 35°C (95°F) and 38°C (100.4°F). The time of freeze-drying is reduced by 30-50% as compared with the freeze-drying of kefir containing about 12% total solids.

Food Value: The final product contains 3-4% moisture. After reconstitution with tap water to 12% total solids the product possesses all the microbiological and biochemical characteristics expected of kefir.

Related Products: Freeze-dried preparations.

Reference: Gouliaev, V. L., and N. G. Alexeiev. 1982. Intensification of freeze-drying process of culture milk products. *Proceedings 21st International Dairy Congress* **1**(1):282. Moscow: Mir Publishers.

KEFIR-BUTTERMILK (En); Babeurre au Kéfir (Fr); Kefir-Buttermilch (De).

Short Description: Different countries; traditional product. Related to cultured buttermilk but prepared from skim milk with kefir cultures instead of buttermilk cultures.

Reference: Mann, E. 1984. Kefir and kumys (in French). *Revue Laiterie Francaise* **429:**38-39.

KEFIR-CULTURED MILK (En); Lait fermenté "kéfir" (Fr); Kefir-Sauermilch (De).

Short Description: Germany, Switzerland, Canada; nontraditional product; cow's milk; set type mainly.

History: There are several commercial products available under the name kefir, but instead of being true kefirs they are only cultured milks.

Microbiology: The microflora in these products includes added baker's yeast, and a cream culture (*Streptococcus lactis* and/or *S. lactis* subsp. *cremoris* in combination with *Leuconostoc mesenteroides* subsp. *cremoris* and/or *S. lactis* subsp. *diacetilactis*) or a yoghurt culture (*S. thermophilus* and *Lactobacillus delbrueckii* subsp. *bulgaricus*), none of which originated from kefir.

Manufacture: The addition and development of yeast must be carefully controlled (maximum 1×10^5 to 3×10^5/ml). By doing so, carbon dioxide and, therefore, package bulging is limited; also whey separation in this set-type coagulum would then be prevented.

Related Products: Kefirlike products.

References:

Duitschaever, C. L. 1985. Formulation of a Mixed Culture for the Production of Kefir. Guelph, Canada: University of Guelph.

Kurmann, J. A. 1985. *Fermented Milks* (in French). Grangeneuve, Switzerland: Agriculture Institute.

KEFIRLIKE PRODUCTS (En); Produits ressemblant au kéfir (Fr); Kefirartige Produkte (De).

Short Description: Various countries; nontraditional products; cow's milk; direct consumption; stirred beverage; various flavors, sometimes similar to cultured buttermilk.

Microbiology: Difficulties in preparing a good-quality kefir culture have led to attempts to replace kefir grains with single-use starters. These consist of pure microorganisms, usually isolated from the kefir grains. The variable composition of these starters, sometimes containing nearly only lactic acid streptococci, results in the manufacture of kefir of varying organoleptic properties frequently devoid of characteristic kefir flavor. New methods subsequently substantiated the claim that improved quality kefir can be produced by using single-use starters. Some examples are mentioned below.

Manufacture: A method developed in Poland (Kramkowska et al. 1982) refers to the preparation of a single-use culture for the production of kefir starter. It consists of mixing kefir grains with neutralized kefir in a mixer, freeze-drying, and then mixing the solids with a dried yeast preparation in such a proportion as to obtain 10^4/g yeast cells upon vivifying. Reportedly, the quality of kefir produced is similar to that obtained when using kefir grains.

A method developed in Sweden (Pettersson 1984) refers to the preparation of freeze-dried concentrated kefir starter for use as a bulk starter milk inoculum. It consists of separate cultivation, concentration, freeze-drying of the various kefir microorganisms with subsequent mixing of the dried culture at a desired ratio, for example, lactic acid streptococci:citric acid fermenting streptococci:lactobacilli:yeasts, approx. 75:24:0.5:0.1. Kefir is then made by incubating pasteurized milk with 1% of a liquid bulk starter at 22°C (71.6°F) for 18-22 h and subsequent cooling, stirring, and packaging. Reportedly, starter activity and kefir quality are uniform. It also allows for changing the starter composition whenever needed or desired.

Another method developed in Germany refers to the use of two types of deep-frozen concentrated starter for making kefir. One type, consisting of kefir lactic acid streptococci and lactobacilli, is used in the first stage of kefir manufacture to ferment pasteurized milk at 24-27°C (75.2-80.6°F) for 16-18 h until a pH of 4.40 is obtained, followed by cooling. The other type, consisting of yeasts and lactobacilli, is propagated in milk at 31-33°C (87.8-91.4°F) for 22 h until a pH of 3.70 is obtained, followed by cooling. This latter culture is added to the fermented milk at a rate of 0.05% (in summer) or 0.5% (in winter) and held 24 h at 5°C (41°F). A shelf life of 24 days is claimed for the final products (Rašić 1987).

The use of starters consisting of pure microorganisms in the manufacture of kefir has potential in the creation of a variety of products with kefirlike flavor.

References:

Kramkowska, A., K. Kornacki, B. Bauman, and D. Fesnak. 1982. Method of production of a single-use kefir culture. *Proceedings 21st International Dairy Congress* **1**(1):304. Moscow: Mir Publishers.

Pettersson, H. E. 1984. Freeze-dried concentrated starter for kefir (Poster No. 16). *IDF/FIL Bulletin Document 179.* Brussels: International Dairy Federation.

Rašić, J. Lj. 1987. Other products. In *Milk—The Vital Force,* ed. Organizing Committee of the 22nd International Dairy Congress, pp. 673-682. Dordrecht: D. Reidel.

KESÄVELLI (En, Fr, De); denotes summer (sour) milk soup.

Short Description: Finland; traditional product; buckwheat or oatmeal is added immediately prior to consumption (*see* Milk-cereal preparations).

Reference: Martiny, B. 1907, p. 63.

KESHK (En, Fr, De); Keschk.

Short Description: Middle East, eastern Asia to Turkestan; traditional whey-drained product; ewe's, goat's, cow's, and mixed milks; consumed with bread; dried product, salted; agreeable sour salty taste and aroma, dark colored.

Manufacture: A fermented milk product homemade from buttermilk, which is put into a cloth bag to remove the whey. The curd is salted and shaped into balls and dried in the sun.

Food Value: Before consumption, the balls are moistened with water. They are eaten with bread.

Related Products: Kishk, kashk.

References:

Fleischmann, W., and H. Weigmann. 1932, p. 371.
Siegenthaler, E. 1968. Two procedures for cheesemaking in the tropics and emerging countries. *Milchwissenschaft* **23:**623-629.

KHERAN (En, Fr, De); sometimes written karan or heran.

Short Description: USSR (Siberia, Tartar people), probably originated in Tibetan monasteries (lamaseries); traditional fermented milk food; type of milk not known, probably milk from various species.

History: Kheran owes its name to the Tartar conqueror Timur, also known as Tamerlane or Tamburlaine (1336-1405), who had his capital at Samarkand and ruled over vast territories in central Asia and a large part of India.

Microbiology: Microflora is not well known. According to FIL-IDF (1983) the milk is coagulated by *Streptococcus thermophilus* Kheran.

Manufacture: Original kheran contained a very small amount (a few grams per liter) of an extract of apples (sour varieties or not-quite-ripe apples).

Food Value: Claimed to restore strength eroded by fatigue (Sheb-Kheran) and to cure illness caused by digestive upsets and liver disorders.

Related Product: Kheran®.

References:

FIL-IDF Dictionary. 1983, p. 129.
Kopytine, N. 1956. "Kheran," fermented milk of the Siberian Tartars (in French). *Annales de Gembloux* (Belgium) **62**(1):1-50.

KHERAN® (En, Fr, De); trade name.

Short Description: Belgium (prepared by Kopytine); nontraditional commercial product; cow's milk; stirred, filled into bottles; creamy color, consistency is firm and compact without syneresis; pH 4.6-4.7; good keeping quality; the taste is very mild and sweet, resembling slightly sour cream.

Microbiology: The starter culture consists of *Streptococcus thermophilus* Kheran, isolated from human feces using a special procedure, and *Lactobacillus delbrueckii* subsp. *bulgaricus,* a slow-acid variant (considered by Kopytine to be active in the intestine).

Manufacture: Kheran® is made from pasteurized or sterilized homogenized milk that is coagulated exclusively through the controlled lactic acid fermentation by means of special lactic cultures. The incubation is at 42-45°C (107.6-113°F) for 2.5 h until an acidity of 82.5°Th (0.74% titratable acidity) is produced. The final product is yoghurtlike.

Related Product: Kheran.

Reference: Kopytine, N. 1956. "Kheran," fermented milk of the Siberian Tartars (in French). *Annales de Gembloux* (Belgium) **62**(1):1-50.

KHOORMOG (En, Fr, De); Khormog.

Short Description: Mongolia; traditional fermented milk product; made from various milks (cow, yak, ewe, camel) but not from mare's milk; lactic and alcoholic fermentation; utilized mainly for its alcohol content; the manufacture is similar to that of airag; a related fermented milk is umdaa.

References:

Accolas, J. P., J. P. Defontaines, and F. Aubin. 1978. Milk and dairy products in the Mongolian Republic (in French). *Le Lait* **58:**284.

FIL-IDF Dictionary. 1983, pp. 129, 220.

KISELA VARENIKA (En, Fr, De); denotes boiled milk which has been allowed to turn sour.

Short Description: Yugoslavia (Bosnia, Croatia); traditional product; cow's, ewe's, or mixed milks; national name for plain set yoghurt or soured milk home-made from boiled milk fermented with empirical starter culture; direct consumption; firm consistency; acidic taste and agreeable aroma.

Reference: Rašić, J. Lj. 1982. Special products. *Proceedings 21st International Dairy Congress* **2:**151. Moscow: Mir Publishers.

KISELO MLEKO (En, Fr, De); denotes sour milk.

Short Description: Yugoslavia (central and eastern areas); traditional product; ewe's, cow's, or mixed milks; used for direct consumption; set; acidic, agreeable taste and aroma.

History: Kisělo mleko is the national name for the plain set yoghurt made in homes from boiled ewe's, cow's, or mixed milks.

Microbiology: The starter culture consists of *Streptococcus thermophilus* and *Lactobacillus delbrueckii* subsp. *bulgaricus.*

Manufacture: The milk is inoculated with 2-3% of the previous day's yoghurt, called *maya,* and incubated in special containers, often earthenware cups, until coagulation, and then cooled. Commercial manufacture involves the use of processed milk, modern equipment, and pure yoghurt cultures.

Related Products: Kiselo mljako, yoghurt.

Reference: Vitković, D., and J. Lj. Rašić. 1963. Let's look at Yugoslavia. *Dairy Industries* **4:**324-328.

KISELO MLEKO, SLANO (En, Fr, De); denotes sour milk, salty.

Short Description: Yugoslavia (Macedonia); traditional products; ewe's or cow's milk or a mixture of both; soft, compact.

Microbiology: The starter culture consists of *Streptococcus thermophilus* and *Lactobacillus delbrueckii* subsp. *bulgaricus.*

Manufacture: In one specific description, cow's milk with Danish yoghurt culture was incubated for 3 h at 43°C (109.4°F), cooled to 5°C (41°F), put in cloth bags 24 h later, was then salted and allowed to drain for 48 h and salted again to desired taste.

Food Value: The composition of the final product is 16.5% fat, 10.6% protein, 2.67% common salt.

Reference: Lazarevska, D., T. Cizbanovski, and N. Kapac-Parkaceva. 1975. Composition, characteristics and organoleptic quality of salted soured milk (in Serbo-Croatian). *Mlekarstvo* **25:**58-64. Cited in *Dairy Science Abstracts,* 1976, **38**(719):75.

KISELO MLJAKO (En, Fr, De); denotes sour milk.

Short Description: Bulgaria; traditional product; ewe's, cow's, buffalo's, or mixed milks; used for direct consumption; set products; acidic taste and aroma.

History: Kiselo mljako is the national name for the plain set homemade yoghurt made from boiled milk.

Microbiology: The starter culture consists of *Streptococcus thermophilus* and *Lactobacillus delbrueckii* subsp. *bulgaricus.*

Manufacture: The milk is inoculated with 1.5-2% of the previous day's yoghurt, called *podkvasa,* incubated in special containers until coagulation, and then cooled. Commercial manufacture involves the use of processed milk, pure yoghurt culture, controlled fermentation, and cooling.

Related Products: Kiselo mleko, yoghurt.

Reference: Nikolov, N. 1962. *Bulgarian Yoghurt and Other Milk Products* (in Bulgarian). Sofia: Zemizdat.

KISHFA (En, Fr, De).

Short Description: Iraq; ewe's milk; fermented milk product.

Reference: Khalaf. S. H., A. Y. Shareef. 1985. The bacteriological quality of kishfa and yoghurt in Mosul city. *Food Microbiology* **2:**241-242. Cited in *Dairy Science Abstracts,* 1986, **48**(2727):321.

KISHK (En, Fr, De); Kish.

Short Description: Middle East/Egypt; traditional product, key product; a mixture of concentrated sour buttermilk and boiled ground wheat kernels;

fermented dry pieces; direct consumption after moistening with water; acidic, salted taste and sour aroma.

Microbiology: The starter culture consists of lactic acid bacteria, principally lactobacilli (*see* Laban zeer).

Manufacture: The product is traditionally made from concentrated salted sour buttermilk, called laban zeer, and sometimes from concentrated salted whole milk soured with the addition of boiled, dried, and ground wheat kernels. The wheat grains are boiled in water until they are soft, then they are spread on straw mats, dried in the sun, and crushed using a stone hand mill. The desired amount of crushed wheat, after seed coat removal, is put in an earthenware container and is moistened with slightly salted boiling water. Laban zeer, diluted with water or raw milk, is then added until a weak, smooth paste is obtained. The container is covered and the contents allowed to ferment with lactic acid bacteria, principally lactobacilli, for 24 h or more. The fermented mass is mixed well, sometimes spiced with heat-treated powdered black pepper, and then shaped into small, round or irregular pieces. These are placed on mats and sun-dried for several days.

Kishk is stored in mud chambers similar to those used for storing wheat grains.

Food Value: The final product contains 5.4-6.9% moisture and 11.2-14.5% protein, in addition to carbohydrate, fat, ash, and common salt.

Related Products: Trahanas, tarhana.

References:

El-Gendy, S. M. 1983. Fermented foods of Egypt and the Middle East. *Journal of Food Protection* **46:**358-367.

Farr, D. 1980/81. Traditional fermented foods: a great potential for the future. *Nestlé Research News.* Switzerland.

KIS YOGHURT (En); Kis yaourt (Fr); Kis Joghurt (De); denotes winter yoghurt (consumed during the winter months, prepared in spring or summer).

Short Description: Turkey (towns in the Mediterranean region); traditional domestic, concentrated fermented milk; usually goat's milk, but also made from ewe's and cow's milks; heat-treated after culturing.

Microbiology: The starter culture consists of *Streptococcus thermophilus* and *Lactobacillus delbrueckii* subsp. *bulgaricus;* destruction of culture bacteria after boiling.

Manufacture: There are two different procedures for making kis yoghurt. The first one depends on boiling fresh yoghurt, the second procedure consists of pouring yoghurt into cloth sacks suspended in a clean place to remove whey. The curd is then transferred to metallic vessels and mixed with water, followed by heating the mixture. The heat treatment of the two procedures involves boiling for 1-1.5 h with stirring. Approximately 1% salt is added. The yoghurt is then cooled at room temperature. After cooling, the product is transferred into suitable containers and paraffin, olive oil, or liquefied animal fat is used to cover the surface of the yoghurt to prevent contamination and mold growth. The container is covered and stored in a cold place. Under these conditions, the product can be stored for 6-7 months. The final product has a creamy and sandy texture.

Food Value: The product can be consumed at breakfast like white soft cheese after first mixing with herbs or spices and olive oil; it can also be substituted for fresh yoghurt when preparing certain dishes. In the villages, a family usually makes approximately 20 kg of this product every year to cover its needs. The average composition is 32.18% total solids, 11.21% protein, 9.65% fat, 5.69% lactose, 4.24% salt. The acidity is 357.5°Th (3.22% titratable acidity).

Related Products: Concentrated fermented milks.

This entry was contributed by Prof. Hasan Yaygin,
Izmir, Turkey.

KJÄLDER MJÖLK, Kjaeldermelk (in Norwegian); Lait de cave (Fr); Kellermilch (De); the word means cellar milk.

Short Description: Norway; traditional fermented milk product with good keeping quality; the fermentation involves hydrolysis of the lactose by lactic acid bacteria with subsequent conversion of the hexoses by *Torula* yeast species. The final product or culture contains about 0.5% ethanol and has 220°Th or more (2.0% or more titratable acidity).

Reference: *FIL-IDF Dictionary.* 1983, p. 129.

KLILA (En, Fr, De).

Short Description: Southern Algeria; traditional product; ewe's milk; dried pellets obtained from leben.

References:

Harrati, E. 1974. Studies on Algerian lben and klila (in French). Thesis, University of Caen, France.
Wolf, C. 1932. The milk in the colonies (in French) *Le Lait* **12:**225.

KOJURTNAK (En, Fr, De).

Short Description: USSR (Kazakhstan and Kirghizia); traditional product; ewe's and goat's milk; domestic manufacture.

Manufacture: The product is made by shepherds who put milk into a wooden or leather bowl and then boil it by adding a red-hot stone. The cooked milk is poured into the stomach obtained from a wether and is allowed to ferment after mixing with kumiss in a 1:2 ratio. The taste is airanlike.

Reference: Rudenko, S. I. 1969. Studies about the nomads (in German). In *Földes,* L., p. 29.

KOKKELI (En, Fr, De).

Short Description: East Finland; traditional homemade whey-drained fermented milk product; the milk is fermented and kept warm in an oven to facilitate whey separation; the kokkeli is consumed as such or is diluted with milk and drunk as a beverage; it is a concentrated fermented milk.

Reference: Grotenfelt, G. 1966. Bidrag till kännedomen of mjölkhushallningens utveckling i Finland (in Swedish). *Den allmänna, finska landbruksutställningen.* Kuopio: K. Malmström, p. 7. Cited in Forsén, R. pp. 10-11.

KOLDSKÅL (En, Fr, De), denotes refreshing preparation in a bowl or a cup ready for consumption in summer.

Short Description: Denmark; nontraditional buttermilk preparation; cow's milk.

Manufacture: Koldskål is prepared as a beverage and a dessert. To prepare the beverage, UHT-treated sweet buttermilk (0.3-1.0% fat) is fermented with mesophilic starter cultures and then mixed with sugar and concentrated lemon juice. To prepare the dessert, lemon peel, ground wheat, and usually seasoned crackers or biscuits are added to buttermilk in a bowl.

Reference: *Tetra Pak Review.* 1984. **60:**54.

KOLOMENSKI (En, Fr, De).

Short Description: USSR; nontraditional product; cow's milk; fermented lowfat or skim milk; beverage; creamy viscosity; agreeable taste and aroma.

Microbiology: The starter culture consists of *Lactobacillus delbrueckii* subsp. *bulgaricus* and *Streptococcus lactis* subsp. *diacetilactis* (ratio 1:3).

Manufacture: The product is made from heat-treated, homogenized milk (either 2.5% or 1.0% fat) or skim milk. The milk is fermented with the above-mentioned culture for 3-4 h and then cooled.

Food Value: The product is consumed plain, sweetened with sucrose, or flavored with strawberry syrup. It may be fortified with vitamin C.

References:

Zobkova, Z. S., E. A. Bogdanova, and I. I. Kočegina. 1978. New types of milk products (in Russian) *Molochnaya Promyshlennost* **4:**16-17. Cited in *Dairy Science Abstracts,* 1978, **40**(7126):753.

Zobkova, Z. S., I. M. Padaryan, L. G. Mytnik, K. G. Egorova, and G. L. Plish. 1978. Method of obtaining a "Kolomenski" type soured milk beverage (in Russian) *USSR Patent* 626 750. Cited in *Dairy Science Abstracts,* 1979, **41**(7496):823-833.

KÖÖRZIK (En, Fr, De).

Short Description: USSR (Kalmyk Republic), Mongolian origin; traditional beverage; cow's and/or mare's milk; alcohol-milk beverage for infants; made by mixing araq (distilled from airan) with fresh cow's milk.

Reference: Bergmann, J. 1804/1805. *Nomadic Incursions Among the Kalmyks in the Years 1802 and 1803,* 4 parts (in German) Riga. Cited in Takamiya, T. 1978, p. 102.

KOROT (En, Fr, De).

Short Description: USSR (originally Bashkirian region of eastern Russia); nontraditional whey-drained product; cow's skim milk.

Microbiology: The starter culture consists of separate cultures of lactic yeast and *Lactobacillus acidophilus* and *Lactobacillus delbrueckii* subsp. *bulgaricus.*

Manufacture: Skim milk is pasteurized, cooled to 30-45°C (86-113°F) and cultured first with "lactic" yeast at 30-32°C (86-89.6°F) for 6 h and then with 8% of an acidophilus culture at 38-40°C (100.4-104.0°F) until the acidity is 230-250°Th (2.07-2.25% titratable acidity). Subsequently, *Lactobacillus delbrueckii* subsp. *bulgaricus* culture is added at a rate of 8-10% and the incubation is continued at 40-45°C (104-113°F) to an acidity of 350-400°Th. The product is stirred constantly. It is then heated to 92-95°C (197.6-203°F) with vigorous stirring, cooled, drained, pressed to contain 55-58% moisture, and passed through a mill. After the addition of an acidophilus culture (1-2%) and 0.01% ascorbic acid, salt and spices or sugar are incorporated and the product is packed in glass or ceramic containers or foil tubes or is dried to about 15% moisture, milled, and pressed into brickettes or tablets.

Food Value: It is used in soups or is dissolved in water to make a refreshing beverage.

Reference: Kirdoda, I. 1963. Bashkir cheese "Korot" (in Russian). *Molochnaya Promyshlennost'* **24:**23-24. Cited in *Dairy Science Abstracts,* 1963, **25**(1332):190.

KOS (En, Fr, De).

Short Description: Albania; traditional product; cow's or ewe's milk; national name for plain set yoghurt.

Reference: Nikolov, N. 1962. *Bulgarian Yoghurt and Other Fermented Milk Products* (in Bulgarian). Sofia: Zemizdat.

KOSAMA-MILK (En); Lait "Kosama" (Fr); Kosama-Milch (De); trade name; a word created to mean colloidal, sour skim milk, from the German: *Ko* (Kolloidale/colloidal), *sa* (saure/fermented), *ma* (Magermilch/skim milk).

Short Description: Germany (Rhine/Ruhr region); nontraditional product; skim cow's milk; a cultured buttermilk.

Reference: Mohr, J., W. Koenen. 1958. *Butter* (in German), pp. 389-390. Hildesheim: Th. Mann K. G.

KOSHER MILK PRODUCTS (En), Produits laitier "Kosher" (Fr), Kosher Milchprodukte (De); denotes that the food has been prepared according to or is served ritually appropriate according to Jewish law.

Short Description: Israel; Jewish law; traditional kind of preparation; milk produced by animals that are ruminants and have cloven hooves, for example, cows, sheep, goats, deer. The milk of mares, camels, and others, is not permitted and is therefore not kosher; kosher does not mean "pure," "hygienic," or "unadulterated," although it is frequently perceived as that by some people.

History: Milk and milk products are mentioned more than fifty times in the Bible.[1,2] Nearly half of the time it is in reference to the abundance and beauty of the promised land of Israel, which is described as "a land flowing with milk and honey".[3] The milk most commonly mentioned in the Bible is that of sheep and goats,[4] but cow's milk was also known.[5] Milk is considered one of the finest of foods[6] and is used as a term of abundance,[7] as a standard of whiteness[8] and, with honey, as a standard of sweetness.[9] Milk was sometimes used for its calming if not soporific qualities.[10]

The halakha (Jewish law) limits the consumption of animals and animal products, including milk, to those derived from ritually slaughtered kosher[11] animals. Milk is one of the few exceptions to the general prohibition of ingesting animal food products derived from living animals.[12] There is no requirement in Jewish law regarding the fat, protein, or water content of milk or its derivatives, provided there is no admixture of milk or other products (including vitamins) derived from nonkosher animals. The absence of such undesired additives is guaranteed by Rabbinical supervision, although strict secular governmental supervision is also acceptable according to many religious authorities.[13] The rennet used to coagulate milk, as well as any other enzymes used in the preparation of cheese, must come from ritually slaughtered kosher animals. The introduction of fungal enzymes relieves to some extent the need for kosher enzymes.

During Passover, Jews refrain from eating or deriving benefit from *chametz*. The latter includes grain-flour-containing products (e.g., bread, cakes, crackers, cereals, etc.) for which the traditional precautions preventing leavening have not been taken. For Passover use, traditional Jews buy only milk obtained with Rabbinical supervision, which guarantees that even inadvertent admixture of *chametz* has been prevented. Of importance to Jewish dairymen is the fact that milk is forbidden for consumption if the cows were fed *chametz* feed during Passover. Hence, in the modern State of Israel nongrain feed is used for dairy cows during this holiday.

Food Value: Perhaps the most intricate of Jewish dietary laws centers around the separation of meat and milk. Three times we find in the Scriptures the prohibition, "Thous shalt not seethe a kid in its mother's milk."[14] According to Rabbinical tradition,[15] these three verses refer to three distinct prohibitions: (1) not to cook meat and milk together, (2) not to eat such a mixture; and (3) not to derive any benefit from such a mixture. "Milk" includes all kosher dairy products, while "meat" traditionally includes Kosher animal and poultry flesh, to the exclusion of fish. "Halakha" further ordained that the separation of meat from milk be as complete as possible. Thus, separate utensils, pots, dishes and cutlery must be used for dairy and meat foods. While meat and milk may not be eaten together at the same meal, imitation milk, as well as nondairy coffee whiteners and margarine, may be used.

Overall, it should be noted that the Jewish dietary laws are quite complex and extensive and the reader is referred to several advanced texts.[1b, 1c]

Notes and References:

1. For general discussions in English of milk and milk products in Jewish tradition and law, see: (a) *Encyclopedia Judaica,* 1971, vol. 5, Cheese, p. 371; vol. 6, Dietary Laws, p. 40; vol. 11, Milk, p. 1578; Jerusalem: Keter Publishing House Ltd. (b) I. Grunfeld, 1975, *The Jewish Dietary Laws,* vol. 1. N.Y.: The Soncino Press. (c) J. Cohn, 1970, *The Royal Table,* New York, N.Y: Feldheim Publishers.
2. S. Mandelkern, 1971, *Concordantiae,* 9th ed. Tel Aviv: Schocken Press. See *chalav, chema,* and *gevina.*
3. See for example, Exodus 3:8, 17; Numbers 13:27; Deuteronomy 6:3; Jeremiah 11:5.
4. For example, Exodus 23:19; Deuteronomy 32:14; Proverbs 27:27.
5. Isaiah 7:21-22.
6. Deuteronomy 32:14; Isaiah 55:1.
7. Joel 4:18; Isaiah 60:16.
8. Genesis 49:12; Lamentations 4:7.
9. Song of Songs 4:11.
10. Judges 4:19.
11. The word *kosher* means prepared in accordance with Jewish dietary laws. Because of the strict supervision required for kosher products, kosher in common parlance has come to mean proper, pure, unadulterated.
12. Babylonian Talmud, Bechorot 6 b.
13. M. Feinstein. 1959. Responsa 48 and 49. In *Responsa Igrot Moshe,* Yoreh Deah, New York: Balshon.

14. Exodus 23:19, 34:26; Deuteronomy 14:21.
15. Babylonian Talmud, Hullin 115 b.

This entry was contributed by A. A. Frimer,
Department of Chemistry, Bar-Ilan University,
Ramt Gan, Israel.

KRAMERIJ (En, Fr, De).

Short Description: Netherlands; traditional homemade preparation; a cooked mixture of skim milk and true buttermilk (no living bacteria).

Reference: Martiny, B. 1907, p. 66.

KUBAN (En, Fr, De); Kouban.

Short Description: USSR (district Don); traditional beverage similar to kumiss and kefir; lactic acid and alcoholic fermentation.

Microbiology: The microflora consists of a lactic acid-producing streptococcus much like *Streptococcus hollandicus,* a lactic acid-producing, rod-shaped bacterium much like *Lactobacillus delbrueckii* subsp. *bulgaricus,* and three types of yeast (*Mycoderma, Torula lactis* and other undefined yeasts).

Related Products: Kefirlike products, kumiss, and kumisslike products.

References:

Bogdanov, V. M. 1934. An investigation of kuban, fermented milk. *Journal of Dairy Research* **5:**153-159.
FIL-IDF Dictionary. 1983, p. 129.

KUMISS (En, De); Koumis (Fr); other spellings are kumys, kumyss, koumiss.

Short Description: USSR, Mongolia, Tibet; traditional product, a key product; mare's milk; direct consumption; a frothy beverage; grayish-white; liquid homogeneous consistency; acid, refreshing taste and characteristic aroma.

History: Kumiss probably originated in the central Asian steppes and derives its name from the ancient Kuman tribes or Kumyks. Herodotus in the fifth century B.C. described how the Scythians made kumiss from raw mare's milk. Marco Polo, the first European traveler to the Far East in the thirteenth century mentioned kumiss as a favorite drink of the Mongols. Kumiss is now produced in limited quantities, mainly in the USSR, for therapeutic purposes. Small quantities are also made in Germany.

Microbiology: The starter culture consists of *Lactobacillus delbrueckii* subsp. *bulgaricus* and lactose-fermenting yeasts.

Manufacture: The original manufacture of kumiss involved the use of raw mare's milk and an empirical starter culture derived from previously produced kumiss. Now kumiss is made from carefully pasteurized mare's milk, although other kinds of milk are sometimes used. Mare's milk has lower concentrations of fat, protein, and ash than cow's milk, but it has a higher lactose content. The protein of mare's milk contains less casein than cow's milk, but it has more whey protein. Hence, in fermented mare's milk the proteins coagulate without changing the texture of the product. Kumiss is made from pasteurized mare's milk by adding about 30% of the above-mentioned starter culture, then allowing it to ferment at 26-28°C (78.8-82.4°F) for 2-4 h. For the first 30-60 min of the fermentation the milk is stirred, followed by intermittent stirring. After fermentation, the kumiss is bottled (while stirring), capped, and stored for ripening at 4-6°C (39.2-42.8°F) for 1-3 days. The major products of fermentation are lactic acid, ethanol, and carbon dioxide. Three types of kumiss are made: (1) weak (ripened one day), acidity 60-80°Th (0.54-0.72% titratable acidity), ethanol content 0.7-1.0%; (2) medium (ripened 2 days), acidity 81-105°Th (0.73-0.95% titratable acidity), ethanol 1.0-1.7%; and (3) strong (ripened 3 days), acidity 106-120°Th (0.96-1.08% titratable acidity), ethanol 1.75-2.50%.

A new proposed method for manufacturing kumiss involves inoculating pasteurized mare's milk with 10-20% of the above-mentioned starter culture, mixing well for 15-20 min and fermenting at 25-26°C (77-78.8°F) for 3-5 h. During the first hour, the milk is stirred 3-4 times for 1-2 min and after 2-4 h the stirring is done for 30-60 min. After fermentation, the kumiss is bottled, capped, and stored at refrigerator temperature. The final product has an acidity of 100-140°Th (0.90-1.26% titratable acidity) and 0.5-2.5% ethanol (Koroleva 1988).

Food Value: Kumiss is regarded as a dietetic product, and it is also used in the treatment of tuberculosis and gastrointestinal disorders (Berlin 1962). A therapeutic dose would be 1.5 l/day for adults and 0.4-0.8 l/day for children 8-15 years old.

Related Products: Kumiss from cow's milk, kumiss from camel's milk, chal.

References:

Berlin, P. J. 1962. Kumiss. *Annual Bulletin IDF,* pp. 4-10. Brussels: International Dairy Federation.

Koroleva, N. S. 1982. Special products. *Proceedings 21st International Dairy Congress* **2:**146-151. Moscow: Mir Publishers.

Koroleva, N. S. 1988. Kefir and kumiss. *IDF Bulletin 227* (Fermented Milks. Science and Technology), pp. 35-40. Brussels: International Dairy Federation.

Pedersen, C. S. 1971. *Microbiology of Food Fermentations.* Westport, Conn.: AVI Publishing Company.

KUMISS FROM COW'S MILK (En); Koumiss de lait de vache (Fr); Kumiss aus Kuhmilch (De).

Short Description: USSR; nontraditional product; cow's milk; a frothy beverage; direct consumption; homogeneous consistency; refreshing mild to acid taste, yeasty aroma.

History: Limited availability of mare's milk has led to the development of kumiss from cow's milk. The technology of manufacture is based on a modified cow's milk simulating the composition of mare's milk.

Microbiology: The starter culture consists of *Lactobacillus delbrueckii* subsp. *bulgaricus, L. acidophilus* and *Saccharomyces lactis.*

Manufacture: The manufacture of kumiss from cow's milk involves using a pasteurized mixture of whole milk (34.6%), skim milk (0.8%), and cheese whey (64.6%). It is inoculated with 20% of a starter culture and ascorbic acid is added at that time at a rate of 0.2 g/kg. The mixture is fermented at 28-30°C (82.4-86°F) for 3-4 h, with continuous stirring until an acidity of 75-80°Th (0.67-0.72% titratable acidity) is obtained, then cooled to 16-18°C (60.8-64.4°F) while stirring for 1-2 h. The product is bottled, capped, and allowed to ripen at 6-8°C (42.8-46.4°F) for 1-3 days (Koroleva 1988).

Food Value: The final product (about 1.5% fat) contains (1) in the case of weak kumiss (ripening period of one day), 95°Th (0.85% titratable acidity) and less than 0.6% ethanol; (2) in the case of medium kumiss (ripening period for 2 days), 110°Th (about 1% titratable acidity) and 1.1% ethanol; and (3) in the case of strong kumiss (ripening for 3 days), 130°Th (1.17% titratable acidity) and 1.6% ethanol. It is claimed that kumiss from cow's milk possesses similar dietetic and therapeutic properties as kumiss from mare's milk.

References:

Koroleva, N. S. 1988. Kefir and kumiss. *IDF Bulletin 227* (Fermented Milks. Science and Technology), pp. 35-40. Brussels: International Dairy Federation.

Puhan, Z., and P. Gallmann. 1980. Ultrafiltration in the manufacture of kumys. *North European Dairy Journal* **8-9:**220-224.

KURT (En, Fr, De).

Short Description: USSR (Kazakhstan); whey-drained traditional national product; cow's milk, ewe's milk, or goat's milk; fermented semidried or freeze-dried product; clean, acid taste; salted or unsalted.

Traditional Preparation

Microbiology: The starter culture consisting of lactic acid streptococci and *Lactobacillus delbrueckii* subsp. *bulgaricus* is added at a rate of 5%.

Manufacture: The product is made from low-fat milk, which is pasteurized at 80-85°C (176-185°F) for 10-20 sec, then cooled to 32-34°C (89.6-93.2°F) and inoculated with the above-mentioned starter culture. The milk is incubated until the acidity reaches 75-76°Th (0.67-0.68% titratable acidity). Then the coagulum is heated to 38-42°C (100.4-107.6°F) and held at this temperature for 20-30 min to facilitate whey separation. After whey drainage the curd is put in cloth bags, 7-9 kg at a time, and pressed until the moisture content reaches 76-80%, usually within 3-5 h. After pressing, the curd is molded into pieces of irregular shapes weighing 25-60 g, with or without prior salting, and then air-dried at 35-40°C (95-104°F) until the moisture content is 17% or less.

Food Value: The final product contains not more than 17% moisture, 12% fat in the dry matter, and a maximum of 2.4% salt. It may be reconstituted with water into a beverage.

Related Product: Kurut.

Reference: Bogdanova, G. J., and E. A. Bogdanova. 1974. *New Whole Milk Products of Improved Quality* (in Russian). Moscow: Pishchevoĭ Promy̆shlennosti.

Freeze-dried Product

Microbiology: The starter culture consists of *Streptococcus lactis* subsp. *diacetilactis, Lactobacillus delbrueckii* subsp. *bulgaricus* and *L. helveticus.*

Manufacture: The milk is first standardized in such a way that the final product will have 15% fat in the dry matter. Then it is pasteurized at about 90°C (194°F) and concentrated to 40-42% total solids. It is then incubated for 20 h at 30°C (86°F), or 6-7 h at 40°C (104°F), after inoculating with a starter containing the above-mentioned bacteria. Thereafter, the product is freeze-dried.

Food Value: The final product contains about 2% moisture, 28% lactose, and 5.6% total nitrogen. It can be readily reconstituted into a liquid by adding water.

Reference: Kraevaya, N. N., V. F. Rozdova, N. P. Zakharova, and Ya. I. Kostin. 1982. Development of technology and study of quality of the cultured milk product kurt (in Russian). In *Sbornik Nauchnykh Trudov. Tekhnologiya i Tekhnika Syrodeliya* Moscow, pp. 7-14. Cited in *Dairy Science Abstracts,* 1985, **47**(1825):210.

KURUNGA (En, Fr, De); Kourunga.

Short Description: USSR (eastern Asia); traditional kumisslike beverage from cow's milk; dietetic and therapeutic uses (e.g., patients with tuberculosis); acidic refreshing taste, yeasty flavor.

Microbiology: The starter culture consists mainly of lactic acid streptococci, lactobacilli, and lactose-fermenting yeasts. Strains isolated from natural kurunga samples at different stages of ripening were mainly lactic streptococci (*Streptococcus lactis, S. lactis* subsp. *diacetilactis, S. thermophilus* and *S. lactis* subsp. *cremoris*) and *Lactobacillus delbrueckii* subsp. *bulgaricus* and *L. acidophilus.* Isolated yeasts belong to the genera *Candida* and *Torulopsis.*

It has been suggested to make kurunga from pasteurized milk using a mixed culture consisting of lactic acid bacteria (*S. lactis, L. acidophilus* and *L. delbrueckii* subsp. *bulgaricus,* with the two latter ones having high inhibitory activity) and *Candida* yeast.

Manufacture, domestic: The product is traditionally made in the home from raw whole or skim milk. The milk is fermented in wooden containers at

25-30°C (77-86°F) using as a starter 20% of kurunga previously made. Whenever some of the fermented milk is removed from the container more fresh milk is added.

Manufacture, industrial: The above-mentioned lactic cultures and the yeast are added separately, and in a 2.0-2.5:1 ratio, to pasteurized milk. On reaching an acidity of 85-90°Th (0.76-0.81% titratable acidity) the product is agitated for 10-15 min with simultaneous aeration, and then it is cooled.

Food Value: The final product has an acidity of 190°Th (1.7% titratable acidity) and contains 1.0% ethanol and small quantities of carbon dioxide. Kurunga has been used by Mongolian people for the production of a milk wine called tarassun.

Related Products: Kefir, kumiss.

References:

Bulgadaeva, R. V., G. B. Lev, and T. G. Dubovik. 1978. In *Selection of Microorganisms* (in Russian). Ed. A. G. Grinevich, pp. 58-61. Irkutsk: Mezhuvuzovskii Sbornik. Cited in *Dairy Science Abstracts,* 1983, **45**(515):57.

Khundanov, L. E., and V. A. Krotova. 1949-1951. Fermented milk "kurunga" (in Russian) *Priroda* **38:**59-61. Cited in *Dairy Science Abstracts,* 1950, **11:**195; *Priroda* **39:**52-53. Cited in *Dairy Science Abstracts,* 1952, **14:**956.

Lev, G. B., and G. M. Patkul. 1976. Method of producing "kurunga" cultured milk beverage (in Russian) *USSR Patent* 651 776. Cited in *Dairy Science Abstracts,* 1980, **42**(4197):492.

Patskul, R. M., G. G. Gonchikov, and G. B. Lev. 1977. Inhibitory activity of lactic acid microflora in kurunga in relation to coliform bacteria (in Russian). *Biologiya Mikroorganizmov i ik'h Ispol'zovanie v Naradnom Khozyaistve (Irkutsk)* pp. 83-89. Cited in *Dairy Science Abstracts,* 1979, **41**(5975):673.

KURUT (En, Fr, De); karut (Asian USSR, Pakistan), quurut (Afghanistan), khurud (Mongolia).

Short Description: Middle East, Turkey, Soviet Asia (Kazakhstan), Afghanistan, Mongolia; traditional fermented milk product; goat's, ewe's cow's milk; hard, dried balls of fermented milk or milk curd.

Food Value: Original way of preserving milk in hot climates, widely used by nomadic peoples.

Related Products: Semidried buttermilk: kurut, Turkish; quurut, Afghanistan (Bachmann 1985).
Dried fermented milks: dried fermented milk products; kurt.

References:

Bachmann, M. R. 1985. The dairy technologist's contribution to food preservation in the Third World (in German). *Swiss Food* **7:**18-19.
Campbell-Platt, G. 1987, p. 109.
FIL-IDF Dictionary 1983, p. 130.
Miaki, T. 1980. Mongolian nomadic culture and animal production.IV. (in Japanese). *Animal Husbandry* **34:**319-394. Cited in *Dairy Science Abstracts,* 1982, **44**(1871):214.

KURUT, TURKISH (En); Kurut Turc (Fr); Türkischer Kurut (De).

Short Description: Turkey; whey-drained traditional dried yoghurt buttermilk; goat's, ewe's, cow's milk; used in soups and other food preparations, also crumbled and used as cheese.

Microbiology: The starter culture consists of yoghurt microorganisms (*Streptococcus thermophilus* and *Lactobacillus delbrueckii* subsp. *bulgaricus*).

Manufacture: Kurut is usually made in villages and small towns. The final product looks like very hard cheese and can be stored for long periods at room temperature. In Turkish villages it is customary to make butter from natural yoghurt. The resulting buttermilk is used to make kurut. The buttermilk is transferred into cloth bags, which are suspended in a clean place in order to remove the serum (whey). By tying the bags and pressing them with a stone weight more whey is removed. Then salt is added and mixed into the curd. This final product is divided into portions of 20-60 grams and air-dried for 1-2 weeks.

Food Value: Analyses have shown that this product contained an average 80% total solids, 52% protein, 11% fat and 9.1% salt.

This entry was contributed by Hassan Yaygin, Izmir, Turkey.

KUSHUK (En, Fr, De); Kushik.

Short Description: Iraq (northern part); traditional product; fermented curd mixed with wheat flour and herbs; prior to consumption, the product, in the shape of dried balls, is dispersed in water.

Microbiology: The starter culture consists mainly of yoghurt bacteria and some other microorganisms.

Manufacture: A fermented food prepared by allowing 1 part of dried parboiled whole wheat meal and 2 parts of yoghurt to ferment together for a week. To this fermented mass is added the curd from an equal volume of milk and further fermentation is sustained for 4-5 days. The product is then sun-dried and pulverized. It can be stored for a relatively long period. The estimate is that 20-30 kg are consumed per person annually in that area of Iraq.

Related Products: Kishk, fermented milk vegetable mixtures.

References:

FIL-IDF Dictionary. 1983, p. 130.

Platt, B. S. 1964. Biological ennoblement: Improvement of the nutritive value of foods and dietary regimens by biological agencies. *Food Technology* **18:**662-670. Cited by Hesseltine, C. W. 1965. *Mycologia* **57:**189.

KVAS, NEW MILK BEVERAGE (En); Nouvelle boisson au kvas (Fr); Neues Kwas-Getränk (De).

Short Description: USSR; nontraditional product; whey; beverage with a lactic and an alcoholic fermentation.

Microbiology: The starter culture consists of lactic acid bacteria and baker's yeast.

Manufacture: Rennet whey fermented to an acidity of 60°Th (0.54% titratable acidity) is pasteurized at 90-95°C (194-203°F) for 20 min, cooled to 20-22°C (68-71.6°F) and filtered. Separately, sugar is dissolved in hot whey and the syrup so obtained (after pasteurizing, filtering, and cooling) is mixed with raspberry or strawberry syrup. Two-thirds of this mixture is added to the cooled whey, along with a suspension of baker's yeast. After fermentation at 20°C (68°F) for 20-24 h and cooling to 2-8°C (35.6-46.4°F), the remaining third of the mixed syrup is added. The resulting mixture is then held at 2-8°C (35.6-46.4°F) for 24 h, decanted from the sediment and filtered. The finished product can be kept for 2 days when stored at about 8°C (46.4°F).

Related Product: Kvas from whey.

Reference: Kalmysh, V. S. 1983. The new milk kvas beverage (in Russian) *Tovarovedenie* **16:**49-51. Cited in *Dairy Science Abstracts,* 1987, **49**(4217):485.

KVAS FROM MILK, BULGARIAN (En), Kvas Bulgare du lait (Fr); Bulgarische Milk-Kwas (De).

Short Description: Bulgaria; traditional, popular summer beverage; ewe's milk; yoghurt-based (*see* Kiselo mljako).

Reference: Kvatchkoff, I. 1937. Studies on Bulgarian fermented ewes' milk (in French) *Le Lait* **17:**472-488.

KVAS FROM WHEY, POLISH (En); Kvas de lactosérum Polonais (Fr); Polnischer Molkenkwas (De).

Short Description: Poland; nontraditional product; deproteinized whey-fermented drink; consumed as a beverage; greenish-yellow color, sour taste, characteristic flavor.

Microbiology: Culture of thermophilic lactic acid bacteria (first stage of fermentation); yeast culture (second stage of fermentation).

Manufacture: The product is made from sweet cheese whey that has been separated to remove residual butterfat and heat-deproteinized. Then the whey is cooled, inoculated with 5% of the bacterial culture and incubated at 42°C (107.6°F) for 2 h. After incubation the whey is inoculated with the yeast culture, colored with caramel, bottled, and held at 8°C (46.4°F) for 40 h to promote yeast activity. The final product has an acidity of about 112°Th (1.0% titratable lactic acid).

Related Product: Kvas from whey, Russian.

Reference: Holsinger, V. H., L. P. Posati, and E. D. DeVilbiss. 1974. Whey beverages: a review. *Journal of Dairy Science* **57:**849-859.

KVAS FROM WHEY, RUSSIAN (En), Kvas de lactosérum, Russe (Fr); Russischer Molkenkwas (De); kvas denotes sour, fermented.

Short Description: USSR and other countries; nontraditional product; deproteinized whey; flavored fermented whey drink; consumed as a refreshing beverage; acidic, agreeable taste and aroma (brown color).

Microbiology: The starter culture consisting of baker's yeast is mixed in at a rate of 0.02%. The yeast culture is prepared by adding 200 g fresh baker's yeast and 50-60 g sugar syrup to 4 liters pasteurized whey at 30°C (86°F). The whey is incubated at that temperature for 2-3 h and then cooled until used.

Manufacture: The product is made from sweet cheese whey or acid whey obtained in tvorog (quarg) manufacture. After separating from whey residual butterfat, it is deproteinized, usually after acidification, by heating at 95-97°C (203-206.6°F) for 1-2 h and then cooling to 25-30°C (77-86°F). The coagulated whey proteins are removed by filtration and the clear supernatant is processed further. About 1.5% of pasteurized sugar syrup is added to the defatted, deproteinized whey, which is then inoculated with a yeast culture. The incubation is at 25-30°C (77-86°F) for 12-15 h until the acidity reaches 80-90°Th (0.72-0.81% titratable acidity). Then another 2.5% sugar syrup is added to impart sweetness, and 1.0% caramel coloring is added. Fruit flavoring may also be used. After cooling to 6°C (42.8°F), the product is bottled and refrigerated until consumed.

Food Value: Kvas made from whey has an acidity of 90-100°Th (0.81-0.90% titratable acidity) and contains 0.4-1.0% ethanol.

Related Products: Whey kvas; whey cultured products; kvas from whey, Polish.

Reference: Bogdanova, G. J., and E. A. Bogdanova. 1974. *New Whole Milk Products of Improved Quality* (in Russian). Moscow: Pishchevoĭ Promȳshlennosti.

L

LABAN. *See* Leben.

LABAN HAMID (En, Fr, De).

Short Description: Egypt; traditional buttermilk beverage; *see* Leben.

Reference: Morcos, S. R. 1977. Egyptian fermented foods. *Symposium on Indigenous Fermented Foods.* Bangkok, Thailand. Cited by Steinkraus, K. H. 1983. pp. 258-290.

LABAN KERBAH (En, Fr, De).

Short Description: Upper Egypt; traditional product; buttermilk obtained by churning laben in skin bags called kerbah; utilized for the preparation of laban zeer, laban khad, and laban hamid.

References:

El-Gendy, S. M. 1983. Fermented foods of Egypt and the Middle East. *Journal of Food Protection* **46:**358-367.
Steinkraus, K. H. 1983, pp. 257-258 (about the skin bag kerbah).

LABAN KHAD (En, Fr, De).

Short Description: Upper Egypt; buttermilk beverage; utilized for the preparation of laban zeer, laban hamid, and laban kerbah.

Manufacture: Milk is poured into a skin bag (called kerbah) and left to sour for a period determined by previous experience. Air is blown into the kerbah before closing it tightly. The bag is then shaken until the fat globules coalesce (churning process). After removal of the fat (butter granules), the remaining fluid is called laban khad.

Reference: Abd-El-Malek, Y., and M. Demerdash. 1970. Studies on the microbiology of some fermented milk in Egypt. I. Sour milk. *Proceedings 2nd Conference Microbiology,* Cairo, pp. 151-152.

This entry was contributed by S. A. Abou-Donia, Alexandria, Egypt.

LABAN MATRAD (En, Fr, De); laban matared. Matared signifies the earthenware pots used to incubate and store the milk.

Short Description: Lower Egypt; traditional fermented milk (cream is skimmed off after coagulation); for type of milk utilized see Laban and Laban rayeb.

This entry was contributed by S. A. Abou-Donia, Alexandria, Egypt.

LABAN MUNAKKAH (En, Fr, De).

Short Description: Saudi Arabia; nontraditional product; cow's, goat's, or mixed milks; national type of flavored set yoghurt; industrially manufactured; direct consumption; pleasant taste and aroma.

Reference: Salji, J. P. 1986. Fermented dairy products of Saudi Arabia. *Cultured Dairy Products Journal* **21**(3):6-7.

LABAN RAYEB (En, Fr, De).

Short Description: Lower Egypt; traditional product; the cream layer is skimmed off after coagulation, for types of milk utilized; slightly acid taste with an aroma resembling that of buttermilk.

Microbiology: *Streptococcus lactis, Kluyveromyces marxianus* subsp. *marxianus* and coliforms are the predominant types of bacteria present in laban rayeb. *Lactobacillus casei* is also occasionally encountered, especially in old samples.

Manufacture: For household manufacture, fresh milk is poured into a 6-liter earthenware pot (called matared). Before a newly made matared is used, the pores in the walls are closed by soaking the inside of the pot with cottonseed oil, olive oil, or egg yolk beaten with oil; this process is followed by baking in a kiln. Milk is poured into the matared and left undisturbed one day or several days at approximately 20 to 25°C (68 to 77°F) until the cream has risen and the milk has coagulated. The cream layer is removed and beaten by hand to make butter. The remaining curd, known as laban rayeb or laban matared, is made into arish cheese, which is either consumed fresh or is pickled (brined). After each operation the matareds are thoroughly washed, dried in hot ovens for a period of 2 h, and then left to cool before being refilled with fresh milk.

Related Products: Laban, laban matrad, goubasha.

Reference: Demeroash, M. A. 1960. Studies on the microbiology of fermented milks common in Egypt. Thesis, University of Cairo. Faculty of Agriculture, Cairo. Cited by Steinkraus, K. H. 1983, pp. 257 and 284.

LABAN ZEER (En, Fr, De), denotes laban stored in an amphora-shaped earthenware vessel (called zeer).

Short Description: Egypt; traditional product; cow's, goat's, ewe's, buffalo or mixed milks or, respectively, their buttermilks; concentrated sour buttermilk used for the manufacture of kishk, sometimes used in a salad or to make beverages after dilution with water; semisolid consistency, acid and salty taste.

Microbiology: Lactic acid streptococci, mainly *Streptococcus lactis, Leuconostoc* spp. and mesophilic lactobacilli such as *Lactobacillus casei, L. plantarum* and *L. brevis,* dominate in the product (*see* kishk).

Manufacture: The product is made at home from spontaneously soured milk that is churned. After removing the butter granules, the remainder is called sour buttermilk in Lower Egypt and laban khad or laban kerbah in Upper Egypt. This product is stored in earthenware containers named zeer. During storage some whey permeates through the porous container walls thus concentrating the product. After each addition of a new lot of sour buttermilk some salt is added to the contents of each zeer. The finished product has a pH value of 3.5-3.8.

Related Products: Laban khad and laban hamid.

References:

Abou-Donia, S. A. 1984. Egyptian fresh fermented milk products. *New Zealand Journal of Science and Technology* **19:**7-18.

El-Gendy, S. M. 1983. Fermented foods in Egypt and the Middle East. *Journal of Food Protection* **46:**358-367.

LABEN (En, Fr, De).

Short Description: Saudi Arabia; nontraditional product; plain yoghurt beverage.

Microbiology: Yoghurt starter culture.

Related Products: Leben and laban.

References:

Al-Shaikhli, J. S. 1980. A study of fermented milks in the Riyadh area. Incidence of coliform contamination (in English and Arabic). In *Fourth Symposium, Riyadh University Press* **68:**68. Cited in *Dairy Science Abstracts,* 1981, **43**(925):120.

Salji, J. P., S. R. Saadi, and A. Mashhadi. 1987. Shelf life of plain liquid yoghurt manufactured in Saudi Arabia. *Journal of Food Protection* **50:**123-126. Cited in *Dairy Science Abstracts,* 1987, **49**(4578):523.

LABEN RAIB (En, Fr, De).

Short Description: Saudi Arabia; yoghurtlike fermented milk product; *see* laban rayeb.

Reference: Kosikowski, F. V. 1977, p. 70.

LABNEH (En, Fr, De); various spellings in different countries or regions: labnah (Saudi Arabia), labneh (Lebanon, Israel, Arab sector, Bangladesh), labaneh (Jordan).

Short Description: Middle East; traditional whey-drained product, a key product; cow's, goat's, ewe's, buffalo or mixed milks; concentrated plain yoghurt; important part of the winter supply; semisolid consistency, agreeable taste and aroma.

Microbiology: A yoghurt culture (*Streptococcus thermophilus* and *Lactobacillus delbrueckii* subsp. *bulgaricus*).

Manufacture: Labneh or lebneh is usually made from surplus summer milk. The heat-treated milk is fermented at 42-45°C (107.6-113.0°F) for 2-2.5 h using 2% of a yoghurt culture, with subsequent cooling and stirring. The stirred yoghurt is placed in a hanging cloth bag so that the whey separates out. To facilitate whey drainage the bags can also be gently pressed by piling them on top of each other. The finished product is packaged and cooled.

Food Value: The composition of labneh may vary depending on the kind of milk used and the rate of concentration. The final product contains 21-26% total solids, 8-11% fat and 1.6-2.5%, or even more, lactic acid. Labneh can also be processed into other types of products (*see* Labneh anbaris and Gibneh-labneh).

Related Products: Torba or tulum made in Turkey, and tan (or than) made in Armenia. A similar product made in Egypt is called laban zeer.

References:

Gording, S. 1980. Milking animals and fermented milks of the Middle East and their contribution to man's welfare. *Journal of Dairy Science* **63:**1031-1038.

Hofi, M. A. 1988. Labneh (concentrated yoghurt) from ultrafiltered milk. *Scandinavian Dairy Industry* **1:**50-52.

Tamime, A. Y., and R. K. Robinson. 1978. Some aspects of the production of a concentrated yoghurt (labneh) popular in the Middle East. *Milchwissenschaft* **33:**209-212.

LABNEH ANBARIS (En, Fr, De).

Short Description: Middle East; traditional whey-drained product; cow's, goat's, ewe's, buffalo or mixed milks; direct consumption; partially dried products; agreeable taste and aroma.

Microbiology: Yoghurt microorganisms.

Manufacture: The product is homemade from a concentrated yoghurt (labneh) by shaping the curd into balls and then placing these in the sun for partial drying. The balls are packaged in glass jars and covered with olive oil (gibneh-labneh).

Reference: Tamime, A. Y., and R. K. Robinson. 1978. Some aspects of the production of a concentrated yoghurt (labneh) popular in the Middle East. *Milchwissenschaft* **33:**209-212.

LACCILLIA (En, Fr, De); trade name.

Short Description: United Kingdom; nontraditional product, pharmaceutical preparation.

Microbiology: Freeze-dried preparation containing more than 10^8/g of viable *Lactobacillus acidophilus.*

Therapeutic Value: Used for gastrointestinal disorders.

Reference: Reuter, G. 1969. Composition of bacterial cultures used for therapeutic purposes (in German). *Arzneimittel-Forschung* **19:**103-109.

LAC CONCRETUM (En, Fr, De); denotes coagulated milk.

Short Description: Areas north of the Roman Empire; traditional product; prepared in antiquity.

History: Important and popular soured milk food consumed by the peoples of ancient Scythia (now part of USSR) and Germanic tribes.

Microbiology: Spontaneously soured milk microorganisms.

Related Products: Dickmilch (thick milk, set milk) and clabbered milk.

Reference: Tacitus. Germaniae 23. Cited by Fleischmann, W. and H. Weigmann. 1932, p. 370.

LACTANA B (En, Fr, De); trade name.

Short Description: Germany (Federal Republic); nontraditional dried baby food that contains bifidobacteria (pharmaceutical preparation).

Microbiology: Containing viable bifidobacteria.

Food Value: Lactana-B (since 1954) is a partly adapted bifidus food for infants that is claimed to impart a bifidum flora in the large intestine, within 2-4 days, similar to that produced with breast-feeding.

Per 100 g, the powder contains 22 g fat, 13.3 g protein, 60 g carbohydrate, 2.90 g mineral salts; the product is enriched with vitamins A, D_3, and C, as well as iron. A 100-g portion has 2082 kJ (491 kcal). Packages available in the market come in sizes of 125, 400 and 1000 g.

Reference: Rašić, J. Lj., and J. A. Kurmann. 1983, pp. 138-139.

LACTANA-STRAINED FRUIT (En); Lactana aux fruits (Fr); Lactana mit Früchten (De).

Short Description: Germany (Federal Republic); nontraditional dried product, pharmaceutical preparation; fruit milk; used for feeding babies older than 3-4 months.

Microbiology: Dried preparation containing viable bifidobacteria.

Food Value: Per 100 g, the powder contains 9.1 g fat, 12.5 g protein, 70.0 g carbohydrate, 2.8 g minerals, 0.9 g fiber and vitamins A, D_3, E, B_1, B_6, C, and niacin, as well as calcium pantothenate.

Reference: Rašić, J. Lj., and J. A. Kurmann. 1983, pp. 137-138.

LACTINEX (En, Fr, De); trade name.

Short Description: United States; nontraditional product; cow's milk; pharmaceutical preparation.

Microbiology: Freeze-dried preparation containing large numbers of viable *Lactobacillus acidophilus* and *L. delbrueckii* subsp. *bulgaricus.*

Therapeutic Value: Used to treat skin diseases and oral infections.

References:

Rašić, J. Lj., and J. A. Kurmann. 1978, pp. 135, 349.

Robins-Browne, R. M., and M. M. Levine. 1981. The fate of ingested lactobacilli in the proximal small intestine. *American Journal of Clinical Nutrition* **34:**514-519.

LACTOBACILLI-CONTAINING FERMENTED FRESH MILKS (En); Laits fermentés avec des lactobacilles (Fr); Mit Laktobazillen-Kulturen fermentierte Milch (De).

Short Description: Various countries; nontraditional products and pharmaceutical preparations.

History: As a result of Metchnikoff's theory and other research findings with lactobacilli, these organisms are considered important contributors to health and agents in the treatment of intestinal disorders and other ailments.

Products: *Lactobacillus delbrueckii* subsp. *bulgaricus* products.

Biolactin, which is used in the treatment of purulent inflammations, lesions, and burns. It is contained in a paste prepared with *L. delbrueckii* subsp. *bulgaricus* (Mincev, P. 1930. *Veterinariya Sbirka* **5:**6. Cited by Rašić, J. Lj., and J. A. Kurmann, 1978, p. 135).

Bulgaricus cultured buttermilk (see there); Bulgaricus milk (see there); Bulgaricus milk for babies (Mayer, J. B. 1948. Development of a new milk for babies with *B. bifidum* I. The bulgaricus milk (in German). *Zeitschrift für Kinderheilkunde* **65:**293-318.

Bulgaricum, L. B. (see there).

Lactobacillus casei products, such as yakult, that contain an intestinal strain (from Shirota) called *Lactobacillus casei* var. *rhamnosus.*

Lactobacillus acidophilus products, including acidophilus milk (see there); acidophilus products (see there); and Ribolac "Zyma," a pharmaceutical preparation containing *Lactobacillus acidophilus.*

Mixed lactobacilli products, such as lactinex, that contain *Lactobacillus delbrueckii* subsp. *bulgaricus* and *Lactobacillus acidophilus* (see there) and are used for eczemas and oral infections (Rašić, J. Lj., and J. A. Kurmann. 1978, p. 135).

LACTO-BACILLINE (En, Fr, De); denotes containing organisms of the genus *Lactobacillus.*

Short Description: USSR, Europe (especially France); nontraditional products.

History, Pharmaceutical Preparation: Composed of a culture of two species (mainly *Lactobacillus delbrueckii* subsp. *bulgaricus*) proposed by I. I. Metchnikoff for the souring of milk or for internal use to suppress putrefactive digestion in the intestine; now only of historical interest in that it was prepared by the society named Le Fement (rue Denfert Rochereau 77, Paris, France) when Metchnikoff was living in Paris in 1906.

History, Sour Milk: Lacto-Bacilline is another name for sour milk or yoghurt that has been produced through the fermentation of milk with the above-mentioned cultures (Metchnikoff clabber).

References:

Great Soviet Encyclopedia. 1973. A translation of the third edition, vol. 14, p. 162. New York: Macmillan.

Leva, J. 1973. Examination of effect of lactobacilline and yoghurt (in German). *Berliner Klinische Wochenschrift,* Number 19.

LACTOBACILLUS MILK FOR BIOTHERAPY OF INFANTILE DIARRHEA (En); Biotherapy infantile de la diarrhée par du lait fermenté aux lactobacilles (Fr); Lactobacillus-Milch für die Biotherapie von Durchfallserkrankungen bei Kindern (De).

Short Description: Argentina; nontraditional fermented milk; cow's milk.

Microbiology: The strains used were originally selected on the basis of their high antibacterial and enzymatic activity. In one particular study the lactobacilli used were *Lactobacillus casei* and *L. acidophilus.*

Manufacture: Partially fermented milk was produced in an Argentinian children's hospital lactation room through the addition of concentrated microorganisms (10^{10} bacteria/ml) to commercial pasteurized milk. The mixture was incubated at 37°C (98.6°F) for 4 h. The final concentration of microorganisms in the fermented milk was 10^8 bacteria/ml.

Therapeutic Value: The basis for treating certain gastrointestinal disorders with lactobacillus milk is the well-known ability of lactobacilli to produce lactic acid and other inhibitory substances that affect the normal growth of pathogenic enterobacteria. Lactobacilli also reduce the reduction-oxidation potential and regulate the ecology of microorganisms in the intestinal tract. Lactobacilli also display proteolytic activity and stimulate the immune system.

Treatments with fermented milk were 91% effective and no adverse results were observed. The objective in using lactobacillus milk is to provide a medicinal food that is harmless, highly nutritious, and extremely efficient in the prevention and treatment of infantile diarrhea, the main cause of mortality among infants between 1 month and 3 years of age, especially in Third World countries.

This entry was contributed by Guillermo Oliver
and Silvia N. Gonzale, Argentina.

LACTOFIL (En, Fr, De); denotes sour milk.

Short Description: Sweden (developed by Starnert in the early 1950s, southern part of Sweden); nontraditional product; cow's milk; concentrated fermented milk; direct consumption; firm, pasty consistency; pleasant taste and aroma.

Microbiology: Cream culture consisting of *Streptococcus lactis* subsp. *cremoris, S. lactis* subsp. *lactis, S. lactis* subsp. *diacetylactis* and *Leuconostoc mesenteroides* subsp. *cremoris.*

Manufacture: Skim milk is pasteurized at 90°C (194°F) and then fermented at 20°C (68°F) after the addition of 1-2% starter culture. After 20-24 h, the coagulated milk is gently stirred and heated to 35°C (95°F) for 3 h. The whey is drained off, removing about 50 liters of whey per 100 liters of milk, and then cream (40% fat) is added to give the product a fat content of 5%. This mixture is homogenized at 25-30°C (77-86°F) under a pressure of 50-100 bar, cooled to 10-12°C (50-54°F) and packaged in retail containers. Shelf life can be expected to be about two weeks at 0-5°C (32-41°F).

Food Value: When manufactured by ultrafiltration, the final product contains 6.5%, protein, 3.5% fat and 4% carbohydrate; the caloric value is 310 kJ (74.04 kcal)/100 g. The final product contains 14-15% total solids.

Related Products: Ymer, milk protein paste, concentrated fermented milks.

References:

Bertelsen, E., B. Ljungren, and N. Mattsson. 1961. Manufacture of laktofil—a concentrated cultured milk product (in Swedish). *Meddelande Svenska Mejeriernas Riksforening* **68.** Cited in *Dairy Science Abstracts,* 1962, **24:**977.

Starnert, A. 1978. Lactofil-a product in time (in Swedish). *Nordisk Mejeriindustri* **5:**185. Cited in *Dairy Science Abstracts,* 1978, **40**(5581):579.

LACTOPRIV (En, Fr, De); trade name.

Short Description: Germany; nontraditional product; contains vegetable matter; pharmaceutical product.

Microbiology: Freeze-dried preparation containing viable bifidobacteria.

Therapeutic Value: Used to treat disturbed balances in the gut microflora, especially after antibiotic therapy and irradiation; also recommended for lactase-deficient persons.

100 g of powder contains 17.35 g vegetable protein, 12.30 g vegetable fat, 64.0 g carbohydrate, 4.0 g salts, 1.19 g fiber, 6.70 mg saccharin; fortified with iron and vitamins A, D_3 and C.

Reference: Rašić, J. Lj., and J. A. Kurmann. 1983, p. 140.

LACTOROL (En, Fr, De); trade name.

Short Description: Poland; nontraditional product; cow's milk; fermented milk.

Microbiology: Cream culture consisting of *Streptococcus lactis* subsp. *cremoris* and/or *S. lactis* subsp. *lactis, S. lactis* subsp. *diacetylactis* and/or *Leuconostoc mesenteroides* subsp. *cremoris.*

Manufacture: Pasteurized milk with a 2% fat content is fermented by the addition of starter culture until 75-100°Th (0.76-0.90% titratable acidity) lactic acid has been developed.

Related Product: Cultured buttermilk.

Reference: Lang, F., and A. Lang. 1971. Fermented dairy beverages and specialties in Poland. *Milk Industry* **69:**13-15.

LACTOSAT (En, Fr, De).

Short Description: Sweden; nontraditional product; cow's milk; a natural alternative to antibiotics for domestic animals.

Microbiology: Yoghurt culture consisting of *Streptococcus thermophilus* and *Lactobacillus delbrueckii* subsp. *bulgaricus.*

Manufacture: Spray-drying of yoghurt.

Food Value: The product contains 96% total solids, 37% milk protein, 44% lactose, 6% lactic acid, 9% salts; the pH is 4.5 and the product contains 500 million or more viable lactic acid bacteria per g.

Related Product: Yoghurt.

Reference: Rašić, J. J., and J. A. Kurmann. 1978, p. 346.

LACTRONE (En, Fr, De); trade name.

Short Description: Germany; nontraditional whey beverages, usually diluted before drinking; with citrus flavor.

Microbiology: Combined fermentation of whey with lactobacilli and a kefir culture.

Related Product: Milone.

Reference: Schulz, M. E. 1941/42. *German Patent* 731 521. Cited by Schulz, M. E. 1965. pp. 663-664.

LACTULOSE-ENRICHED FERMENTED FRESH MILK PRODUCTS (En); Produits laitiers fermentés enrichit avec de la lactulose (Fr); Mit Laktulose angereicherte Sauermilchprodukte (De).

Short Description: Various countries; nontraditional products (pharmaceutical preparations) and traditional products; mainly cow's milk.

Microbiology: Lactulose (β-galactosidofructose) has been reported to be a bifidogenic factor (in vivo). It is ineffective in vitro as a growth factor.

Food Value: Lactulose is a disaccharide composed of one molecule each of galactose and fructose, but it is not present in a free state in human milk. It is only upon heating milk that a part of the lactose is converted to lactulose. In sterilized liquid formula foods lactulose amounts to 2-5% of the total carbohydrate content. Its inclusion in the diet of formula-fed infants at a daily rate of 1.0-1.5 g per kg body weight stimulates growth of bifidobacteria in the large intestine within 24-96 h.

Products: The following products are known to contain lactulose: Eugalein Töpfer, yoghurt, and infant food.

References:

Morinaga Milk Industry Co., Ltd. 1988. *Bibliography of Bifidobacterium and Lactulose,* Rev. ed. No.1-83, 5-Chome Higashihara, Zama-City, Kanagawa-Pref., Japan 228: Morinaga Milk Industry Co., Ltd.

Olano, A., S. J. Lopez-Covarrubias, M. Ramos, and J. A. Suarez. 1986. The use of lactulose in the manufacture of low-lactose yogurt. *Biotechnology Letters* **8:**451-452. Cited in *Dairy Science Abstracts,* 1986, **48**(5561):659.

Rašić, J. J., and J. A. Kurmann. 1983, pp. 43-45, 72-86, 96-100.

LAPTE-AKRU (En, Fr, De); from *lapte acru* (in Rumanian), meaning sour milk.

Short Description: USSR (Moldavia); national, traditional product; cow's milk and cream; stirred or set type; plain or flavored; smooth, creamy consistency; acidic and agreeable aroma.

Microbiology: A culture of *Streptococcus thermophilus.*

Manufacture: The product is made from a cream-milk mixture that is standardized to 10% fat, homogenized, pasteurized at 95-98°C (203.0-208.4°F) for 3-5 h and cooled to 38-40°C (100.4-104.0°F). This mixture is inoculated with a 5% culture, mixed well, and incubated until firmly coagulated, followed by stirring, cooling, and packaging. *Lapte-akru* can be sweetened with 5% sugar and flavored with vanilla or flavored with fruit syrup at a rate of 12%.

Food Value: The final product has a maximum acidity of 110°Th (1.0% titratable acidity) and about 10% fat.

Related Product: *Streptococcus thermophilus* fermented milks.

Reference: Bogdanova, G. J., and E. A. Bogdanova. 1974. *New Whole Milk Products of Improved Quality* (in Russian). Moscow: Pishchevoĭ Promȳshlennosti.

LASSI (En, Fr, De); also called chas or matha.

Short Description: India; traditional beverage; buttermilk of churned dahi, diluted with water and consumed sweet or salted.

Manufacture: Lassi is a whipped or heavily agitated liquid product made from buttermilk of churned dahi. It may or may not be sweetened and it may or may not be diluted with water to adjust the solids content as stipulated by regulations.

Related Product: Dahi.

References:

Bhandari, V. 1985. Effect of some processing variables on acid development in lassi from skim milk prepared by the continuous agitation method. *Indian Journal of Animal Sciences* **55:**293-295. Cited in *Dairy Science Abstracts,* 1985, **55:**293-295.

Sukumar, De 1980. *Outlines of Dairy Technology,* p. 463. Delhi: Oxford University Press.

LÄTTFIL (En, Fr, De); denotes lightly fermented milk.

Short Description: Scandinavia (Sweden); traditional product; cow's milk; fermented low-fat milk drink; stirred product; pleasant taste and aroma; smooth body.

Microbiology: A mixed culture consisting of *Streptococcus lactis* (a), *S. lactis* subsp. *cremoris* (b), *S. lactis* subsp. *diacetylactis* (c) and *Leuconostoc mesenteroides* subsp. *cremoris* (d) at a ratio 85:15 (a + b + c = 85; d = 15).

Manufacture: It is made from low-fat milk (0.5% fat), homogenized and pasteurized at 90-91°C (194.0-195.8°F) for 3 min, cooled, and inoculated with 1-2% of the above culture. The milk is incubated at 20-21°C (68.0-69.8°F) for 20-24 h until coagulation, then stirred, cooled, and packaged.

Food Value: A good-quality Lättfil has an acidity of 90-95°Th (0.81-0.85% titratable acidity), a pH of 4.5, and a count of about 9×10^8 cells (colony-forming units) per ml.

Related Product: Lättfil is also called lowfat buttermilk and is closely related to the cultured buttermilk of North America.

References:

Alm, L. 1982. The effect of fermentation on nutrients in milk and some properties of fermented liquid milk products. Thesis, University of Stockholm, Sweden.

Bertelsen, E., and B. Olsson. 1977. Quality of liquid milk products in 1976 (in Swedish). *Svensk Meijeritidningen* **69**:9-11. Cited in *Dairy Science Abstracts,* 1978, **40**(693):85.

LEBEN (En, Fr, De); various ways of spelling in different countries or regions: laban (Egypt, Lebanon, Iraq), laben (Saudi Arabia), lben (Morocco), lebben (Israel), leben (Lebanon, Iraq, Jordan), liban (northern Iraq).

Short Description: Middle East (Egypt, Lebanon, Iraq, etc.); old traditional product, key product, a fermented milk; mainly cow's, goat's, ewe's, buffalo, or mixed milks; direct consumption or with added spices; set or stirred type; pleasant taste and aroma; naturally soured milk, skimmed before or after coagulation, sometimes made from whole milk.

Microbiology: The microflora has been shown to consist of *Streptococcus thermophilus, Lactobacillus delbrueckii* subsp. *bulgaricus, L. acidophilus, Leuconostoc lactis, Kluyveromyces marxianus* subsp. *marxianus, Saccharomyces cerevisiae.*

Manufacture: Methods of preparation, microflora, and fermentation processes of leben vary greatly. Originally the product was characterized by lactic acid fermentation and slight alcoholic fermentation, and the latter could vary considerably. Today, leben shows only lactic acid fermentation. It is very much a yoghurtlike product.

The manufacture of homemade Lebanese laban has been described as follows (Baroudi and Collins 1976): Boil milk for 1 min, cool it to 50°C (45°C after inoculation). Inoculate the milk with 2.5 to 3.0% laban saved from a previous batch but not older than 3 weeks (refrigerated storage). Allow the temperature to drop to a room temperature of 26°C (78.8°F) during the incubation of about 9 h, followed by another 3 to 5 h of incubation at room temperature prior to refrigeration.

Related Products: Laban is also the generic name for different products (see specific names); yoghurt.

References:

Baroudi, A. A. G., and E. B. Collins. 1976. Microorganisms and characteristics of laban. *Journal of Dairy Science* **59:**200-202. Cited in *Dairy Science Abstracts,* 1976, **38**(4425):470.

El-Gendy, S. M. 1983. Fermented foods in Egypt and the Middle East. *Journal of Food Protection* **46:**358-367. Cited in *Dairy Science Abstracts,* 1983, **45**(6192):646.

Rist, E., and J. Khoury. 1902. Studies on the fermented milk "leben" from Egypt (in French). *Annales de l'Institut Pasteur* **16:**65-84.

Rosell, J. M. 1939. Milk, some methods in its utilization (in French). *Le Lait* **19:**238-239.

LEBENIÉ (En, Fr, De).

Short Description: Israel; nontraditional product; fermented milk mixed with quarg (white fresh cheese) and homogenized (according to Kun 1953).

Reference: Kun, A. 1953. *Deutsche Molkerei-Zeitung* **74:**1630. Cited in Schulz, M. E. 1965, p. 668.

LEGUME-CONTAINING FERMENTED FRESH MILK PRODUCTS (En); Produits laitier fermentés additionnés de légumes (Fr); Sauermilchprodukte mit Gemüse-Zusätzen (De).

Short Description: Various countries; nontraditional products; mainly buffalo's and cow's milk.

Peanut (Groundnut) Addition

Manufacture: A peanut-extended buffalo milk (*see* Miltone). Peanut flour (meal) is used as the additive to fermented milks.

Soybean Addition, Product Formula 1 (According to Yamanaka et al., 1970)

Microbiology: Use of *Streptococcus thermophilus* and *Lactobacillus delbrueckii* subsp. *bulgaricus* as the lactic starters.

Manufacture: A sour milk beverage consisting of soybean protein, milk, and sucrose. Addition of an amino acid mixture (alanine, arginine, aspartic acid, methionine, glycine, lysine, and sodium glutamate) was claimed to mask the beany flavor of the product.

Soybean Addition, Product Formula 2 (According to Kanda et al., 1976)

Microbiology: *L. acidophilus* starter culture microorganisms.

Manufacture: After addition of *L. acidophilus* starter culture, incubation at 37°C (98.6°F) for 24 h.

Food Value: A fermented acidophilus product consisting of soybean extract (500 ml), whey solids (10 g), sucrose (25 g) and flavoring (e.g., lemon).

Soybean Addition, Product Formula 3 (According to Dimov et al., 1982)

Microbiology: Yoghurt starter microorganisms.

Manufacture: Mixture of high-acid yoghurt and heat-treated soybean extract at a ratio of 1:1.

Food Value: Designated as dietetic fermented product from soybean extract and cow's milk.

References:

Dimov, N., O. Djondjorova, and A. Kojev. 1982. Dietetic cultured milk product from soy and cow milk (in French). *Proceedings 21st International Dairy Congress* **1**(1):297. Moscow: Mir Publishers.

Kanda, H., H. L. Wang, C. W. Hesseltine, and K. Warner. 1976. Yoghurt production by Lactobacillus fermentation of soybean milk. *Process Biochemistry* **11**(4):23-25.

Yamanaka, Y., O. Okamura, and W. Hasegawa. 1970. Method of preparing a sour milk beverage. *U.S. Patent* 3,535,177.

LEMONADE (En); Limonade (Fr); Limonade (De).

Short Description: Various countries; nontraditional product; the generally accepted definition for lemonade is "sweetened lemon juice and water." However, lemonade can also be made from fermented whey.

Products: Whey lemonade (called molkine), whey cultured products, whey products.

LEUCONOSTOC-CONTAINING FERMENTED FRESH MILK PRODUCTS (En); Produits laitiers fermentés contenant de Leuconostoc microorganisms (Fr); Sauermilch mit Leuconostoc-Mikroorganismen (De).

Short Description: Various countries; mostly nontraditional products; several milk types (cow's, goat's, buffalo milk, etc.)

Microbiology: *Leuconostoc* is a genus of gram-positive bacteria; several species occur in a variety of dairy products and also in fermented vegetables (e.g., sauerkraut). Leuconostocs are heterofermentative lactic acid bacteria that produce small quantities of lactic acid. There is also carbon dioxide production. Important species of the genus *Leuconostoc* are *Leuconostoc mesenteroides, L. mesenteroides* subsp. *dextranicum, L. mesenteroides* subsp. *cremoris* and *L. lactis.*

Products: Specific Leuconostoc-containing fermented fresh milk products are products prepared with cream cultures (containing *Leuconostoc*), such as cultured buttermilk and natural buttermilk, and kefir and kefirlike products.

Reference: Devoyod, J. J., and F. Poullain. 1988. The Leuconostocs, properties: their role in dairy technology (in French). *Le Lait* **68:**249-279.

LIBAN (En, Fr, De), liban arbeel, indicates liban (name for Iraqi soured milk) from Arbeel, northern Iraq.

Short Description: Iraq; sour milk or sour milk beverage; goat's, ewe's, buffalo's milk (diluted with water, and with the addition of sugar, salt, and ice).

Related Product: Leben.

Reference: Hamdi, Y. A. 1983. Iraqi liban arbeel. *Symposium on Indigenous Fermented Foods, Bangkok.* Cited by Steinkraus, K. H. 1983, p. 260.

LIFE START ORIGINAL (En, Fr, De); trade name.

Short Description: United States (California); nontraditional product; cow's milk; freeze-dried preparation; dietary adjunct in feeding babies (pharmaceutical preparation).

Microbiology: The product contains about 2×10^9 cells of viable *Bifidobacterium infantis* per gram of product.

Therapeutic Value: The product has beneficial effects on the gut microflora and plays a protective role against gastrointestinal disorders and infections.

Reference: Nakaya, R. 1984. Role of Bifidobacterium in enteric infection. *Bifidobacteria Microflora* **3:**3–9.

LIFE START TWO (En, Fr, De); trade name.

Short Description: United States (California); nontraditional product; cow's milk; freeze-dried pharmaceutical preparation.

Microbiology: The product contains about 2×10^9 cells of viable *Bifidobacterium bifidum* per gram of product.

Therapeutic Value: Used as a dietary adjunct against an imbalance of the gut microflora and in the treatment of gastrointestinal and liver disorders.

Reference Anand, S. K., R. A. Srinivasan, and K. L. Rao. 1985. Antibacterial activity associated with Bifidobacterium bifidum. II. *Cultured Dairy Products Journal* **20**(1):21,23. Cited in *Dairy Science Abstracts,* 1985, **47**(2834):320.

LIOBIF (En, Fr, De); trade name.

Short Description: Yugoslavia; nontraditional product; cow's milk; freeze-dried pharmaceutical preparation.

Microbiology: The product contains large numbers of viable *Bifidobacterium bifidum*—active in acid production, slightly acetic acid flavored—high survival rate at gastric pH.

Therapeutic Value: Used in the treatment of enteric infections in babies.

Reference: Lipinska, E. 1978. Survival of bifidobacteria under unfavourable physiological conditions. *Proceedings 20th International Dairy Congress, Paris* **E:**528-529. Brussels: International Dairy Federation.

LLAMA MILK FERMENTED FRESH PRODUCTS (En); Produits laitiers fermentés à partir de lait de lama (Fr); Sauermilchprodukte aus Lama-Milch (De).

Short Description: South America; nontraditional products; llama milk; llamas are any of a group of South American mammals related to the camel but smaller and without humps. Llamas are used as beasts of burden and as a source of wool, meat, and milk.

Food Value: The average composition of llama milk is as follows: 86.55% water, 3.90% protein, 3.15% fat, 5.60% lactose and 0.80% ash.

Products: There is no evidence that traditional fermented milk products are made from llama milk or the milk of vicunas and alpacas, which are related to llamas, but spontaneously fermented milks have most likely been produced.

Reference: Lampert, L. M. 1970. *Modern Dairy Products,* p. 11. New York: Chemical Publishing Company, Inc.

LO (En, Fr, De); means sour milk.

Short Description: China, Kan-su province (northwest China); traditional fermented milk, yoghurtlike; buffalo's, cow's, camel's, ewe's, mare's, and yak milk; very viscous fermented milk.

Manufacture: The milk is concentrated by boiling and is then cooled and incubated with product made the previous day or earlier. Because of its high viscosity, lo can be put on a leaf instead of in a cup or dish when consumed.

Related Product: Yoghurt.

Reference: Hermanns, M. 1969. *The Nomads of Tibet* (in German). Vienna, p. 66. Cited by Takamya, T. 1978, pp. 98-99 and 289.

LONG-LIFE YOGHURT (En); Yaourt de longue conservation (Fr); Joghurt mit langer Haltbarkeitsdauer (De).

Short Description: Various countries; cow's milk, usually; yoghurt with prolonged storage life; heat-treated after culturing.

Microbiology: To avoid misunderstanding, and since the product does not contain viable yoghurt bacteria or viable bacteria in sufficient numbers, the label yoghurt should not be granted to such products, according to traditionalists.

Manufacture: Among the different methods for prolonging the storage life of yoghurt, aseptic manufacture and heat treatment after fermentation are the most important. The first refers to the technology of manufacturing yoghurt using aseptic techniques and/or methods of aseptic acidification. The second refers to methods of heat treatment of yoghurt to destroy contaminants and the majority, if not all, of culture bacteria.

Related Products: Heat treatments have found wide application in the manufacture of flavored drinks (*see* Beverages) and in the preparation of various sauces and culinary preparations based on yoghurt (*see* Cooked yoghurt).

Reference: Rašić, J. Lj., and J. A. Kurmann. 1978, pp. 283-293.

LONG MILK (En); Lait filant (Fr); Langmilch (De).

Short Description: Scandinavia (Sweden, Norway, Finland); traditional product; cow's milk; stirred product; direct consumption; ropy consistency, long, slimy strings, and high viscosity; mild acid and agreeable flavor.

History: Also known as langfil, long milk is one of the oldest fermented milk products made in Sweden. It was already described in the eighteenth century by the botanist Linné under the name *Sätmjölk.* Originally the product was made by adding to milk and incubating it with parts of the plant butterwort (*Pinguicula vulgaris,* L.). During incubation lactic acid bacteria are shed from the plant and, consequently, acidify the milk and produce ropiness.

Microbiology: A mixed culture consisting of *Streptococcus lactis* var. *longi* (capsule-forming strains; not considered as subspecies) and *Leuconostoc mesenteroides* subsp. *cremoris.*

Manufacture: Modern manufacture involves the use of pure cultures for controlled acid fermentation. The product is made from milk standardized to 3.0% fat, pasteurized at 90-91°C (194.0-195.8°F) for 3 min and cooled to 18°C (64.4°F). It is inoculated with 2% of the above culture, incubated at 17-18°C (62.6-64.4°F) until coagulation, then stirred, cooled, and packaged.

Food Value: The final product has an acidity of 90-95°Th (0.81-0.85% titratable acidity), a pH of 4.5, and a count of about 7×10^8 cells (colony-forming units) per milliliter.

Related Products: Tätte mjölk, viili, langfil, pitkäpimä (Finnish).

References:

Alm, L. 1982. The effect of fermentation on nutrients in milk and some properties of fermented liquid milk products. Thesis, University of Stockholm, Sweden.
Forsén, R. 1966, p. 76.

LONG WHEY (En), Lactosérum filant (Fr); Fadenziehende Molke (De); Lange Wei, denotes ropy whey.

Short Description: The Netherlands; ropy whey obtained by fermentation with hydrocolloid-forming lactic acid bacteria, in earlier times used in the manufacture of Edam cheese (*Streptococcus hollandicus* resp., a ropy strain of *S. lactis* subsp. *cremoris*).

Reference: Martiny, B. 1907, p. 137.

LOW-(CONTENT) FERMENTED MILK PRODUCTS (En); Produits laitiers fermentés avec une teneur réduite (Fr); Sauermilchprodukte mit reduzierten Gehaltsbestandteilen (De).

Short Description: Various countries; nontraditional products; mainly cow's milk.

Products: Fermented milk products whose name is preceded by the term *low-* are products that have been modified or have been formulated in such a way that one or more components have been reduced, for example low-calorie (energy-reduced), low-fat (skimmed or fat-reduced), low-carbohydrate (lactose-reduced), low-sodium (unsalted or sodium-reduced).

LOW-SODIUM YOGHURT (En); Yaourt allégé en sodium (Fr); Joghurt mit vermindertem Natriumgehalt (De).

Short Description: Several industrialized countries; cow's milk; dietetic product.

Manufacture: Yoghurt prepared from low-sodium skim milk or partially skimmed milk (e.g., "pennac" milk contains 0.23 g sodium/liter as compared with 1.50 g/liter for ordinary milk).

Food Value: Dietetic use by patients suffering from high blood pressure and cardiovascular and kidney diseases who are advised to be on a low-sodium diet.

Related Products: Yoghurt, yoghurt preparations.

References:
Rašić, J. Lj., and J. A. Kurmann. 1978, pp. 116, 355.
Rialland, J. P., and J. P. Barbier. 1984. Acidulated decationized milk. *United States Patent* 4,460,616. Cited in *Dairy Science Abstracts,* 1985, **42**(1841):211.

LÜNEBEST (En, Fr, De); trade name.

Short Description: Germany; nontraditional product; cow's milk; used for direct consumption; cultured milk called *Lünebest special-joghurt.*

Microbiology: A mixed culture consisting of *Lactobacillus acidophilus, Bifidobacterium bifidum* and yoghurt culture bacteria. Usually made as a low-fat fruit-flavored product and having firm body and specific flavor.

Related Products: Acidophilus-bifidus-yoghurt.

Reference: Rašić, J. Lj., and J. A. Kurmann. 1983, pp. 132, 216.

LYNTYCA (En, Fr, De); Lyntca.

Short Description: Poland; traditional fermented whey beverage; ewe's milk; whey from Brinza cheese; alcoholic fermentation; homemade product.

Reference: Hetcho, N., Basel, Switzerland, personal communication.

LYOBIFIDUS (En, Fr, De); trade name.

Short Description: France; nontraditional product, pharmaceutical preparation; freeze-dried.

Microbiology: The product contains a minimum of 10^6 viable *Bifidobacterium bifidum* cells per gram of product.

Therapeutic Value: Used for diarrhea.

Reference: Schneegans, E., A. Haarscher, A. Lutz, J. Levi-Silage, and J. Schmittbühl. 1966. Contribution to the study of Bifidobacterium bifidum. Implantation experiments in the presence of nutrients and of pathogenic Escherichia coli (in French). *Sem. Hôp. Paris* **42**(26/27):457-462.

LYUBITELSKII (En, Fr, De); Lyubitel'skii, denotes favorite.

Short Description: USSR; nontraditional product; cow's milk fermented beverage; liquid, slightly viscous consistency; acidic taste and pleasant aroma.

Microbiology: Cultures of *Streptococcus thermophilus* and *S. lactis* subsp. *diacetylactis;* in some cases also *S. lactis.* When combined cultures are used, they should preferably also contain *S. lactis* strains. When separate cultures are used *S. thermophilus* is added to the milk first, and the ratio of *S. lactis* subsp. *diacetylactis* to *S. thermophilus* is 4:1.

Manufacture: The product is made from skim milk, pasteurized, cooled and inoculated with 2-3% of the above culture. Incubation is carried out in two stages, with the inoculum first added to ⅕-⅙ of the milk volume at 25-26°C (77.0-78.8°F). The resultant mixture is then added to the remaining milk at 37-38°C (98.6-100.4°F) so that the final temperature is 35-36°C (95.0-96.8°F) at which incubation is continued to an acidity of 85-90°Th (0.76-0.90% titratable acidity).

Related Products: Lactic acid streptococci fermented milks.

References:

Koroleva, N. S., L. A. Bannikova, I. N. Pyatnitsyna, S. B. Zadoyana, G. V. Guzikova, V. I. Klimova, L. M. D'yakova, and E. A. Bogdanova. 1978. Method of obtaining "lyubitel'skii" soured milk (in Russian). *USSR Patent* 608,513. Cited in *Dairy Science Abstracts,* 1979, **41**(6506):729.

Zobkova, Z. S., E. A. Boganova, and I. I. Kochergina. 1978. New types of whole milk products (in Russian). *Molochnaya Promȳshlennost'* **4:**16-17. (Ru).

M

MAAS (En, Fr, De).

Short Description: South Africa; cultured milk or cultured buttermilk; whole, low-fat, or skimmed milk.

Microbiology: Commerical butter starters or mixtures of mesophilic lactic streptococci are used.

Reference: Bolstridge, M. C., and G. Roth. 1985. Enterotoxigenicity of strains of *Staphylococcus aureus* isolated from milk and milk products. *South African Journal of Dairy Technology* **17**:91-95. Cited in *Dairy Science Abstracts, 1987,* **49**(301):33-34.

MADEER (En, Fr, De).

Short Description: Middle East, especially in the desert regions of Saudi Arabia; traditional product; goat's and/or ewe's milk; dried, fermented milk product, very much like a hard, dry cheese in consistency; heat-treated after culturing.

Manufacture: Madeer from goat's milk is prepared as follows: Fresh whole milk is fermented overnight at ambient temperature, followed by churning 1-2 hours for fat separation. The buttermilk obtained is boiled 3-4 hours, then the viscous product is cooled and the semisolid mass is molded by hand.

Pieces are sun-dried (direct sunlight) at 50-60°C (122-140°F) for 2-4 days (Sawaya et al. 1984).

Food Value: Madeer forms a major component of the family diet of desert dwellers and plays a significant role in the nutritional well-being of these people. When fresh milk is in short supply, stored madeer is stirred into water and eaten like yoghurt, or it may be dissolved in hot water to make soups.

Related Product: Oggt.

References:

El-Erian, A. F. M. 1979. Studies on oggt. *Proceedings, Saudi Biological Society* **3:**7-13.
Sawaya, W. N., J. P. Salji, M. Ayaz, and J. K. Khalil. 1984. The chemical composition and nutritive value of madeer. *Ecology of Food and Nutrition* **55:**29-37.

MAFI (En, Fr, De).

Short Description: Lesotho (South Africa); yoghurt-type fermented milk.

Reference: Campbell-Platt, G. 1987, p. 232.

MAGOU (En, Fr, De).

Short Description: South Africa (Bantu); fermented traditional vegetable beverage; possibility of combination with milk is recognized.

Microbiology: The starter culture consists of *Lactobacillus delbrueckii.* The *L. delbrueckii* was adapted to grow in magou by subculturing in a medium containing 8% maize (corn) meal in water, 2% dried skim milk, 0.1% K_2HPO_4, 1% wheat flour extract, and 1% yeast extract. The content of yeast extract was reduced at each transfer so that after the seventh transfer the culture could grow in its absence.

Manufacture: Magou is made from 8% of corn flour (maize meal) in water, heated at 121°C (249.8°F) for 15 min and incubated at 50°C (122°F) in the presence of 0.5-1% unheated wheat bran. Improvement in flavor and consistency is achieved when 0.5-2% dried skim milk is added. By fermentation with the above-mentioned starter culture a pH of 4.3-4.5 was reached in about 8 h at 50°C (122°F). A combination of magou with milk is suggested.

References:
Campbell-Platt, G. 1987, p. 123.
Kriel, J. B., and L. A. Wood. 1977. Certain aspects regarding the manufacture of magou (in Afrikaans). *South African Journal of Dairy Technology* **9**(4):155-158. Cited in *Dairy Science Abstracts* 1978, **40**(4194):464.
Schweigart, F., and S. A. Fellingham. 1963. A study of fermentation in the production of Mahewu, an indigenous sour maize beverage of Southern Africa. *Milchwissenschaft* **18:**241-246.

MAĬSKIĬ (En, Fr, De).

Short Description: USSR; nontraditional cultured milk beverage; cow's milk.

Microbiology: The starter culture consisting of lactic streptococci and *Lactobacillus acidophilus* (4:1 ratio) is mixed in at a rate of 3%.

Manufacture: Maĭskiĭ is made from a mixture of equal portions of cheese whey and skim milk, with or without 3% sugar. The mixture is incubated at 28-30°C (82.4-86.0°F), after inoculation with the above-mentioned culture, until a firm coagulum is formed and the acidity reaches 90°Th (0.81% titratable acidity). The final product is kept for 36 h at 6 ± 2°C (42.8 ± 36.6°F) before leaving the factory.

Food Value: The total solids content is about 10.5% for the plain variety and about 13.5% for the sweetened product.

Reference: Molochnikov, V. V., and L. S. Trufanova. 1985. Maĭskiĭ—a cultured milk beverage (in Russian). *Molochnaya Promȳshlennost'.* **4:**16-18. Cited in *Dairy Science Abstracts,* 1986, **48**(1293):153.

MAJA (En, Fr, De); Maya.

Short Description: Serbia (Yugoslavia), Turkey, and the Balkan area; starter culture for traditional products.

Microbiology: The name signifies ferment (culture).

Wild culture(s) for homemade products. National name for starter culture in Serbia (Yugoslavia), Turkey, and other countries. To make homemade sour milk preparations some of the sour milk from a previously made batch is used to inoculate milk (the culture and the fermentation are not well defined).

Industrial starter culture(s). The name is used to designate industrially prepared starter cultures.
Related designations for cultures. Rob, Roba, Podkvasa, Zakvaska (USSR).

References:
Martiny, B. 1907, p. 75.
Fleischmann, W., and H. Weigmann. 1932, p. 371.
Földes, L. 1969, pp. 864 and 873.

MALA (En, Fr, De); a term obtained by contracting *maziva lala.*

Short Description: Kenya; nontraditional product.

Microbiology: The starter culture consists of *Streptococcus lactis* subsp. *cremoris, S. lactis* subsp. *diacetilactis* and *Leuconostoc mesenteroides* subsp. *cremoris.* The starter culture is mild in taste, with some CO_2 gas production, good viscosity.

Manufacture: Milk is heated to 85°C (185°F) for 30 min, then cooled and after inoculation incubated at room temperature (20-40°C/68-104°F). Optional flavoring with mango juice or passion fruit juice (12-10%), addition of sugar (10-12%), and stabilization (0.16% pectin, 0.9% gelatin, or 0.6% sodium caseinate). Keeping quality at 20-40°C (68-104°F) is about 2-4 days.

Food Value: The final product contains 11.5% total solids, 3.5% fat, 8.5-9.0% nonfat solids, and has 135°Th (1.2% titratable acidity).

Related Product: Maziva lala.

Reference: Wenger, U. 1988. Where milk products are stored at 30°C/86°F (in German). *Schweizerische Milchzeitung* **114**(32):5.

MALYSH (En, Fr, De); trade name.

Short Description: USSR; nontraditional product; cow's milk; acidophilus baby food.

Reference: Koroleva, N. S., V. F. Semenikhina, L. N. Ivanova, I. V. Oleneva, M. B. Sundukova, and E. A. Khorkova. 1982. Milk products cultured with Lactobacillus acidophilus and bifidobacteria for infants (in Russian). *Molochnaya Promȳshlennost'.* **6:**17-20.

MALYUTKA (En, Fr, De); trade name.

Short Description: USSR; nontraditional product; cow's milk; acidophilus baby food.

Reference: Nabukhotnyi, T. L., S. A. Cherevko, F. I. Samigullina, and A. I. Grushko. 1983. Use of adapted propiono-acidophilic "Malyutka" and "Malysh" formulas in complex treatment of acute gastrointestinal diseases of infants (in Russian). *Voprosy Pitaniya* **6:**27-30. Cited in *Dairy Science Abstracts,* 1985, **47**(2122):243.

MARE'S MILK FERMENTED MILK PRODUCTS (En); Produits laitier fermentés a partir de lait de jumaux (Fr); Produkte aus Stutenmilch (De).

Short Description: USSR, mainly Mongolia; traditional products; mare's milk.

History: The utilization of mare's milk for the manufacture of fermented milk products is very old and several products are known: kumiss (known by the nomadic tribes of earliest antiquity), karakosmos, araka.

Food Value: The milk of horses has the following chemical composition: 11.2% total solids, 1.9% fat, 1.3% casein, 1.2% whey proteins, 6.2% lactose, 0.5% minerals (Jenness and Sloan 1970). Fermented mare's milk products are believed to have therapeutic and dietetic properties (in the USSR these products are used in treating tuberculosis).

References:

Accolas, J. P., J. P. Deffontaines, and F. Aubin. 1975. Rural activities in the Mongolian Republic (in French). *Etudes Mongoles* (Paris) **6:**7-98.

Jenness, R., and R. E. Sloan. 1970. The composition of milks of various species: a review. *Dairy Science Abstracts* **32:**610.

MARRIOTT'S ACIDIFIED BABY MILK FORMULA (En); Aliment pour bébé selon Marriott (Fr); Säuglingsnahrungsmittel nach Marriott (De).

Short Description: United States; nontraditional baby food; cow's milk; a precursor of pelargon, which is a product acidified with lactic acid; first made by W. M. Marriott (born 1885), an American pediatrician working in St. Louis, Missouri; acidified milk product (*see* Acidified baby milk formula).

Reference: Feer, E. 1937. Sour milk products in the diet of various nations and in modern medicine (in German). In *Volume Jubilaire en l'Honneur de Monsieur Louis E. C. Dapples,* pp. 202-209. Vevey, Switzerland: Nestlé S.A.

MAST (En, Fr, De); Maast; signifies coagulated milk.

Short Description: Iran, Iraq, and Afghanistan; traditional, national type of yoghurt; the milk of ewes, goats, or cows; firm consistency and acidic flavor.

References:
Földes, L. 1969, p. 854.
Steinkraus, K. H. 1983, p. 249.

MATTHA (En, Fr, De).

Short Description: India (common to all regions); traditional, refreshing beverage; dahi (fermented milk); mattha is highly flavorful and mildly to highly acidic.

Manufacture: After churning of dahi (previously mixed with cold water), grains of makkan (butter granules) that appear on the top of the fluid buttermilk are scooped out and patted into a smooth compact mass. The remaining fluid is called mattha. It is essentially a rural domestic product.

Food Value: The final product contains about 0.5% fat and 6-7% milk solids.

Related Product: Buttermilk, natural.

Reference: Bandyopadhyay, A. K., and B. N. Mathur. 1987. Indian milk products: a compendium. In *Dairy India,* ed. P. R. Gupta, pp. 211-218. Delhi: Dairy India, Priyadarshini Vihar.

MA TUNG (En, Fr, De).

Short Description: China; kumiss-type.

Reference: Campbell-Platt, G. 1987. pp. 108, 127.

MATZOON (En); mazun (Fr, De); matsun, matsoni, maconi.

Short Description: Of Armenian origin; Georgia, Caucasus (USSR); traditional product; the milk of ewes, goats, buffalo, or cows or mixtures thereof; yoghurtlike product traditionally made from boiled milk and an undefined starter culture; firm consistency and acidic flavor.

Microbiology: Traditional product made with undefined starter culture consisting of thermophilic and mesophilic lactic streptococci and thermophilic lactobacilli, and often with yeasts. Starter culture with defined microflora: proposed *Streptococcus thermophilus* and *Lactobacillus delbrueckii* subsp. *bulgaricus.*

Related Product: Yoghurt.

References:

Awetikow, G., and A. Magakjan. 1961. Accelerated production method for the fermented milk "Mazun" (in Russian). *Molochnaya Promȳshlennost'* **22:**14-16. Cited in *Milchwissenschaft* (in German). 1962, **17:**108.

Düggeli, M. 1905. Bacteriological researches about Armenia Matzoon (in French). *Le Lait,* p. 279.

Koroleva, N. S. 1975. *Microbiology of Whole Milk Products* (in Russian). Moscow: Pishchevoĭ Promȳshlennosti.

MAZIWA LALA (En, Fr, De); signifies milk that is "sleeping."

Short Description: East Africa, Kenya; traditional spontaneously soured milk beverage prepared by nomads or homemade; related, but industrially prepared, is a cultured buttermilk (mala).

Reference: Shalo, P. L., and K. K. Hamsen. 1973. Maziwa lala—a soured milk (in French). *World Animal Review* **5:**33.

MEGADOPHILUS (En, Fr, De); trade name.

Short Description: California United States; nontraditional product; freeze-dried; pharmaceutical preparation.

Microbiology: Freeze-dried preparation containing 5×10^9/g of viable *Lactobacillus acidophilus.* Selected strain of *L. acidophilus* demonstrated to have antibiotic activity against many pathogens (Shahani et al., 1977).

Therapeutic Use: Used as a dietary adjunct against an imbalance of the intestinal microflora and in the treatment of gastrointestinal disorders.

Reference: Shahani, K. M., J. P. Vakil, and A. Kilara. 1977. Natural antibiotic activity of *Lactobacillus acidophilus* and *L. bulgaricus.* II. Isolation of acidophilin from *L. acidophilus. Cultured Dairy Products Journal* **12:**8.

MESOPHILIC FERMENTED FRESH MILK PRODUCTS (En); Produits laitiers fermentés mesophiles (Fr); Mesophile Sauermilchprodukte (De).

Short Description: Various countries; mainly traditional and some nontraditional products; cow's milk.

Microbiology: Mesophiles are any organisms having an optimum growth temperature in the range 20-40°C (68-104°F). Fermented milks prepared with mesophilic starter cultures use incubation temperatures mainly between 10/15-20/30°C (50/59-68/86°F).

Products: Kurmann (1984) has categorized subgroups of fermented milks.

1. Lactic acid fermentation and slime production: Scandinavian fermented milks, e.g. long milk, viili
2. Lactic acid fermentation without slime production using cream cultures: fermented milks prepared with cream cultures; cultured buttermilk (*see* Cultured buttermilk)
3. Lactic acid fermentation, with increased total solids content (concentrated fermented milks): domestic products, e.g. cellar milk; industrial products, e.g. lactofil, skyr, ymer
4. Mixed lactic acid and alcoholic fermentation (*see* Kumiss and similar products, Kurunga, Kefir, Airan)
5. Various products (*see* Streptococcus lactis milk)

References:

Driessen, F. M., and Z. Puhan. 1988. Technology of mesophilic fermented milk. *IDF Bulletin 227* (Fermented Milks. Science and Technology), pp. 75-81. Brussels: International Dairy Federation.

Kurmann, J. A. 1984. Production of fermented milk in the world. II. Aspects of the production of fermented milks. *IDF Bulletin 179* (Fermented Milks) p. 26. Brussels: International Dairy Federation.

METCHNIKOV PROSTOKVASHA (En, Fr, De); denotes sour milk according to Metchnikov, the famous Russian biologist.

Short Description: USSR; nontraditional product; cow's milk and cream; yoghurtlike product; firm consistency and pleasant flavor.

Microbiology: The starter culture consisting of *Streptococcus thermophilus* and *Lactobacillus delbrueckii* subsp. *bulgaricus* (ratio 4:1) is added and mixed in at a rate of 5%.

Manufacture: The product is made from a mixture of milk and cream, standardized to 6.1% fat, then homogenized and pasteurized at 85-90°C (185-194°F) for 10-15 min and cooled to 40-45°C (104-113°F). The mixture is inoculated with the above-mentioned culture, mixed well, packaged into retail containers, and incubated for 2-4 h until firmly coagulated and then cooled.

Food Value: The final product contains 6% fat and has an acidity of 110°Th (about 1% titratable acidity).

Related Product: Rjazhenka.

Reference: Koroleva, N. S. 1975. *Microbiology of Whole Milk Products* (in Russian). Moscow: Pishchevoǐ Promȳshlennosti.

MICIURATU (En, Fr, De); Mezzoradu.

Short Description: Italy (Sicily); traditional fermented milk beverage; related product gioddu.

Reference: *FIL-IDF Dictionary.* 1983, p. 145.

MIDDLE EAST FERMENTED FRESH MILK PRODUCTS (En); Produits laitiers fermentés du Moyen-Orient (Fr); Sauermilchprodukte des Mittleren Ostens (De).

Short Description: Middle East region; important traditional and several nontraditional products; different types of milk (buffalo, goat, cow).

Products: The Middle East is an important source of several interesting fermented milk products. Numerous products are prepared in Egypt, Iran, Iraq, Saudi Arabia, Sudan, and Turkey (see Appendix B).

The products originating in these regions are not always confined to well-defined geographical boundaries but can also be found in adjacent territories.

MILD (LOW-ACID)CULTURED FRESH MILK PRODUCTS (En); Produits laitiers fermentés doux (Fr); Mildgesäuerte Sauermilchprodukte (De).

Short Description: Various countries; nontraditional products; cow's milk mainly; prepared with thermophilic cultures.

Microbiology: The starter culture termed *MSK* (German: Mildsäuernde Kultur) consists of *Lactobacillus acidophilus, Bifidobacterium longum,* and *Streptococcus thermophilus.* It is available commercially in concentrated and deep-frozen form. There are different types of MSK cultures: MSK or MSK/R cultures for yoghurt and cultured milk; MSK/V culture for yoghurt with more viscosity; MSK/Q culture for quarg or curdled milk made by the thermo-technique or for concentrated fermented milks.

Manufacture: The manufacture of mild cultured fresh milk products requires specific technical facilities, but superior dietary benefits are claimed. The critical factor in manufacturing these products is selecting an appropriate culture. One such example is MSK cultures. These cultures, when correctly used, enable the manufacture of mildly acid and aromatic cultured milk products without notable postacidification and without the addition of other cultures. However, when other cultures are added, for example, some yoghurt culture, it is possible to make products of different tastes, flavors, and viscosities.

References:

Hunger, W. 1985. Fermented milk products prepared with "mild acidifying cultures (in German). *Deutsche Molkerei-Zeitung* **106:**826-833.

Laboratorium Wiesby GmbH and Co., D-2260 Niebüll, P.O. Box 1366, Germany (promotional literature).

MILK BRANDY (En); Alcohol distillé de lait fermentés (Fr); Alkohol aus distillierten Sauermilchprodukten (De).

Short Description: Various countries; traditional products; the milk of mares, cows goats, or ewes.

Manufacture: Milk brandy is made by distillation of certain fermented milks, especially those after intensive lactic and strong alcoholic fermentations (e.g., kumiss, kefir, airan).

Related Products: The best known brandy products are aker, araq, arsa, and chorsa.

MILK-CEREAL PREPARATIONS (En), Préparations de lait-cereal (Fr), Milch-Getreide Zubereitungen (De).

Short Description: Various countries; mainly traditional and some non-traditional products (muesli, gua-nai); milk and cereals.

Products: Milk-cereal preparations include

- Subsequent addition of nonfermented cereals: crowdies (oatmeal), fura (millet), honey clabbered milk (flour), kesävelli (buckwheat or oatmeal), muesli (various mixed cereals and fruits)
- Mixed milk-cereal fermentations: busa (rice), gua-nai (rice), jalebi (refined flour), rabadi (maize flour or corn meal)

MILK "CHAMPAGNE" (En); "Champagne du lait" (Fr); "Milch-Champagner" (De); denotes champagne-style fermented fresh milk beverage.

Short Description: Eastern Europe; nontraditional products; skim milk; alcoholic, gassy beverage.

Manufacture: In Bulgaria milk champagne is made by the fermentation of milk with kefir and kumiss starters, with the addition of sugar, honey, and botanical extracts (Kvatchkoff 1937).

In Poland milk champagne is made by acidifying pasteurized skim milk with a cream starter culture to 55-87.5°Th (0.49-0.78% titratable acidity). Lemon essence or lemon peel, 2% sugar, and other flavoring ingredients and yeast starter are also added. Fermentation is carried out in bottles at 18°C (64.4°F) for 12 h (Lang and Lang 1971).

References:

Kvatchkoff, I. 1937. Studies on Bulgarian fermented ewes' milk (in French). *Le Lait* **17**:475.

Lang, F., and A. Lang. 1971. Fermented dairy beverages and specialties in Poland. *Milk Industry* **69**(2):13-15.

MILK-FISH MEAL HYDROLYSATE, FERMENTED (En); Lait-hydrolysat de farine de poisson, fermentés (Fr); Milch-Fischmehl-Hydrolysat, vergoren (De).

Short Description: Japan; nontraditional product; milk and a fish meal hydrolysate; no commercial products are known to exist.

History: In regions where fish is widely available and considered an important food (e.g., Japan) research has been conducted to combine milk with a fish meal hydrolysate.

Manufacture: The growth of lactic acid bacteria in such mixtures is stimulated, but firmness of the curd is slightly diminished.

Reference: Yuguchi, H. 1981. Studies on the utilization of fish meal hydrolysate for fermented milk products. VII. Correlation of the chemical composition of the fish meal hydrolysate with the stimulatory activity for the growth of lactic acid bacteria and for the decrease of curd tension of fermented milk products (in Japanese). *Japanese Journal of Dairy and Food Science* **30**:95-104. Cited in *Dairy Science Abstracts,* 1982, **44**(6915):761.

MILK-FRUIT PREPARATIONS, FERMENTED (En); Preparations de lait-fruit fermentés (Fr); Milch-Frucht-Zubereitungen (De).

Short Description: Various countries; mainly nontraditional products; cow's milk and fruit.

Products: Known fermented milk-fruit preparations include those preparations with the subsequent addition of nonfermented fruit, for example, fruit yoghurt and fru-fru, and mixed milk-fruit preparations, such as fruit-jellied yoghurt and niyoghurt (coconut juice).

MILK-MEAT EXTRACT MIXTURES, FERMENTED (En); Mixture fermentés de lait et extrait de viande (Fr); Gesäuerte Milch-Fleischextrakt-Produkte (De).

Short Description: Germany; nontraditional products; not available.

History: Around 1900 attempts were made to ferment mixtures of milk and meat extracts (similar to Liebig's extract). However, no such commercially made products were sold.

Reference: Martiny, B. 1907, p. 81.

MILK-PLANT MIXTURES, FERMENTED (En); Mixture lait-plantes fermentés (Fr); Gesäuerte Milch-Pflanzenzubereitung (De).

Short Description: Various countries; traditional and nontraditional products; milk and edible plants.

Products: Fermented milk-plant mixtures are composed of three product groups: milk-cereal preparations, milk-fruit preparations, and milk-vegetable preparations.

Milk-plant mixtures can be divided into two groups: (1) Those with the subsequent addition of nonfermented cereals, for example, kesävelli or those with the subsequent addition of nonfermented cereals and fruits, for example, muesli and (2) Mixed milk-cereal fermentations, for example, gua-nai, jalebi, niyoghurt, rabadi, and trahana.

Reference: Kurmann, J. A., and J. Lj. Rašić. 1988. Technology of fermented special products. *Bulletin IDF No. 227* (Fermented Milks. Science and Technology), pp. 101-114. Brussels: International Dairy Federation.

MILK-VEGETABLE MIXTURES, FERMENTED (En); Mixtures de lait-végétaux, fermentés (Fr); Gesäuerte Milch-Vegetabilien-Mischung (De).

Short Description: Various countries; traditional and nontraditional products; milk and vegetables.

Manufacture: Vegetables are plants or parts of plants cultivated for food. Some foods that are botanically fruits, such as tomatoes and cucumbers, and seeds, such as peas and beans, are included with the vegetables. The technology of fermenting milk-vegetable mixtures is still in the stage of early research and development. Such products would be an excellent source of low-cost protein and high nutritional value. They might also be used for special or therapeutic purposes. Instead of whole milk, skim milk, natural buttermilk, or whey may be used as the milk component. Usually soybean extract is the vegetable component. The beany flavor of soy extract may be reduced both by its appropriate preparation and by the subsequent lactic acid fermentation, which further improves the product's acceptability.

Fermented milk-vegetable mixtures may be made either by mixing separately cultured skim milk or cultured sweet buttermilk with the heat-treated soy extract in a suitable proportion or by fermenting soy extract containing milk solids or whey solids with lactic cultures. Two examples illustrate these methods. In the first, skim milk yoghurt of high acidity is mixed with the heat-treated soy extract (in a ratio 2:1 to 1:1) and then cooled. In the second, the heat-treated soy extract containing 2% whey solids, 5% sucrose, 1% gelatin, and 5% lemon flavoring is fermented with 5% of an acidophilus culture at 37°C (98.6°F) and then cooled. Further advances in the manufacturing technology of fermented milk-vegetable mixtures will undoubtedly be made in the future.

Products: Two categories of milk-vegetable products are known and described: (1) those with the subsequent addition of nonfermented vegetables, for example, apero yoghurt (tomatoes); and (2) mixed milk-vegetable and legume extract fermentations, for example, milk-soy-extract fermentations, buffalo milk-peanut extract (miltone).

Reference: Rao, D. R., S. R. Pulusani, and C. B. Chawan. 1986. Fermented soybean milk and other fermented legume milk products. In *Legume-Based Fermented Foods,* ed. N. R. Reddy, M. D. Pierson, and D. K. Salunke, pp. 119-134. Boca Raton, Fla.: CRC Press.

MIL-MIL (En, Fr, De); trade name.

Short Description: Japan; nontraditional fermented milk beverage.

Microbiology: The starter culture consists of *Bifidobacterium bifidum, Bifidobacterium breve,* and *Lactobacillus acidophilus.*

Manufacture: A manufacturing process for this fermented milk has been developed by the Yakult Honsha Company in Japan, where it is commercially produced. Mil-mil is made by fermenting standardized, heat-treated milk with the above-mentioned starter. The product is sweetened with small quantities of glucose and/or fructose and is colored with carrot juice. Currently it is available in three different flavors.

Food Value: Carrot juice provides provitamin A. The product may be used as a soup as well.

References:

Anonymous. 1978. Soups and a health drink in Tetra Briks in Japan. *Packaging* **49:**37. Cited in *Dairy Science Abstracts,* 1979, **41:**406.

Hori, S. 1983. Characteristics of the bifidus yoghurt "Miru-Miru E; Mil-mil E (in Japanese). *New Food Industry* **25:**25-32. Cited in *Dairy Science Abstracts,* 1984, **46**(3718):424.

MILONE (En, Fr, De); trade name.

Short Description: Germany; nontraditional product; whey obtained from cow's milk with plant extracts and sweeteners added; whey beverage.

Microbiology: The starter culture consists of lactic acid bacteria (species not specified) and a lactose-fermenting yeast.

Manufacture: Whey is fermented to 110°Th (1% titratable acidity) then mixed (1:1 ratio) with an infusion derived from a mixture of plants. The tannins extracted from the plants and the whey proteins coprecipitate and are removed. The remaining liquid is inoculated with a lactose-fermenting yeast. Upon incubation approximately 0.8% ethanol is produced. The final product is sweetened with nonnutritive sweeteners, filtered, and bottled with carbon dioxide.

Related Products: Whey cultured products (alcoholic beverages).

Reference: Schulz, M. E., and K. Fackelmeyer. Fermented beverage from plant extract and whey (in German). *Milchwissenschaft* **3:**165-174.

MILTONE (En, Fr, De).

Short Description: Southern India; nontraditional product; a mixture of milk or sometimes fermented milk and peanut meal (called miltone or peanut milk). Since 80-90% of the milk in India is buffalo milk, this is a buffalo milk product.

History: Miltone is a vegetable toned milk developed by the Central Food Technological Research Institute (CFTRI) in Mysore in Southern India where peanuts are grown. This product is made to boost the solids content of milk, especially of diluted (toned) milk. Miltone has a peanutlike taste. In the home, the flour of other cereals is sometimes added to milk.

Food Value: Miltone contains 11.5% total solids, 4% proteins, 2% fat, 2.5% lactose, and 2.5% glucose and malt dextrin (Dairy India 1983).

References:

Krishnaswamy, M. A., J. D. Patel, S. Dhanaraj, and V. S. Govindarajan. 1971. Curd from miltone (vegetable toned milk). *Journal of Food Science and Technology* **8:**41-46.

Swaminathan, M., and H. A. B. Parpia. 1967. Milk substrates based on oilseeds and nuts. *World Review Nutrition and Diet* **8:**184.

MISTI DAHI (En, Fr, De).

Short Description: India; traditional fermented milk; partially concentrated buffalo milk (18% milk solids) and sweetened with 14% sucrose; the commercial importance of misti dahi is increasing, and therefore the technology for producing it industrially is being developed.

Reference: Ghosh, J., and G. S. Rajorhia. 1990. Selection of starter culture for production of indigenous Indian fermented milk product (Misti dahi). *Lait* (Lyon) **70**(2):147-154.

MLADOST (En, Fr, De); trade name.

Short Description: Bulgaria; nontraditional product; cultured milk beverage; cow's milk.

Microbiology: The starter culture consists of non-specified lactic acid bacteria (probably thermophilic; yoghurtlike).

Manufacture: In the manufacture of mladost, milk is pasteurized at 90-95°C (194-203°F) for 5-10 min. It is then pumped into tanks or vats where starter culture is added at the appropriate temperature. Coagulation is completed in 3 h. The coagulum, with an acidity of 85-90°Th (0.76-0.81 titratable acidity) and a pH of 4.3-4.4, is cooled to 20°C (68°F), in the summer to 10°C (50°F), and packaged.

References:

Dinev, V. 1974. Technology, microflora and storage of "Mladost" cultured cows milk (in Bulgarian). *Sbirka* **7:**29-30. Cited in *Dairy Science Abstracts,* 1976, **38:**575.

Mladenov, M. G., V. Aleksiev, P. Todorov, and M. Bachiiska. 1984. Microbiological study of "Mladost" cultured milk (in Bulgarian). *Veterinarnomeditsinski Nauki* **21:**103. Cited in *Dairy Science Abstracts,* 1986, **46**(6200):704.

MOLKOSAN (En, Fr, De); trade name derived from the German word for whey (*Molke*) and the Latin word for health (*sana*).

Short Description: Germany; nontraditional product; concentrated whey, fermented, containing L (+)-lactic acid, rich in minerals; pharmaceutical preparation.

Food Value: Diluted with tap water or mineral water, Molkosan is made into a beverage, usually sweetened with honey. It is suitable for preparing sauces, especially low-calorie sauces, muesli, and other foods, and for cosmetic purposes such as skin applications and as a bath water additive.

Reference: Prospectus from 1988. An A. Vogel product, Bioforce SA, CH-9325. Roggwil/TG, Switzerland.

MOLODOST (En, Fr, De).

Short Description: USSR; nontraditional product; cow's milk; direct consumption; yoghurtlike product; firm consistency; acidic and cooked flavor.

Microbiology: The starter culture consisting of *Streptococcus thermophilus* and *Lactobacillus delbrueckii* subsp. *bulgaricus* (ratio 4:1) is mixed in at a rate of 5%.

Manufacture: The product is made from pasteurized skim milk fortified with 1.5% skim milk powder. The mix is heat-treated at 95-99°C (203-210.2°F) for 2-3 h, cooled to 40-43°C (104-109.4°F) and inoculated with the above-mentioned culture, then incubated to coagulation and cooled.

Food Value: The final product has an acidity of 100-130°Th (0.9-1.17% titratable acidity) and contains at least 9.5% nonfat solids. It may be sweetened with 5% sucrose.

Related Products: Mladost (Bulgaria), varenets.

Reference: Bogdanova, G. J., and E. A. Bogdanova. 1974. *New Whole Milk Products of Improved Quality* (in Russian). Moscow: Pishchevoĭ Promȳshlennosti.

MONGOLIAN FERMENTED FRESH MILK PRODUCTS (En); Produits laitiers fermentés Mongoles (Fr); Mongolische Sauermilchprodukte (De).

Short Description: Mongolian Republic; traditional and nontraditional products; different milk types (also mare's milk).

Manufacture: Accolas et al. (1975) give an excellent survey of the main manufacturing processes for Mongolian milk products.

Products: Airaq (araki), acidophilin, bjaslag, kefir, khoormog, prostokvasha, smetana, tarag, tarassun (kurunga), tsutsugi, umdaa, xurud.

Reference: Accolas, J. P., J. P. Deffontaines, and F. Aubin. 1975. Rural activities in the Mongolian Republic (in French). *Etudes Mongoles* (Paris) **6:**7-98.

MOSKOWSKI (En, Fr, De).

Short Description: USSR; nontraditional product; low-fat cultured milk beverage; homogeneous, creamy consistency, and clean lactic flavor.

Microbiology: The starter culture consisting of *Lactobacillus acidophilus* and containing both mucilaginous and nonmucilaginous strains (ratio 1:4) is mixed in at a rate of 1%.

Manufacture: The product is made from a mixture of skim milk, whole milk, 30% fat cream, dried whole milk, and dried skim milk by incubation at 40°C (104°F) with the above starter culture.

Related Product: Acidophilus milk.

Reference: Butin, V. I., Z. S. Zobkova, I. M. Goldina, and S. E. Mogilevskii. 1975. Cultured milk beverage Moskovskii with increased protein content (in Russian). *Trudy Vsesoyznyi Nauchno-Issledovatelskii Institut Molochnoi Promyshlennosti* **36:**56-68, 91. Cited in *Dairy Science Abstracts,* 1976, **38**(3373):365.

MOUSSI MILK PRODUCTS, FRESH FERMENTED (En); Produits laitiers fermentés mousseux (Fr); Geschäumte Sauermilchprodukte (De).

Short Description: Various countries; nontraditional products; cow's milk.

Manufacture: Some stirred yoghurts are aerated in order to create products suitable as sour milk desserts. The yoghurt is usually heat-treated after culturing, so it does not contain viable bacteria. Various manufacturing procedures are suggested by suppliers of hydrocolloid stabilizers.

Reference: Westphal, U. 1985. A continuous aerator for whipping dessert products (in German). *Deutsche Molkerei-Zeitung* **106:**890, 892, 894.

MULTICURDLED FERMENTED MILK PRODUCTS (En); Multifermentations de produits laitiers fermentés (Fr); Mehrfachgärungen von Sauermilchprodukten (De).

Short Description: United States; nontraditional product; cow's milk and colostrum; no such commercial product is currently available; little has been reported about multicurdled fermented products.

Microbiology: The starter culture consists of *Streptococcus lactis, Leuconostoc mesenteroides* subsp. *cremoris, Lactobacillus delbrueckii* subsp. *bulgaricus, S. thermophilus, L. acidophilus,* and *Bifidobacterium bifidum.*

Manufacture: First the milk is incubated with cultures of lactic acid-forming bacteria (e.g., *S. lactis, L. mesenteroides* subsp. *cremoris*) at 21°C (70°F) to hydrolyze approximately 20% of the lactose into glucose and galactose. This incubation is followed by a second fermentation with *L. delbrueckii* subsp. *bulgaricus* and *S. thermophilus* at 41-46°C (105-114°F). After cooling up to 10% colostrum milk is added and the mixture may be aged for 24 h before it is subjected to a third fermentation with strains of *L. acidophilus* or *B. bifidum* at 26.9°C (98°F).

Food Value: Claims are made for a buttermilk- or yoghurt-like food product of high nutritional-therapeutic value and low lactose and increased lactase content.

Reference: Roberts, J. G. 1977. Multi-curdled milk product and process for the preparation thereof. *United States Patent* 4,034,115. Cited in *Dairy Science Abstracts,* 1978, **40:**266.

MURSIK (En, Fr, De).

Short Description: Kenya; traditional product; cow's milk.

Manufacture: Sour milk is ripened for 4-14 days in containers called *kalebasse.* The containers have previously been treated with special herbs.

Reference: Klupsch, H. J. 1984. *Fermented Milk Products, Beverages and Desserts* (in German). Gelsenkirchen-Buer: Th.Mann Verlag.

N

NEPALESE FERMENTED FRESH MILK PRODUCTS (En); Produits laitiers fermentés du Nepal (Fr); Nepalesische Sauermilchprodukte (De).

Short Description: Nepal; mostly traditional products; buffalo, jhopa, and yak milk.

History: Milk in Nepal is obtained not only from dairy cattle, and particularly from water buffalo as in neighboring India, but also from female jhopa and yak. Jhopa are traditional to the area and have been domesticated for a long time and much attention is given to pure blood lines. When a jhopa is bred with another species, the result is a yak, which is sterile (similar to a mule produced by a horse and a donkey). Yaks are sturdy work animals, but are also valued for their milk and wool. Since they are considered "inferior" to cows, their meat is also consumed, which is not the case with cows in that area. It is estimated that of all the milk produced in Nepal 20-25% is consumed in the fermented form.

Products: The major fermented milk products are dahi (a yoghurt type), sho (yoghurt), shomar (obtained by mixing the whey of buttermilk cheese with milk and allowing further fermentation), shosim (naturally fermented milk curd), thara (yoghurt buttermilk from churned yoghurt), and zurpi or churpi (dried buttermilk).

References:

Shrestha, M. P. 1989. Personal communication. Kathmandu, Nepal: Institute of Medicine, Technical University.

Tokita, F., A. Hosono, T. Ishida, F. Takahashi, and H. Otaini. 1980. Variety and manufacturing methods of native milk products in Nepal. *Journal of the Faculty of Agriculture, Shinshu University* **17**(2):117-127. Cited in *Dairy Science Abstracts,* 1984, **46**(1936):235.

NIYOGHURT (En, Fr, De).

Short Description: Philippines; nontraditional product; various milk-plant mixtures (coconut milk, cow's dried skim milk).

Microbiology: Yoghurt starter culture.

Manufacture: Homogenization and pasteurization of a blend of skim milk and coconut milk, followed by the addition of sugar, the above-mentioned starter culture, and pineapple. Incubation is for 3 h at 46°C (114.8°F).

Food Value: 1.9% fat, 4.4% protein and 1.1% minerals.

Related Products: Milk-plant mixtures, fermented; milk-vegetable mixtures, fermented.

References:

Davide, C. L. 1986. Cadtri cheese and niyoghurt made of skimmilk and cocomilk. *Research at Los Banos* **5**(2):28-29. Cited in *Dairy Science Abstracts,* 1988, **50**(2976):332.

N. N. 1985. Niyoghurt vs. yoghurt. *Cocommunity Newsletter* **15**(9):8. Cited in *Dairy Science Abstracts,* 1987, **49**(4971):568.

NONO (En, Fr, De); hausa, fulani milk.

Short Description: Northern Nigeria; probably a traditional product; goat milk; a local yoghurt-type commonly consumed in areas of West Central Africa.

Microbiology: Undefined microflora (spontaneous fermentation) with a wide variety of bacteria among which nonhemolytic *Streptococcus* and *Lactobacillus* were abundant.

Manufacture, Domestic: Nono is prepared by allowing raw whole milk to ferment naturally in a closed 1-liter vessel. Cream is skimmed off.

References:

Atanda, O. O., and M. J. Ikenebomeh. 1988. Changes in the acidity and lactic acid content of "Nono," a Nigerian cultured milk product. *Letters in Applied Microbiology* **6**(6):137-138. Cited in *Dairy Science Abstracts,* 1988, **50**(6551):735.

Odunfa, S. A. 1985. African fermented foods. In *Microbiology of Fermented Foods,* ed. B. J. B. Wood, vol. 2, pp. 185. London, New York: Elsevier.

NOVAYA (En, Fr, De).

Short Description: USSR; nontraditional product; cow's milk; pasty consistency.

Microbiology: *Lactobacillus acidophilus* starter culture bacteria.

Food Value: With 8 or 4% fat, or fat-free and flavored with fruit syrups and extracts.

Related Product: Acidophilus paste.

Reference: Zobkova, Z. S., E. A. Bogdanova, and I. I. Kochergina. 1978. New types of whole milk products (in Russian). *Molochnaya Promȳshlennost'* **4:**16-17. Cited in *Dairy Science Abstracts,* 1978, **40**(7127):753.

NOVINKA (En, Fr, De); denotes novelty.

Short Description: USSR; nontraditional beverage; sweet buttermilk (byproduct of unripened cream butter manufacture); stirred, acidic taste, and agreeable flavor.

Microbiology: The starter is composed of *Streptococcus lactis* subsp. *cremoris, S. lactis* subsp. *lactis,* and *S. lactis* biovar. *acetonicus.*

Manufacture: Novinka is made from sweet buttermilk blended with whole milk, which is pasteurized, then concentrated to 13% total solids, and standardized with milk to 1.5% fat. Fermentation is with a nonspecified amount of starter at 22-25°C (71.6-77.0°F) until coagulation (acidity of 80-90°Th/0.72-0.81% titratable acidity). The soured product is then cooled and bottled.

Food Value: 13-15% total solids, 1-2% fat.

References:

Gushchina, I. M., A. K. Maksimova, and A. I. Nikulina. 1979. Method of obtaining "Novinka" cultured milk beverage (in Russian). *USSR Patent* 682,213.

Zobkova, E. C., E. A. Bogdanova, and I. I. Kochergina. 1978. New kinds of fermented milk products (in Russian). *Molochnaya Promȳshlennost'* **4:**16-17.

O

OFILUS (En, Fr, De); trade name.

Short Description: France; nontraditional product since 1987; cow's milk; set type.

Microbiology: *Ofilus nature:* containing *Streptococcus thermophilus, Lactobacillus acidophilus,* and *Bifidobacterium bifidum.*
Ofilus double douceur: containing *S. lactis* subsp. *lactis, S. lactis* subsp. *cremoris, B. bifidum,* and *L. acidophilus.*

Products: *Ofilus nature,* 3.6% fat and *Ofilus double douceur* (double-mild Ofilus), 10% fat.

OGGT, COOKED (En, Fr, De); Oggt cuite (Fr); Erhitzter Oggt (De); Oggtt, Okt.

Short Description: Saudi Arabia (bedouins), Arabian peninsula; traditional (homemade) product; goat and camel milk; heat-treated and sun-dried.

Microbiology: Spontaneous milk fermentations microorganisms.

Manufacture, Homemade: Milk is fermented spontaneously for 1-2 days and is then churned. The residual buttermilk is boiled while stirring until it thickens. The thick paste is allowed to cool to about 86-95°F (30-35°C) and then shaped by hand into small cakes, which are pressed on canvas fabric and sun-dried.

Food Value: The yellowish-white final product consists of irregularly shaped pieces, 5-9 cm in diameter and 0.8-2.4 cm thick. It has no viable microorganisms; useful as a long-storage and emergency food.

References:

Al-Ruquaie, I. M., H. M. M. El-Nakhal, and A. N. Wahdan. 1987. Improvement in the quality of the dried fermented milk product oggtt. *Journal of Dairy Research* **54:**429-435.

El-Erian, A. F. M. 1979. Studies on oggtt. In *Proceedings of the 3rd Conference of Biological Aspects of Saudi Biological Society,* pp. 7-13. Al-Hassa, Saudi Arabia: King Faisal University.

OGGT, NEW PREPARATION (En); Oggt, preparation nouvelle (Fr); Oggt, neue Zubereitungsart (De); Oggtt, Okt.

Short Description: Saudi Arabia; homemade product; cow's and sheep's whole buttermilk; heat-treated after culturing.

Microbiology: Commercial lactic ferment CH-01 from Chr. Hansen's Laboratories, Denmark, containing mesophilic lactic acid bacteria and *Leuconostoc* sp. (Al-Mohizea et al., 1988). Yoghurt starter cultures microorganisms by Al-Ruquaie et al. (1987).

Manufacture: A flow chart of oggt preparation is given by Al-Mohizea et al. (1988).

Food Value: Depends on the type of milk (or buttermilk) utilized. Sheep milk oggt: 8.1% moisture, 91.9% total solids, 40.5% protein, 15.8% fat, 35.4% lactose, and 8.3% ash. Cow's milk buttermilk oggt: 6.3% moisture, 93.7% total solids, 33.2% protein, 11.7% fat, 7.4% lactose, and 7.7% ash.

References:

Al-Mohizea, I. S., I. H. Abu-Lehia, and M. M. El-Behery. 1988. Acceptability of laboratory made oggt using different types of milk. *Cultured Dairy Products Journal* **23**(3):20-21.

Al-Ruquaie, I. M., H. M. El-Nakhal, and A. N. Wahdan. 1987. Improvement in the quality of the dried fermented milk product oggtt. *Journal of Dairy Research* **54:**429-435.

OGGT, SALTED (En), Oggt salé (Fr), Gesalzener Oggt (De); Oggtt, Okt.

Short Description: Northern Arabian peninsula and Syria, Jordan, and Lebanon; traditional, domestic, whey-drained, dried fermented milk product; goat and sheep milk.

Manufacture: This product is made by fermenting and churning the milk and then adding about 10% salt but not heating. The salted fermented product is filled into cloth containers and allowed to drain for 1 day. The resulting paste is shaped manually into cakes, larger and more regular-shaped than those of cooked oggt. These cakes are also sun-dried. The white oggt pieces are 8-12 cm in diameter and 2-4 cm thick.

Food Value: Both types of oggt can be consumed either dry or after reconstitution with water (the product is then known as *merase*).

Reference: Al-Ruquaie, I. M., H. M. M. El-Nakhal, and A. N. Wahdan. 1987. Improvement in the quality of the dried fermented milk product oggtt. *Journal of Dairy Research* **54:**429-435.

OMAERE (En, Fr, De); Omaire.

Short Description: Central part of South-West Africa, Herero people (a Bantu race); traditional product; cow's milk; kefir-milklike fermented milk; stored in hollowed-out gourds.

Microbiology: Nonspecified lactic acid bacteria and yeasts.

Related Product: Omatuko (agitated omaere).

Reference: Martiny, B. 1907, p. 89.

OMNIFLORA (En, Fr, De); trade name.

Short Description: Germany; nontraditional product, pharmaceutical preparation; freeze-dried.

Microbiology: Freeze-dried preparation containing *Lactobacillus acidophilus, Bifidobacterium longum,* and a saprophytic *Escherichia coli.*

Therapeutic Use: Used for normalization of disordered intestinal flora in humans.

Reference: Reuter, G. 1969. Composition and administration of bacterial cultures for therapeutic purposes (in German). *Arzneimittelforschung* **19:**103-109.

OPLAGT MILK (En); Lait acidifié spontané (Fr); Gesäuerte Dickmilch (De); oplagt maelk, op-lagt maelk, denotes clabbered milk.

Short Description: Denmark; traditional whey-drained product (classifiable as lying between ropy milk and quarg); cow's milk.

Microbiology: Spontaneous milk fermentation microorganisms; slime-producing bacteria.

Manufacture: Spontaneous souring of raw milk with growth of slime-producing bacteria; separation of whey, after cutting the coagulum by passage through cheese cloth.

Related Product: Clabbered milk.

References:

Martiny, B. 1907, p. 89.

Skovgaard, N. 1981. *Hygiene of Milk Production* (in Danish), p. 137. Copenhagen, Denmark: Carl Fr. Mortensen A/S.

OSOBYI (En, Fr, De); trade name.

Short Description: USSR; traditional product, a modified kefir (low-fat, protein-enriched); cow's milk; *see* Kefir.

Reference: Inozemtseva, V. F., I. M. Padaryan, and G. S. Lavrenova. 1979. A new type of dietetic cultured-milk drink (in Russian). *Molochnaya Promȳshlennost'* **49:**30. Cited in *Dairy Science Abstracts,* 1982, **44**(4561):513.

OUZBECK (En, Fr, De); denotes geographic origin.

Short Description: USSR (central Asia, Uzbek region); traditional product; *see* Katyk.

Reference: Abdourazakova, S., E. Guinzbourg, and T. Astakhova. 1965. Study of the microflora of the "coagulated milk ouzbeck" (in French). *Molochnaya Promȳshlennost'* (Moscow) **10**:41. Cited in *Le Lait.* 1967, **47**:85.

OXYGALA (En, Fr, De); denotes soured.

Short Description: Roman Empire; traditional product, whey-drained; sheepmilk; must have been fairly solid, curdlike with good keeping quality.

History: The preparation is described by Columella, a Roman writer on agriculture (approximately A.D. 60).

Microbiology: Spontaneous milk fermentation microorganisms.

Manufacture: According to Columella a bunch of aromatic herbs and onions is added to milk in a jar, which is then closed. After five days the separated whey is drawn off. Three days after that the whey is drawn off again and the herbs are removed and replaced by condiments and leeks. Two days later whey is again removed, salt is added, and the contents of the jar is mixed. The jar is kept closed until the product is consumed.

Oxygala is considered a spontaneously fermented milk product, which is quarglike.

Related Products: Concentrated fermented milks.

References:

André, J. 1981. *L'Alimentation de la Cuisine a Rome* (in French). Paris: Les Belles Lettres.

Columella. *J. Junius Moderatus,* 12,8:1-2. Cited in André, J. 1981. *L'Alimentation de la Cuisine a Rome* (in French). Paris: Les Belles Lettres.

P

PADUSA (En, Fr, De).

Short Description: India; preparation of dahi and ghee. Ghee is clarified butterfat obtained by prolonged boiling of cream.

Reference: *FIL-IDF Dictionary.* 1983, p. 162.

PARAG (En, Fr, De).

Short Description: India; nontraditional product; acidified whey beverage.

Microbiology: The starter culture consists of *Lactobacillus acidophilus* and *Streptococcus thermophilus.*

Manufacture: The beverage is made by mixing clarified paneer whey (whey from drained acid curd) and buttermilk in equal parts, heating to 100°C (212°F) for 15-20 min, cooling and incubating with the above-mentioned starter culture at 39°C (102.2°F) for 20-25 hours; then clarifying, adding 10-12% sugar syrup, pineapple flavoring, coloring and, finally pasteurizing.

Reference: Srivastava, M. K., and P. P. Lohani. 1986. Utilization of by-products by dairies—whey and buttermilk. *Indian Dairyman* **38:**389-390. Cited in *Dairy Science Abstracts,* 1987, **49**(875):102.

PARAGHURT (En, Fr, De).

Short Description: Denmark; nontraditional product; pharmaceutical product containing *Streptococcus faecium.*

Reference: Frii-Møller, A., and H. Hey. 1983. Colonization of the intestinal canal with a *Streptococcus faecium* preparation (Paraghurt). *Current Therapeutic Research* **33**(5).

PATENTED FERMENTED FRESH MILK PRODUCTS PROCESSES (En); Produits laitier fermentés brevetés (Fr); Patentierte Sauermilchprodukte (De).

Short Description: Various countries; nontraditional products; mainly cow's milk; product names shown with the symbol ® are registered trademarks.

History: Since about 1900 several processes for the manufacture of fermented milk products have been patented. The period of patent protection varies from country to country; it is for usually 25-50 years.

Products: Some of these products described in this volume include Biogarde®, Bioghurt®, Bifighurt®, and Kheran®.

PAYYODHI (En, Fr, De).

Short Description: India; sweetened dahi, typically from eastern India. Sugar is added before incubation (6-12%). It is customary to flavor the product with saffron; it has a light brown color and a caramellike taste.

Reference: Gupta, P. R. 1983. *Dairy India,* p. 248. Delhi: Priyadarshini.

PEDIOCOCCUS-CONTAINING FERMENTED FRESH MILK PRODUCTS (En); Laits fermentés contenant des Pediococcus (Fr); Pediokokken enthaltende Sauermilchprodukte (De).

Short Description: Czechoslovakia; nontraditional product; cow's milk.

Microbiology: *Pediococcus* is a genus of gram-positive bacteria of the *Streptococcaceae*; several species occur in fermented beverages and vegetable matter. The cells are coccoid. Glucose is degraded homofermentatively with the formation of lactic acid and without gas production. Pediococci belong to the lactic acid bacteria, which are characterized by some resistance to antibiotics.

Manufacture: Pediococcus-containing fermented milk products include biokys, the manufacture of which is discussed separately (*see* Biokys).

PELARGON (En, Fr, De); trade name.

Short Description: Switzerland; baby milk formula.

Microbiology: The starter culture consists of *Streptococcus lactis.*

Manufacture: Pelargon is a formulated product made from whole milk, which is heat-treated, concentrated, inoculated, and incubated with the above-mentioned starter culture to the desired acidity and then cooled. The fermented milk is mixed with heat-treated carbohydrates, such as maltodextrin, maize (corn) or rice starch, and sucrose, and usually vegetable oil, then homogenized and spray-dried; some 20% of the *Streptococcus lactis* cells survive the drying (Grieder 1969).

Food Value: The product is intended for feeding healthy babies older than 2 months.

Reference: Grieder, H. R. 1969. Experience with a new biologically acidified baby food product (in German). *Praxis* **58**(2):1236–1238.

PESKÜTAN (En, Fr, De).

Short Description: Turkey (Anatolia area, villages and small towns); traditional product; sheep skim milk; domestic milk-plant product mixture; heat-treated after culturing.

Microbiology: The starter culture consists of *Streptococcus thermophilus* and *Lactobacillus delbrueckii* subsp. *bulgaricus.* During processing the culture bacteria are destroyed.

Manufacture: Sheep skim milk yoghurt is used to make Peskütan. The yoghurt is first brought to a boil in big vessels under continuous stirring with the addition of flour, crushed wheat, and some salt. After a limited time of boiling the mixture is cooled to room temperature and transferred into earthenware or metal containers. Liquefied animal fat is added to the surface of the product to prevent contamination. If kept in a cold place the product will stay edible for 1-2 years. Whenever some of the product is used, it is carefully removed from underneath the protective fat layer, which often is butterfat.

Food Value: Pesküitan is sold commercially in the market and consumers use it as a cheese spread. A study has shown that this product contains, on the average, 67.3% total solids, 2.6% fat, 3% salt, and has 195°Th (titratable acidity of 1.75%).

Related Products: Cooked yoghurt, concentrated fermented milks.

This entry was contributed by Prof. Dr. Hassan Yaygin, Izmir, Turkey.

PHARMACEUTICAL PREPARATIONS CONTAINING FERMENTED MILK ORGANISMS (En); Produits pharmaceutiques contenant de micro-organisms de lait fermentés (Fr); Pharmazeutische Präparate welche Sauermilch-Mikroorganismen enthalten (De).

Short Description: Various countries; nontraditional products; pharmaceutical preparations.

Microbiology: The major organisms utilized in these preparations are *Lactobacillus acidophilus,* bifidobacteria, and yoghurt bacteria, *Lactobacillus casei.*

Manufacture: The products are manufactured mainly by freeze-drying the bacterial suspensions. Numerous products exist and their function is based on the presence of viable bacteria. Such product categories include

- Freeze-dried preparations. Acidophilus zyma, bacid, bifider, bifidogene, biolactis (capsules), bulgaricum I. B., cerolac B (animal use), enpac, Eugalan Töpfer forte, Euga-lein Töpfer, infloran Berna, lactinex, lactopriv, life start original, life start two, liobif, lyobifidus, megadophilus, omniflora.
- Tablets. Bulgaricum tablets.
- Microencapsulated preparations. The purpose of encapsulating is mainly to circumvent bitter taste. It is possible to encapsulate cultured milk with viable organisms.

PIIMÄ (En, Fr, De); denotes sour milk (thick milk).

Short Description: Finland; sour milk beverage; whole or skim cow's milk; prepared with cream culture microorganisms, sometimes with the addition of ropy strains (Forsén 1963); similar to long milk, which is prepared from skim milk, according to Fleischmann and Weigmann (1932).

References:

Martiny, B. 1907, p. 34.
Fleischmann, W., and H. Weigmann. 1932, p. 385.
Forsén, R. 1963. The influence of the microflora on the Finnish fermented milk beverage "Piimä" (in German). *Milchwissenschaft* **18:**22-25.

PITKÄPIMÄ (En, Fr, De); denotes long milk.

Short Description: Finland (northwestern region); traditional product; fermented ropy milk; related product is long milk.

References:

Martiny, B. 1907, p. 94.
Forsén, R. 1966, p.10.

PODKWASA (En, Fr, De); Podkvasa.

Short Description: Various countries; traditional product; regional name for wild yoghurt culture and starter cultures in Bulgaria (*see* Maja).

References:

Nikolov, N. 1962. *Bulgarian Yoghurt and Other Fermented Milk Products* (in Bulgarian). Sofia: Zemizdat.
FIL-IDF Dictionary. 1983, p. 170.

POLKREM (En, Fr, De); trade name.

Short Description: Poland; nontraditional product; cow's milk; cultured cream; pasty consistency.

Manufacture: Polkrem is made from homogenized mixtures of cream, skim milk concentrated by ultrafiltration, and a whey protein concentrate obtained by whey ultrafiltration.

Food Value: The mixture contains 2.5% fat and 4.8% protein.

Reference: Chojnowski, W., S. Poznanski, W. Bednarski, and I. Poliwko. 1981. Production of cultured beverages from skim milk and whey concentrated by ultrafiltration (in Polish). *Roczniki Institutu Przemyslu Mleczarskiego* **21**:25-33. Cited in *Dairy Science Abstracts,* 1982, **44**(1478):167.

PROBIOTICS (En); Probiotics (Fr); Probiotika (De).

Short Description: Various countries; traditional and nontraditional products; mainly dried or freeze-dried.

History: Probiotic is a word coined by Parker in 1974. It is derived from the Greek and means *for life* and contrasts with the better known *antibiotic* which means *against life.* According to Parker's original definition, probiotics are "organisms and substances which contribute to intestinal microbial balance." The definition was intended to include only microbial culture suspensions and crude microbial culture products. The term *probiotic* is generally used to describe animal feed supplements, but it would also include yoghurt used for human consumption (Fuller 1986). The most commonly used bacteria in probiotics for human use are lactobacilli, bifidobacteria, and streptococci.

References:

Fuller, R. 1986. Probiotics. *Journal of Applied Bacteriology Supplement,* pp. 1S-7S.

Parker, R. B. 1974. Probiotics, the other half of the antibiotics story. *Animal Nutrition and Health* **29**:4-8.

PROGURT (En, Fr, De); trade name.

Short Description: Chile; nontraditional product; cow's milk; protein-enriched cultured product developed by Schacht and Syrazinski (1975); whey-drained.

Microbiology: The starter culture (1) consists of *Streptococcus lactis* subsp. *diacetilactis* and *S. lactis* subsp. *cremoris* (ratio 1:1); or (2) includes the addition of *Lactobacillus acidophilus* and/or *Bifidobacterium bifidum* starter culture.

Manufacture: Progurt is made by fermenting heat-treated (95°C/203°F for several seconds) skim milk with 1-3% of above-mentioned starter (1). When the milk reaches 80-90°Th (0.72-0.81% titratable acidity) after 12-18 h it is gently heated to 40-45°C (104-113°F) to separate the whey, which is partially removed. The cream used to standardize the fat content and 0.5-1% of above-mentioned starter culture (2) is added. Finally, the product is homogenized, cooled, and packaged.

Food Value: Progurt has a pH of 4.4-4.5 and contains approximately 5% fat, 6.2% protein, 3% lactose, and 0.7% ash; no data are available on the survival rate of the culture bacteria.

Reference: Schacht, E., and A. Syrazinski. 1975. Progurt, a new cultured product: its manufacturing technology and dietetic value (in Spanish). *Industria Lechera* **646:**9-11. Cited in *Dairy Science Abstracts,* 1976, **38:**744.

PROHLADA (En, Fr, De).

Short Description: USSR; nontraditional product; beverage; deproteinized whey; liquid consistency; characteristic flavor.

Microbiology: The starter culture consists of *Lactobacillus delbrueckii* subsp. *bulgaricus, L. acidophilus,* and lactose-fermenting yeasts.

Manufacture: Prohlada is made from heat-treated deproteinized whey by fermenting it with the above-mentioned starter culture. The amount of starter and incubation conditions are not specified. The product may be sweetened and/or flavored with fruit (*see* Beverages).

Reference: Zobkova, E. C., E. A. Bogdanova, and I. I. Kochergina. 1978. New kinds of fermented milk products (in Russian). *Molochnaya Promȳshlennost'* **4:**16-17.

PROPIONIC ACID BACTERIA FERMENTED MILK PRODUCTS (En); Laits fermentés contenant de bacteries propioniques (Fr); Propionsäurebakterien enthaltende Sauermilchprodukte (De).

Short Description: Czechoslovakia, USSR; nontraditional product; cow's milk.

Microbiology: *Propionibacterium* is a genus of gram-positive, asporogenous, anaerobic bacteria that have been classified as relatives of the actinomycetes. All species are nonmotile. *Propionibacterium* spp. ferment glucose (and lactic acid) with the formation of propionic, acetic, and other acids. The organisms are highly pleomorphic, that is, individual cells may be cocci or branched or unbranched rods; in stained preparations groups of cells often give the impression of Chinese language characters. Propionibacteria may be found in the alimentary tract of man and animals and in dental plaques. *Propionibacterium freudenreichii* subsp. *shermanii* (formerly *P. shermanii*) and *P. jensenii* (formerly *P. peterssonii*) are important in the manufacture of certain types of cheese.

Manufacture: Propionibacteria are used

- In acidophilus baby foods and in propionic acidophilus milk (Doležalek, and Havlickova 1969)
- As mycostatic ingredients. Whey fermented with propionibacteria is dried and used as a mycostatic agent in breads, pastries, and other bakery products (Pyne and Schwarz 1986)
- For vitamin B_{12} enrichment of kefir (Karlin 1965) of cultured milks (Cerna and Hrabova 1977), and of deproteinized whey (Bielecka et al., 1978). Propionibacteria are known to produce vitamin B_{12} (*see also* Elvit).

References:

Bielecka, M., B. Roczniak, and D. Kornacka. 1978. Use of propionibacteria cultures to prepare vitamin-enriched whey. *Proceedings 20th International Dairy Congress* **D:**1043.

Cerna, J., and H. Hrabova. 1977. A method for the manufacture of cultured milk products enriched with B-group vitamins (in Czech). *Czechoslovak Patent* 174 444. Cited in *Dairy Science Abstracts,* 1979, **41**(2149):243-244.

Doležalek, J., and V. Havlickova. 1969. Possibility of utilization of propionic bacteria in the production of acidophilus milk (in Czech). *Sborník Vysoké Školy Chemicko-Technologické v Praze* **E27:**59-66. Cited in *Dairy Science Abstracts,* 1971, **33**(4230):630.

Karlin, R. 1965. About enrichment of kefir with vitamin B_{12} by addition of *Propinobacterium freudenreichii* subsp. *shermanii. International Journal for Vitamin Research* **35:**358-364.

Pyne, C. H., and R. Schwartz. 1986. Whey fermentation process. *European Patent* EP 0 096 477 B1. Cited in *Dairy Science Abstracts,* 1987, **49**(4582):524.

Singleton, P., and D. Sainsbury. 1987. *Dictionary of Microbiology and Molecular Biology.* New York: John Wiley.

PROSTOKVASHA (En, Fr, De); denotes sour milk.

Short Description: USSR; national product; cow's or buffalo's milk; fermented milk.

Microbiology: The starter culture consisting of *Streptococcus lactis,* and *S. lactis* subsp. *diacetilactis* is mixed in at a rate of 5%. Small amounts of a culture of *Lactobacillus delbrueckii* subsp. *bulgaricus* may be added to promote acid production.

Manufacture: The product is prepared from pasteurized milk and with the above-mentioned starter culture. The incubation is at 38-40°C (100.4-104.0°F) for 5-7 h. The product is very popular because of its clean and delicate flavor and refreshing taste.

References:

FIL-IDF Dictionary. 1983, p. 174.

Koroleva N. S. 1975. *Technical Microbiology of Whole Milk Products* (in Russian). Moscow: Pishchevoĭ Promÿshlennosti.

Lipatov, N. N. 1978. Fermented milks other than yoghurt. *Proceedings, 20th International Dairy Congress. Brussels: International Dairy Federation.*

PUDDINGS (En); Pouding (Fr); Pudding (De).

Short Description: Various countries; nontraditional products.

History: Pudding is a term describing several different desserts, usually cooked. Originally, puddings contained animal fats, meat, fruits, or other ingredients and were pastrylike dishes.

Manufacture: Today puddings are milk-based desserts made with flavorings (for example, chocolate, vanilla) cooked with starch until thickened and then cooled until well set. Numerous variations exist, incorporating such ingredients as cassava root starch (tapioca pudding), coconut milk (Hawaiian haupia pudding), cornmeal (American "Indian pudding") and sweet potatoes in the southern United States. It is only logical to use puddings as a vehicle for fermented milk products also.

Products: Three pudding-based fermented milk products are

Acidophilus pudding. Intermediate-moisture frozen acidophilus pudding.

Buttermilk pudding. Fruit pudding with buttermilk as an ingredient.

Yoghurt pudding. Sweetened fruit-flavored varieties exist, also with whipped egg white. Prepared from fermented low-fat milk with added beaten egg white and flavored with raspberry, cherry, orange, or strawberry concentrate and sugar; 19-20% total solids content, 1% milk fat, and the pH at 3.4-4.0.

References:

Anonymous. 1963. Yoghurt pudding—yoko-cup (in German). *Milchwissenschaft* **18:**309-310.

Kahn, M. L., and K. E. Eapen. 1981. Intermediate-moisture frozen acidophilus pudding. *U.S. Patent* 4,308,287. Cited in *Dairy Science Abstracts,* 1983, **45**(863):105.

Mariani, J. F. 1983. *The Dictionary of American Food and Drink.* New York: Ticknor & Fields.

Q

QUARG MILK (En); Lait au quarg (caillebotte) (Fr); Quark-Milch (De).

Short Description: Germany; nontraditional baby food; cow's milk quarg; dietetic preparation used in the treatment of diarrhea in babies; a cultured milk product.

Microbiology: Mesophilic cream culture starter microorganisms.

Manufacture: There are three steps in domestic preparation:

1. One half liter of fresh boiled and cooled whole milk is inoculated with mesophilic starter bacteria and incubated for 24-28 h. The soured milk is heated to 40°C (104°F). The whey is removed with cheesecloth. The curd residue, the quarg, is stirred. The yield is approximately 125 g quarg. Then 250 g of the whey is boiled and after cooling is vigorously blended into the quarg.
2. 125 g raw milk, 250 g water, and 15 g wheat flour are mixed and boiled for 3-5 minutes with vigorous agitation, followed by cooling.
3. Both mixtures are then combined and sugar is added to obtain a concentration of 8-10%.

Food Value: Treatment of diarrhea in babies as replacement for buttermilk.

Related Products: Concentrated fermented milks.

Reference: Schulz, M. E. 1965, p. 952.

R

RABADI (En, Fr, De).

Buttermilk-Pearl Flour Preparation

Short Description: Northwestern India; traditional preparation; buttermilk from sour cream; nutritious food commonly used by low- and average-income rural populations in the millet-producing region; important staple food item for millions of Indians; consistency of very viscous buttermilk.

Manufacture: Rabadi is prepared by mixing pearl millet flour (*Pennisetum typhoides* Rich.) with buttermilk and allowing the mixture to ferment by keeping it in the open sun for 4-6 hours (during the hot summer months). The final product is sweetened with sugar. It is made and sold by small shops or is made in the home.

Food Value: This traditional method of making rabadi carries with it certain nutritional advantages. The fermentation brings about a partial removal of the antinutrients phytic acid and polyphenols normally occurring in pearl millet, thus enhancing the bioavailability of minerals. Improved digestibility of proteins and carbohydrates is also claimed, and the blending of cereal and milk proteins brings about complementarity of all essential amino acids. After 9 hours of fermentation and concentration of phytic acid and polyphenols are reduced to the same level regardless of fermentation temperature (34, 40, or 50°C/93.2, 104, or 122°F).

References:
Dhankar, N. 1985. Studies on the nutritional value of rabadi—a fermented bajra product. M.S.thesis, Haryana Agricultural University, Hisar, India.
Dhankar, N., and B. M. Chauhan. 1987. Effect of temperature and fermentation time on phytic and polyphenol content of rabadi, a fermented pearl millet food. *Journal of Food Science* **52**(3):828-829.

Buttermilk-Maize Flour Preparation

Short Description: Northern India; traditional preparation; buttermilk.

Microbiology: *Pediococcus acidilactici* (3.6×10^5/g), *Bacillus* sp. (1.1×10^6/g) and *Micrococcus* sp. (7.9×10^5/g) have been isolated from fermented rabadi.

Manufacture: Cooked maize flour (corn meal) is cooled and combined with buttermilk to make rabadi. The mixture is fermented overnight and then consumed. A slight pH change from 6.7 to 6.4 and a volume increase of about 5% occur.

Reference: Ramakrishnan, C. V. 1983. The use of fermented foods in India. *Symposium on Indigenous Fermented Foods,* Bangkok, Thailand. Cited by Steinkraus, K. H. 1983, p. 275.

RAIBI (En, Fr, De).

Short Description: Tunisia; nontraditional product; reconstituted or recombined milk.

Microbiology: Fermented with *Lactobacillus acidophilus* starter culture.

Reference: Draoui, A. 1984. Progress made in the manufacture of fermented milks. Use of reconstituted milks. *IDF Bulletin 179* (Fermented Milks), p. 126. Brussels: International Dairy Federation.

RASOGOLLA (En, Fr, De); Rasgolla.

Short Description: Eastern India (Bengal); traditional preparation, acid-coagulated milk curd; a confection prepared from chhana.

Microbiology: Artificial acidified milk product, without starter culture microorganisms, but possible presence of contaminants.

Manufacture: Sweet, pleasant-tasting preparation from fresh chhana (coagulum prepared by adding acid whey to hot milk). After drainage, the coagulum is kneaded to a creamy consistency with starch or white flour (flavoring is optional) and shaped into small balls that are immediately boiled in sugar syrup and thereafter kept in syrup. As a result, rasogolla has a spongy texture. The sugar solution used to store the product (60%) is usually double the concentration of the boiling solution.

Food Value: Rasogolla contains 55% (max.) moisture, 5% fat, 5% protein, and 45% (max.) sugar.

References:

Goel, V. K. 1970. Studies on the manufacture and packaging of rasgolla. M.Sc.thesis, Panjab University, Chandigadh, India.
Holsinger, V. H. 1974. Nutritious candies. *Cereal Science Today* **19:**316-317.
Sangwan, R. B., and M. M. Sharma. 1984. *Literature on Indian Dairy Products.* New Delhi, India: Metropolitan Book Co., Ltd.

RECOMBINED MILK, FERMENTED (En); Lait recombiné, fermenté (Fr); Rekombinierte, fermentierte Milch (De).

Short Description: Various countries; nontraditional products; mostly cow's milk; used to prepare various fermented milk products in countries and areas where there is a lack of fresh milk.

Recombined products are defined as follows (IDF, 1982): "A recombined product is the milk product resulting from the combining of milk fat and milk solids-non-fat in one or more of their various forms with or without water. This combination must be made as to reestablish the product's specified fat to solids-non-fat ratio and solids to water ratio." There are also three milk components to be mixed together: water, skim milk powder, and anhydrous milk fat (also denoted as anhydrous butterfat, dry butterfat, or anhydrous butteroil, which are products exclusively obtained from butter or cream). Alternatively, the fat and some of the nonfat solids may be supplied by whole milk powder (Gilles and Lawrence 1979).

Manufacture: The recombination may be achieved by mixing powder containing 0% fat with powder containing 26% fat plus water in proportion that will give the desired fat content, or by mixing powdered skim milk containing 0% fat plus water plus anhydrous milk fat (Draoui 1984).

Products: Types of fermented, recombined products are filled milk, fermented; toned milk, fermented; and reconstituted milk, fermented (if made with a low concentration of lactose, this would be of value to lactose-deficient persons according to Abd-El-Gawad et al. 1984); labneh (Abd-El-Salam and El-Alamy 1982); yoghurt (Draoui 1984; Salji et al. 1984); acidophilus drink (Nahaisi and Robinson 1985); and buttermilk.

References:

Abd-El-Salam, M. H., and H. A. El-Alamy. 1982. Production and properties of yoghurt and concentrated (Labneh) from ultrafiltrated recombined milk. *Research Bulletin, Faculty of Agriculture, Ain Shams University,* No. 1803, 11 pp. Cited in *Dairy Science Abstracts,* 1983, **45**(8132):860.

Abd-El-Gawad, I. A., E. S. Girgis, A. M. Mehriz, S. M. K. Anis, and S. N. Amer. 1984. Studies on the production of cultured buttermilk in Egypt. II. Type and ratio of starter cultures. *Annals of Agricultural Science, Mosthtohor* **21:**739-747. Cited in *Dairy Science Abstracts,* 1987, **49**(492):50.

Draoui, A. 1984. Progress made in the manufacture of fermented milks. Use of reconstituted milk. *IDF Bulletin 179* (Fermented Milks), pp. 123-126. Brussels: International Dairy Federation.

Gilles, J., and R. C. Lawrence. 1979. Recombined cultured milk products. *IDF Bulletin Doc. 116,* pp. 28-29. Brussels: International Dairy Federation.

IDF. 1982. Recombination of milk and milk-products. Proceedings of IDF Seminar in Singapore, October 7-10, 1980. *IDF Bulletin Doc. 142.* Brussels: International Dairy Federation.

Nahaisi, M. H., and R. K. Robinson. 1985. Acidophilus drinks: the potential for developing countries. *Dairy Industries International* **50:**16-17. Cited in *Dairy Science Abstracts,* 1986, **48**(2380):281-282.

Salji, J. P., W. N. Sawaya, S. R. Saadi, and W. M. Safi. 1984. The effect of heat treatment on quality and shelf life of plain liquid yoghurt. *Cultured Dairy Products Journal* **19:**10-14. Cited in *Dairy Science Abstracts,* 1984, **46**(7355):834.

Winkelmann, F. 1982. Recombined milk products and imitation milk products in the light of FAO/WHO food standards. *IDF Bulletin Doc. 142,* p. 159. Brussels: International Dairy Federation.

RECONSTITUTED MILK, FERMENTED (En); Lait reconstitué, fermenté (Fr); Rekonstituierte, fermentierte Milch (De).

Short Description: Reconstituted milk products are defined as follows by FAO (Winkelmann 1982): "A reconstituted product is the milk resulting from the addition of water to the dried or condensed form of product in the amount necessary to re-establish the specified water/solids ratio."

The most frequently practiced reconstitution process is that made with skim milk powder (good keeping quality, low cost). Sometimes whole milk powder is used, but it must be free of any oxidized flavor. Sweet whey could be used successfully to replace up to 50% of the water added in the milk reconstitution and subsequent manufacture of zabady (El-Safty and El-Zayat 1984). In all reconstitutions the best available water fit for human consumption should be used. For a related process see Recombined Milk, Fermented.

Manufacture: Reconstituted and recombined milk products ferment well and always show satisfactory acid production. However, they produce lower concentrations of flavor metabolites, such as diacetyl and acetaldehyde, and they are less viscous.

The final quality can be influenced by

- Choice of high-quality raw ingredients (the solubility of a milk powder depends on the manufacturing process)
- Choice of the reconstitution and recombined process
- Total milk solids content (TS) after reconstitution or recombination (usually 10% and sometimes up to 12.5%)
- Heat treatment (pasteurization) during the subsequent manufacturing process with concomitant whey protein denaturation (and viscosity induction)
- Homogenization. Any viscosity defects could be corrected through the use of stabilizers or fruit concentrates, so homogenization is not always necessary.

Food Value: Reconstituted milks become necessary when there are seasonal shortages of fresh milk or when the local demand depends entirely on imported milk powder. Therefore, reconstituted milk is called "milk for millions."

The fermentation process by β-galactosidase-producing bacteria (e.g., yoghurt cultures) allows people with lactase deficiency (lactose intolerants) to digest milk sugar more easily. As a result, consumption of high-quality protein and calcium as provided by milk is assured.

References:

Caric, M., D. Gavaric, and S. Markov. 1985. Production of yoghurt from reconstituted milk (in Serbo-Croatian). *Hrana i. Ishrana* **25**(7/10):153-157. Cited in *Dairy Science Abstracts,* 1986, **48**(2379):281.

El-Safty, M. S., and A. I. El-Zayat. 1984. Physical and chemical properties of zabadi manufactured from skim milk powder reconstituted with sweet whey. *Journal of Dairy Research* **51:**471-475. Cited in *Dairy Science Abstracts,* 1984, **46**(6620):751.

IDF. 1982. Recombination of milk and milk products. Proceedings of IDF Seminar Singapore, October 7-10, 1980, *IDF Bulletin Doc. 142.* Brussels: International Dairy Federation.

Krishna, G. G., B. V. R. Rao, and T. J. Rao. 1984. Yoghurt from whey-based reconstituted milk. *Journal of Food Science and Technology (India)* **21:**48-49. Cited in *Dairy Science Abstracts,* 1985, **47**(646):75.

Nahaisi, M. H., and R. K. Robinson. 1985. Acidophilus drinks: the potential for developing countries. *Dairy Industry International* **50:**16-17. Cited in *Dairy Science Abstracts,* 1986, **48**(2380):281-282.

Newstead, D. F., A. Goldman, and J. G. Zadow. 1979. Recombined milks and creams. *IDF Bulletin Doc. 116,* p. 9. Brussels: International Dairy Federation.

Salji, J. P., A. K. Fawal, S. R. Saadi, A. A. Ismail, and A. Mashhadi. 1985. Effects of processing and compositional parameters on quality of plain liquid yoghurt. *Milchwissenschaft* **40:**734-736. Cited in *Dairy Science Abstracts,* 1986, **48**(2380):281-282.

Winkelmann, F. 1982. Recombined milk products and imitation milk products in the light of FAO/WHO food standards. *IDF Bulletin Doc. 142.* Brussels: International Dairy Federation.

REFORMED YOGHURT (En); Yaourt réformé (Fr); Reform-Joghurt (De).

Short Description: Various countries (*see* Acidophilus milk); nontraditional product; cow's milk.

History: As a result of work on negating Metchnikoff's theory, it was found that *Lactobacillus delbrueckii* subsp. *bulgaricus* is not a human intestinal bacterium but that another lactobacillus, *L. acidophilus,* is a regular component of the human intestinal flora. Therefore, the German scientist Henneberg (1926) recommended preparing acidophilus milk, which he called reformed yoghurt.

Related Product: Acidophilus milk.

Reference: Henneberg, W. 1926/1927. (About *Bacillus acidophilus* milk—reformed yoghurt (in German). *Molkerei-Zeitung* **40:**2633-2635 and **41:**604. Cited by Demeter, K. J. 1941, p. 695.

REINDEER MILK-BASED FERMENTED FRESH MILK PRODUCTS (En); Produits laitiers fermentés à partir de lait de renne (Fr); Sauermilchprodukte aus Renntiermilch (De).

Short Description: Various countries, mainly Scandinavia and Siberia; traditional products; reindeer milk; domestic preparations.

History: Reindeer were one of the more important animal species first utilized by humankind. The domestication (partial or whole) seems to have occurred in prehistoric times, but evidence is lacking. However, it is known that Siberian nomads were milking reindeer by the fifth century B.C. (Zeuner 1967).

A reindeer cow produces milk approximately for 4 months during the summer and is milked three times per day at irregular intervals. Each milking yields about 100-200 g milk and annual milk production amounts to 15-24 liter. Calves are not removed from a herd and also draw milk from their respective dams.

Reindeer are now specifically bred for milk production (human use) in many locations of Scandinavia (Norway, Sweden, Finland) and northern Asia (Siberia, e.g., the Tungang people).

Manufacture: Of the domestic manufacture of fermented milk products and cheeses from reindeer milk by nomads little is known and today it has practically disappeared. Compared to cows' milk production, very little reindeer milk is available today.

The unique chemical composition would give reindeer milk products very distinct organoleptic properties (see Food Value).

Food Value: The average gross composition of reindeer milk from contemporary animals is 33.1% total solids, 16.9% fat, 11.5% protein, 2.8% lactose, and 1.44% minerals.

Products: Martiny (1907) and Fleischmann and Weigmann (1932) have reported on the production of a ropy fermented milk in Finland to which is added frozen herbs, spices, or berries.

References:

Fleischmann, W., and H. Weigmann. 1932, p. 145.
Herre, W. 1955. *The Reindeer as Domestic Animal* (in German), p. 271. Leipzig: Akademische Verlagsgesellschaft.
Martiny, B. 1907, p. 102.
Zeuner, F. E. 1967, pp. 111-112.

RIBOLAC® (En, Fr, De); trade name.

Short Description: Switzerland; nontraditional preparation; freeze-dried, pharmaceutical preparation.

Microbiology: *Lactobacillus acidophilus* (resistant to antibiotics and sulfa drugs).

Therapeutic Value: Usually administered as a preparation with various vitamins (vitamin B_1, 3 mg; vitamin B_2, 4 mg; vitamin P complex/bioflavonoids, 60 mg; vitamin B_3, 4.5 mg; vitamin B_6, 6 mg; vitamin C, 200 mg.

Ribolac® is claimed to arrest the proliferation of pathogenic organisms in the intestine, to prevent intestinal disorders following antibiotic therapy, and, whenever necessary, to restore normal intestinal flora. The minimum number of organisms per capsule is 100,000,000.

RIBOT-MILK (En); Lait ribot (Fr); Ribot-Milch (De).

Short Description: France (region of Bretagne); traditional fermented milk; only the product name and no other information is available.

Reference: Ferialdi, R., and P. Moisan. 1982. Procedure for the manufacture of a fermented milk (in French). *French Patent* 8,105,974.

RIVELLA® (En, Fr, De); trade name.

Short Description: Switzerland; nontraditional soft drink, heat-treated after culturing; cheese whey; clear, sparkling, amber-colored liquid product with a refreshing acidic flavor enjoying considerable consumer acceptance (produced since 1953).

Microbiology: Destruction of fermentation microorganisms by heat-treatment after culturing.

Manufacture: The manufacture of Rivella® involves the following steps:

1. Fresh whey from different sources is collected and standardized; protein is precipitated and removed by filtration or centrifugation.
2. The clarified whey is subjected to a lactic acid fermentation;
3. Additional protein is precipitated and removed by filtration;
4. The fermented whey is condensed in a double-effect falling-film evaporator;
5. A blend of Swiss herbs is added and the concentrate is filtered to remove sugar crystals;
6. The concentrate is diluted with a specified amount of hard water;
7. Following carbonation, the product is filled into brown bottles;

8. After sealing, the bottles are gradually heated from 30-78°C (86-172°F) and then slowly cooled, at which time the product is ready for consumption.

Food Value: The finished beverage contains 9.7% TS (total solids), 0.125% total nitrogen, 35% whey and water, sugar, and natural flavors. The pH is about 3.7. The energy value is 37 kcal/100 ml (155 kJ/dl).

References:

Anonymous. 1960. Rivella—a new form of whey utilization. *Dairy Industry* **25:**113-117.

Marth, E. H. 1970. Fermentation products from whey. In *Byproducts from Milk,* ed. B. H. Webb and E. O. Whittier, pp. 66-67. Westport, Conn.: AVI Publishing Co.

RJAZHENKA (En, Fr, De); Ryazhenka, Riazhenka; denotes sour milk.

Short Description: USSR (Ukraine); traditional national product; cow's milk and cream; yoghurtlike product; set type; firm consistency; sour taste and cooked flavor.

Microbiology: The starter consists of *Streptococcus thermophilus* and *Lactobacillus delbrueckii* subsp. *bulgaricus* (ratio 4:1-5:1 to 40:1-50:1).

Manufacture: Rjazhenka is made from cow's milk with cream admixed to bring the fat content to 4-6%. This mixture is pasteurized at 95°C (203°F) for 2-3 h, then cooled to 45-50°C (113-122°F) and inoculated with 1-5% of the starter. After filling into retail containers the milk is incubated at 43-45°C (109.4-113.0°F) for 4-5 h until coagulated and then cooled. Acidity of the final product ranges between 90 and 120°Th (0.81-1.08% titratable acidity).

Related Products: Yoghurt, metchnikov prostokvasha, varentes.

Reference: Koroleva, N.S. 1975. *Technical Microbiology of Whole Milk Products* (in Russian). Moscow: Pishchevoĭ Promȳshlennosti.

ROB (En, Fr, De); Roba.

Short Description: Egypt, Iraq, Sudan; traditional beverage; cow's, buffalo's, goat's milk.

Microbiology: The microflora is composed of streptobacillus (probably lactobacilli), diplococcus (probably lactic streptococci), and mycoderma (probably *Candida kefir*).

Manufacture: Roba in Egypt is a kefirlike beverage. Milk is boiled, cooled to 40°C (104°F), inoculated with maya (culture), and incubated for 6 hours.

Reference: Kvatchkoff, I. 1937. Studies on Bulgarian fermented milks from ewe's milk (in French). *Le Lait* **17:**475.

RØMMEKOLLE (En, Fr, De); denotes bowl or dish with soured, thickened milk (set milk).

Short Description: Norway; traditional product with distinct cream layer on top.

Reference: Anonymous. 1978. Activities of the Government Research Institute for the Dairy Industry, 1977-1978 (in Danish). *Beredning fra Statens Statens Forsøgsmejeri No. 234.* Cited in *Dairy Science Abstracts,* 1981, **42**(6418):768.

ROPY MILK (En); Lait filant (Fr); Fadenziehende Milch (De).

Short Description: Various countries (mainly Scandinavia; traditional and nontraditional fermented milks; cow's milk (mostly).

Two different types of starters are used in the preparation of ropy milk products:

Mixed Ropy Starter

Microbiology: Ropy milk is the generic term for any fermented milk with slime (mucus) production as a result of growth of certain mesophilic lactic acid-producing bacteria strains (e.g., *Streptococcus lactis, S. lactis* subsp. *cremoris, Leuconostoc* spp.).

Manufacture: In traditional ropy milks a number of phenomena can be observed that are the direct result of microbial growth and the symbiotic relationships among the organisms. The hydrocolloidal substances produced by the cells act as a food stabilizer, thus preventing syneresis and providing a product with a natural and desirable viscosity. These ropy milks have a longer keeping quality than other fermented milks cultivated under similar conditions.

Food Value: Claims are also made for improved immunological properties after consumption of viili. Eaten with a spoon and used on breakfast cereals and fruit in Scandinavia.

Products: This type of starter is used in three Scandinavian ropy milk products, which in the older literature were referred to as tätte or tätte mjölk in Norway; langfil, mainly in Sweden; viili (from whole milk) and piimä (from skim milk) in Finland.

Simple Ropy Starter

Microbiology: Several less well-known fermented ropy milks are all produced with simple ropy starters. Their microflora is different from that of Scandinavian ropy milk.

Products: Products made from simple ropy starters are not commercially important: bulgarian milk; thick-milk, ropy; ropy milk, Bretagne, France (Forsén 1966).

References:

Campbell-Platt, G. 1987, p. 67.
Forsén R. 1986, p. 19.
Macura, D., and P. M. Townsley. 1984. Scandinavian ropy milk. Identification and characterization of endogenous ropy lactic streptococci and their extracellular excretion. *Journal of Dairy Science* **67**:735-744. Cited in *Dairy Science Abstracts,* 1984, **46**(5425):617.

ROUABA (En, Fr, De); denotes acidified milk.

Short Description: Chad (Africa); traditional fermented milk or buttermilk-cereal mixture; cow's milk.

Microbiology: Spontaneous fermentation microorganisms (undefined).

Manufacture: Fresh whole milk is poured into a hollowed-out gourd, inoculated with fermented milk from the previous day, and then incubated for 15-20 hours. After coagulation the gel is stirred with a wooden agitator for 10-15 min and thereafter the whole gourd is shaken for a brief period. Part of the fermented milk may be consumed as such, while the remainder is further agitated in the closed gourd for about 1 h to induce churning (fat separation). The butter is skimmed off. Then the remaining buttermilk is mixed with cooked wheat flour mash (porridge), which may be consumed as a breakfast dish. The keeping quality is approximately one week and is limited by a gradually developing oxidized off-flavor.

This entry was contributed by L. Layotay,
N'Djaména, Chad.

RUSSKIĬ (En, Fr, De).

Short Description: USSR; nontraditional cultured milk drink; cow's milk.

Microbiology: The starter consists of *Streptococcus thermophilus* and *S. acetoinicus* (not classified as species).

Manufacture: The manufacture of this product first involves the preparation of a basic mix that contains 1.5% milk fat and 0.6% added sodium caseinate. This mix is pasteurized at 90-92°C (194.0-197.6°F) for 2-3 min, homogenized at 17.5 MPa, inoculated with 5% of above-mentioned starter culture, and then incubated at 35-37°C (95.0-98.6°F) to an acidity of 80-90°Th (0.72-0.81% titratable acidity). After cooling the product is packaged, at which time a flavoring syrup may be added if so desired (usually at a 10% level).

Related Products: Beverages.

References:

Lavrenova, G. S., V. F. Inozemtseva, I. N. Pyatnitsyna, and E. A. Bogdanova. 1978. Process for the manufacture of milk beverage Russkiĭ with added protein (in Russian). *Molochnaya Promȳshlennost'* **47**:66-69, 95. Cited in *Dairy Science Abstracts,* 1979, **41**(7493):832.

Lavrenova, G. S., V. F. Inozemtseva, and I. N. Pyatnitsyna. 1980. Process for making the cultured milk drink "Russkiĭ (in Russian). *USSR Patent* 731,947. Cited in *Dairy Science Abstracts,* 1981, **43**(1290):174.

S

SAKOULAS (En, Fr, De); denotes strained yoghurt.

Short Description: Greece; traditional product; cow's milk; concentrated fermented milk (whey drained from the yoghurt through sacks or bags).

SALOMAT (En, Fr, De).

Short Description: USSR; nontraditional beverage; whey.

Microbiology: The starter culture consists of a lactose-fermenting yeast (*Saccharomyces lactis*) and a beer yeast (*Saccharomyces carlsbergensis*).

Manufacture: The processing technology includes the following stages: whey separation, whey pasteurization, flavoring ingredients addition, fermentation and, lastly, addition of an infusion made from specially selected herbs.

Reference: Express information service, *Maslodel naya i syrodel naya promyshlenost* (The Butter and Cheese Industry). 1980, **6:**10. Cited by Kravchenki, E. F. 1989. *International Dairy Federation Bulletin* **233:**61-67.

SALTED FERMENTED FRESH MILK PRODUCTS (En); Produits laitiers fermentés sales (Fr); Fermentierte, gesalzene Milchprodukte (De).

Short Description: Various countries; traditional products; the addition of salt.

Manufacture: The addition of salt to fermented milks allows them to be used in soups or as aperitifs because of the salty taste. On the other hand, the addition of sugar provides sweetness and gives the product dessert qualities. Salt is customarily added to some, such as slano mleko, salted sour milk, sosky.

Reference: Lazarevska, D., T. Cizbanovski, and N. Kapac-Parkaceva. 1975. Composition, characteristics and organoleptic quality of salted soured milk (in Serbo-Croatian). *Mlekarstvo* **25:**58-64.

SAMCO'S (En, Fr, De).

Short Description: Nigeria; popular sour cream drink (semifluid cream).

Microbiology: The starter culture consists of mesophilic lactic acid bacteria.

Manufacture: The product is made by allowing pasteurized cream to sour because of action by lactic acid bacteria or similar cultures until it contains more than 0.2% titratable acid (about 22°Th).

Food Value: The product is formulated with milk, the major ingredient of value. It also contains skim milk powder (nonfat milk solids), stabilizers, emulsifiers, and water. The product usually contains 18.5% milk fat and 8.5% milk nonfat solids.

Reference: Ihekoronye, A. I., and P. O. Ngoddy. 1985. *Integral Food Science and Technology for the Tropics.* London: MacMillan.

SAMOKISSELIS (En, Fr, De); Samokisjeliš; denotes spontaneously soured product.

Short Description: Yugoslavia (Gornje Polimlje, Montenegro region); traditional product; concentrated whole milk; soured milk product; prepared for special occasions.

Microbiology: Lactic acid bacteria and yeast.

References:

Barjaktarović, M. 1951. Popular foods and drinks in the Upper Polimlje (in Serbo-Croatian). In *Zbornik radova SAN,* vol. 14, pp. 143-166, Belgrade: Etnografski institut.

Novak, V. 1969. About dairying by the peoples of Yugoslavia (in German). In *Földes,* L. 1969, p. 584.

SAVANYUTEJ (En, Fr, De); Sürütej.

Short Description: Hungary; traditional product; ewe's and cow's milk; homemade with whey separation.

Manufacture, Domestic: Milk is first coagulated with rennet, then boiled milk is added (1:10 parts boiled). This mixture is stored at close to 38°C (97°F). After several hours the mixture is coagulated and the whey is separated and removed. Within several days frothing begins (CO_2 gas production) and the product is ready to be eaten. Shelf life is 2-3 weeks.

Related Products: Clabbered milk, concentrated fermented milk.

Reference: Keszi-Kovacs, L. 1966. Traditional Hungarian dairying (in German). In *Földes,* L. 1969, p. 611.

SAYA® (En, Fr, De); trade name.

Short Description: Germany; nontraditional product; skim milk; coarsely flocculated sour milk beverage with long keeping quality and having a characteristic taste.

Microbiology: The starter culture consists of *Streptococcus lactis* and two strains of *Leuconostoc mesenteroides* subsp. *dextranicum.*

Manufacture: Milk is carbonated with 0.4% carbon dioxide and a protease is also added. The fermentation, at 8-10°C (46.4-50°F), lasts 4-6 weeks. The final acidity is 182.5°Th (1.64% titratable acidity).

Food Value: The product is enriched with vitamins A, B, and C and various amino acids.

Reference: Wehrsarg. 1928. Modern milk therapy during intestinal disturbances and tuberculosis (in German). *Aerztliche Rundschau* (Munich). Cited in *Molkerei-Zeitung.* 1929. **43:**1989-1990.

SCHEMAITSCHJU (En, Fr, De).

Short Description: Poland; nontraditional product; cultured buttermilk; buttermilk containing 1% fat and 11.5% total solids.

Microbiology: The starter culture consists of *Streptococcus lactis, S. lactis* subsp. *cremoris, Leuconostoc mesenteroides* subsp. *cremoris* and/or *S. lactis* subsp. *diacetilactis.*

Manufacture: The product is made with skim milk powder and starter culture added at a rate of 2-5%.

SCHISTON (En, Fr, De).

Short Description: Ancient Rome; traditional products; soured milk invented by physicians during Pliny's time.

Reference: Brothwell, D., and P. Brothwell. 1969. *Food in Antiquity.* London: Thames and Hudson.

SET MILK (En); Lait coagulé (Fr); Dickmilch (De); common name for sour milk, spontaneously soured; means thickened milk, clabbered milk.

Short Description: In earlier times very popular with the German peoples; old traditional product, key product; mainly cow's milk; direct consumption or eaten with bread; mesophilic fermented milk; refreshing beverage or eaten with a spoon (concentrated, whey-drained fermented milks); mild acid flavor, addition of sugar, spices (e.g., cinnamon), or flavorings is optional but customary.

Manufacture, Household Preparation: Very important and popular food prepared in households until about World War II (1939-1945). The manufacture is characterized by a spontaneous coagulation of fresh raw milk, which is put in soup plates or flat dishes (additional contamination with lactic acid bacteria highly probable). Sometimes pieces of plants with milk-coagulating juices, crusts of bread, and, if needed for coagulation, some buttermilk or remainders of sour milk are added. Allowed to stand in a warm place, 15-25°C (60.8-77°F) coagulation would be complete within 24 h. Today, without the necessary raw milk, coagulation is very difficult to achieve, because pasteurized market milk does not have sufficient numbers of lactic acid bacteria in it. Little has been published on this traditional and once very popular product.

Manufacture, Commercial: The product is made from pasteurized skim milk cooled to 20°C (68°F), followed by the addition of 1 g rennet/100 liters milk and 2% starter culture consisting of mesophilic lactic acid bacteria. Formation of a firm coagulum is brought about by both rennet and bacterial activity (usually overnight). Then careful transfer of coagulum (by scooping) is carried out into whey drainage boxes held in a container whose outer jacket is filled with water. The water effectively cools the entire coagulum mass. Rapid cooling ensures that the coagulum does not contract too much, thus ensuring a yield of 50-70%. When thoroughly cooled, the product is packaged. The product is neither a buttermilk nor a fresh cheese (quarg), but a hybrid between the two. Consumers usually add some fresh milk to it and eat it with a spoon. Some varieties are drunk as a beverage. When covered with fresh, cold water, the product can be kept for several days.

Manufacture, Modern Industrial: Today produced industrially only in very small quantities in Germany and Denmark. There is a lack of microbiological studies of the household-prepared products, but it is assumed that the industrial cultures do not correspond to the original mesophilic lactic acid-producing bacterial mixtures of many species. The industrial manufacturing process is like any other preparation of a less or more concentrated mesophilic or thermophilic fermented milk.

Related Products: Klotzmilch, Setzmilch (German); tykmaelk (Danish); rømmekolle (Norwegian); lac concretum (Latin, according to Tacitus).

References:

Anonymous. 1954. "Thickmilk" made from skim milk (in German). Chemisches Institut der Bundesversuchs-und Forschungsantalt für Milchwirtschaft, Kiel. Report of July 1954. Cited in *Milchwissenschaft* (in German). 1954. **9:**273-274.

Klupsch, H.-J. 1984. *Fermented Milk Products—Mixed Beverages and Desserts* (in German). Gelsenkirchen-Buer: Th. Mann.

SHAKHTERSKI (En, Fr, De).

Short Description: USSR; nontraditional product; cow's milk; carbonated fermented milk beverage.

Microbiology: The starter culture consists of special lactic acid bacteria (species not specified).

Manufacture: The beverage is made by incubating skim milk (plus 1% skim milk powder) with a lactic starter, thereafter 10% natural fruit or berry syrup is added, as well as 0.01% ascorbic acid and carbon dioxide.

Food Value: The final product contains 17.2% total solids and has a higher content of free amino acids than skim milk.

Reference: Tylkin, W. B., W. S. Kalmysh, and N. N. Romanskaya. 1977. Properties of the carbonated cultured milk beverage "Shakhterski" (in German). *Lebensmittel-Industrie* **24:**457-458.

SHENINA (En, Fr, De).

Short Description: Iraq (Baghdad region); fermented milk.

Reference: Ishac, Y., N. Kaddouri, and J. Al-Shaikhli. 1970. Studies on fermented milks in the Baghdad area. 2. Survival of pathogens. *Indian Journal of Dairy Science* **23**(4):233-237. Cited in *Dairy Science Abstracts,* 1971, **33**(5933):897.

SHERBET (yoghurt sherbet, frozen confection) (En); Sorbet au yaourt (Fr); Joghurt-Sorbet (De).

Short Description: USA; nontraditional product; cow's milk and other optional edible ingredients such as sugar; frozen product composed of 40-60% skim milk yoghurt and fresh orange, lemon, or pineapple juice; direct consumption or used in making desserts; pleasant taste, refreshing sour flavor.

Reference: Rašić, J. Lj., and J. A. Kurmann. 1978, p. 344.

SHOSIM (En, Fr, De).

Short Description: Nepal; traditional product; spontaneously fermented milk of the Sherpas.

Reference: Tokita, F., A. Hosono, F. Takahashi, T. Ishida, and H. Otani. 1981. Animal products in Nepal. III. The Sherpas and their livestock-farming (in Japanese). *Japanese Journal of Dairy and Food Science* **30:**55-60. Cited in *Dairy Science Abstracts,* 1982, **44**(6616):728.

SHRIKAND (En, Fr, De).

Short Description: India (western part); traditional product mentioned 800-300 B.C. (Prakash 1968); buffalo milk, cow's milk, or mixed milk;

semisoft whole milk product; direct consumption or used to make a dried product called shrikand wadi; pasty consistency; sweet acid taste and agreeable flavor.

Microbiology: The starter culture consists of *Streptococcus lactis* subsp. *lactis* (*see* Chakka, Indian).

Manufacture: The product is made by mixing chakka with sugar and kneading well; usually color and flavor are added. The finished product may be dried by heating in an open pan to make the product called shrikand wadi.

Reference: Prakash, O. 1968. *Food and Drinks in Ancient India,* pp. 63 and 292. Nai-Sarak, Delhi-6: Munshi Ram Manchar Lal, Oriental Booksellers and Publishers.

SHUBAT (En, Fr); Schubat (De); Šubat, Šuvat.

Short Description: USSR (southern Kazakhstan); traditional beverage; fresh camel milk; kumisslike product; microbiology related to that of kumiss; manufacture similar to that of kumiss.

Related Products: Chal, kumiss.

References:

Cherepanova, V. P. 1982. A dietetic and therapeutic product (in Russian). *Zhivotnovodstvo* **10:**63. Cited in *Dairy Science Abstracts,* 1983, **45**(1444):169.

De, S. 1986. *Outlines of Dairy Technology.* Delhi: Oxford University Press.

Nurymbetov, A. 1983. A new branch: dairy husbandry of camels (in Russian). Zhivotnovodstvo **1:**18-19. Cited in *Dairy Science Abstracts,* 1983, **45**(7541):793.

Orlov, V. K., and G. K. Servetnik-Chalaya. 1982. Some physicochemical indices of the fat and the fatty acid composition of the lipids of mares' milk and shubat (in Russian). *Voprosy Pitaniya* **2:**59-61. Cited in *Dairy Science Abstracts,* 1985, **47**(2283):262.

Urazakov, N. U., and Sh. Bainazarov. 1974. Tushchibek—the first clinic history for the treatment of pulmonary tuberculosis with cultured camel milk (in Russian). *Problemy Tuberkuleza* **2:**89-90. Cited in *Dairy Science Abstracts,* 1979, **41**(913):102.

SIBDA (En, Fr, De).

Short Description: Israel; fermented milk that contains fruit juices or fruit preserves.

Reference: Schulz, M. E. 1965, vol. 2, p. 1086.

SILAGE PREPARATIONS (En); Préparations pour l'ensilage (Fr); Zubereitungen für die Silierung (De).

Short Description: Various countries; nontraditional preparations; it has become widespread practice to use specific lactic acid cultures in the ensiling of green cut fodder.

Microbiology: The starter culture consists of mesophilic lactic acid bacteria, mostly lactobacilli of vegetable origin.

Manufacture: Freeze-dried (lyophilized) cultures are normally used for direct inoculation of green cut fodder; sometimes they are first propagated in whey. There are numerous commercial preparations, for example, Biomax SI (*Lactobacillus plantarum*) from Denmark and Siloferm (*Pediococcus acidilactici* and *Lactobacillus plantarum*) from Sweden.

Food Value: Starter culture not only produces a preservative effect, but consumption of starter culture also helps to reestablish and maintain a healthy balance of bacteria in the digestive tract of pigs, calves, and other farm animals. The use of such cultures is also claimed to limit the growth of clostridia in the silage, and it is speculated that this has some significance in dairy cattle as far as milk and cheese quality is concerned (*see* Animal feeds and bacterial concentrates).

References:

Anonymous. Chr. Hansen's Biosystems, 3 Snakt Annae Plads, DK-1250, Copenhagen K. Prospectus.

Marth, E. H. 1970. Fermentation products from whey. In *By-Products from Milk*. 2nd ed., ed. B. H. Webb, and E. O. Whittier, p. 65. Westport, Conn.: AVI Publishing.

SILIVRI YOGHURT (En); Yaourt Siliwri (Fr); Siliwri Joghurt (De); denotes yoghurt from Silivri, a town 90 km from Istanbul.

Short Description: Turkey (first made in Silivri and then produced in other areas); famous traditional yoghurt culture and preparation; ewe's, goat's and buffalo's milk or mixed milk; this yoghurt type contains more dry matter, higher viscosity, and a thicker cream layer than normal yoghurt. These properties provide advantages for transportation under difficult conditions.

Microbiology: The starter culture consists of *Streptococcus thermophilus* and *Lactobacillus delbrueckii* subsp. *bulgaricus*.

Manufacture: Artisanal manufacture involves heating of milk to 85-90°C (185-194°F) over a wood fire while continuously stirring with a special wooden ladle. Heating takes 1-1.5 h and some water is evaporated from the milk. The fat content of the heated milk is between 4-5.5%. The hot milk is then transferred to special vessels placed on a heating table. The milk is always poured in from a high position to cause foaming. Hot wood embers previously prepared are put under the heating table and a second heating is done for 30-40 minutes while a thick cream layer forms on the milk surface. Then the embers and ashes are removed and the milk is cooled to 45°C (113°F). Culture from a one-day-old yoghurt is then added to the milk from 2 or 3 sides by means of a special injector. The vessels are covered with a clean cloth supported by a wooden lattice. Three or four hours of incubation are required to complete the yoghurt making procedure.

Related Product: Yoghurt.

Reference: Izmn, E. R. 1935. Contribution to the knowledge of preparation and composition of Silivri-Yoghurt (in German). *Ankara Yük sek Ziraat Entstitüsu* **11:**49.

This entry was contributed by Prof. Dr. Hassan Yaygin, Bornova-Izmir, Turkey.

SKOFIR (En, Fr, De); trade name.

Short Description: France; nontraditional product; partially skimmed cow's milk; liquid, beverage.

Microbiology: The starter culture consists of lactic acid streptococci and *Lactobacillus acidophilus.*

Manufacture: Pasteurized milk is used and inoculated with the above culture; fermentation at various temperatures.

Reference: Société Meihac, France. 1962. *French Patent* No. 1 299 639 (in French). Cited by Schulz, M. E. 1965, vol. 2, p. 1094.

SKORUP (En, Fr, De).

Short Description: Yugoslavia (Bosnia, Serbia, Montenegro); traditional domestic product; salted and fermented (*see* Kaymak).

References:

Fleischmann, W., and H. Weigmann. 1932, p. 374.
Schulz, M. E. 1965, vol. 1, p. 466.

SKUTA (En, Fr, De); denotes cheese.

Short Description: USSR (Carpathians); sheep or goat whey; beverage with alcoholic fermentation; spontaneous fermentation, lactic acid bacteria, and yeasts.

Related Product: Urda.

Reference: *FIL-IDF Dictionary,* 1983, p. 197.

SKYR (En, Fr, De).

Short Description: Iceland; traditional and very popular product; originally skimmed ewe's milk, in this century cow's milk also used; a very old product; concentrated fermented milk, whey-drained.

Microbiology: The starter culture consists of *Streptococcus thermophilus* and *Lactobacillus delbrueckii* subsp. *bulgaricus.* Fermenting yeasts occur at later stages and in the end product.

Manufacture, Traditional Farmhouse Method: Up to about 1900, skyr was made roughly as follows: Skim milk was heated to 90-100°C (194-212°F) or simply boiled and then cooled to 40°C (104°F). Then some water-diluted skyr from an earlier production was added, approximately 15 g per liter of milk along with cheese rennet, approximately 6 ml per 100 liters. The subsequent souring lasted 4.5 to 5.5 hours until a pH of about 4.7 was reached. The product was then cooled to 18-20°C (64.4-68°F) and left for about 18 hours or to pH 4.2. By pouring the resulting skyr curd into linen bags the whey was allowed to drain through the cloth for about 6 hours, first with the temperature at 19-20°C (66.2-68°F) and then for another 18 hours at 6-8°C (42.8-46.4°F). The total filtering time was about 24 hours. The finished skyr had a pH of 3.8-4.0 and a dry matter content of 17-20%. To make one kilogram of skyr, 5 liters of skim milk were needed.

Manufacture, Modern Industrial Method: The new method is very much like the traditional one, up to the point of filtration (concentration). Instead of filtering through linen bags a quarg separator is used to remove the whey, which is ultrafiltered, and pasteurized protein mass is added to quarg attained after separation (Gudmundsson 1987).

Food Value: Industrially prepared skyr contains 82.5% water, 17.5% dry matter, 13.3% protein, 2.4% lactose, 0.8% minerals, 0.4% fat, 0.3-0.5% alcohol; it also contains acetic acid, acetaldehyde, diacetyl, CO_2, and traces of other substances.

Related Products: Ymer, concentrated fermented milks (whey drained).

References:

Gudmundsson, B. 1987. Skyr. *Scandinavian Dairy Industry* **4:**240-242.
Orla-Jensen, S., and W. Sadler. 1940. Bacteriological examination of the Icelandic sour milk product skyr (in German). *Zentralblatt f. Bakteriologie* **2:**102, 260-261. Cited in *Dairy Science Abstracts,* 1946, **8:**34.

SLAVYANKA (En, Fr, De).

Short Description: USSR; nontraditional culture milk; cheese whey; set or stirred.

Microbiology: The starter culture consisting of streptococci (*Streptococcus lactis, S. lactis* subsp. *cremoris, S. lactis* subsp. *diacetilactis*), *Leuconostoc mesenteroides* subsp. *cremoris,* and *Lactobacillus acidophilus* is mixed in at a rate of 2-3%.

Manufacture: Unsalted cheese whey is concentrated to 12-18% total solids in a vacuum evaporator and then heated to 68-75°C (154.4-158°F) with 10-15 min holding. The concentrate is mixed with pasteurized (e.g., 90-92°C or 194-197.6°F for 2-3 min) skim milk at a total solids ratio of 3-4:1-2. This mixture is fermented with the above-mentioned culture at 28-30°C (82.4-86°F) for 6-8 h until a firm coagulum and an acidity of 100-140°Th (0.90-1.26% titratable acidity) are obtained. To obtain a sweet product, sugar is added (at 3-5% of the weight of the final product) to the skim milk before pasteurization.

Reference: Trufanova, L. S., V. V. Molochnikov, S. N. Karlikanova, P. G. Nesterenko, and V. E. Zhidkov. 1984. *USSR Patent* 1,066,522 A (in Russian). Cited in *Dairy Science Abstracts,* 1985, **47**(648):76.

SMETANKA, CONCENTRATED BUTTERMILK (En); Smetanka, babeurre concentré (Fr); Smetanka, Konzentrierte Buttermilk (De).

Short Description: USSR; nontraditional product.

Microbiology: The starter culture consists of mesophilic lactic acid streptococci.

Manufacture: Buttermilk (7.9% total solids) from sweet-cream butter is ultrafiltered to a total solids content of 13.6-18.1% and then inoculated with lactic streptococci for 16 h at 25°C (77°F). After agitation the product is packaged.

References:

Vyshemirskii, F. A., A. V. Konanykhin, and O. M. Sveshchinskii. 1977. Method for production of the soured milk product Smetanka (in Russian). *USSR Patent* 581,921. Cited in *Dairy Science Abstracts,* 1978, **40**(2880):320.

Vyshemirskii, F. A., N. N. Ozhgikhima, and A. V. Konanykhin. 1979. Ultrafiltration of buttermilk and rational utilization of separate fractions (in Russian). *Promyshlennost* **28:**35-39, 71. Cited in *Dairy Science Abstracts,* 1981, **43**(8080):957.

SMETANKA, CULTURED CREAM (En); Smetanka, crème acidifiée (Fr); Smetanka, Sauerrahm (De).

Short Description: USSR; nontraditional sour cream containing 10% fat; viscous consistency, good balanced flavor.

Microbiology: The starter culture contains *Acetobacter lactis* (no longer considered as species) and *Lactobacillus acidophilus* or *L. casei, Leuconostoc lactis, Leuconostoc mesenteroides* subsp. *dextranicum,* and *Streptococcus lactis* subsp. *cremoris.*

Manufacture: The manufacturing process involves pasteurizing, homogenizing, and cooling cream to fermentation temperature, adding a starter of acid-producing streptococci and flavor-producing leuconostocs. The fermentation is carried out at 26-28°C (78.8-82.4°F) for 6-8 h with periodical mixing during the first 1.5-2 h, then the product is cooled to 15-16°C (59.0-60.8°F). To improve the organoleptic properties, stability of the product and its biological value, 10-15 g dried skim milk is added to the cream (per liter) before pasteurization. The final acidity is 70.0°Th (0.64% titratable acidity).

References:

Obermann, H. 1985. p. 185.

Romanskaya, N. N., G. S. Dyment, R. S. Bashirova, L. D. Tovkachevskaya, S. I. Kochubei, L. A. Mostovaya, and G. Ya. Isaeva. 1980. Process for making cultured milk product "Smetanka" (in Russian). *USSR Patent* 766,565. Cited in *Dairy Science Abstracts,* 1981, **43**(3301):408.

SMY (En, Fr, De).

Short Description: Ancient Egypt; traditional fermented milk product; human or animal milk.

Microbiology: Spontaneously soured milk microorganisms.

Food Value: Ancient physicians of Egypt specifically recommended this product to cure cough.

Reference: Darby, W. J., P. Ghalioungui, and L. Grivetti. 1977. *Food: The Gift of Osiris,* vol. 2, p. 775. London: Academic Press.

SNEZHANKA (En, Fr, De).

Short Description: Bulgaria; nontraditional fermented milk; mixtures of milk from cows and buffalo or cows and sheep; a sweetened yoghurt type.

Microbiology: The starter culture consists of *Streptococcus thermophilus* and *Lactobacillus delbrueckii* subsp. *bulgaricus.*

Manufacture: Milk from cows and buffalo is mixed at a ratio of 80:20 and from cows and sheep at a ratio of 70:30. The fat content in the mixed milk must not be lower than 4.9%. About 6% sugar is added to the milk, which is then treated, cultured, and processed as in yoghurt production. The fermentation proceeds at 42-45°C (107.6-113°F) until an acidity of 90-95°Th (0.81-0.85% titratable acidity) is obtained. The product can be stored at 8-10°C (46.4-50°F).

Related Product: Yoghurt.

Reference: Koroleva, N. S., and M. S. Kondratenko. 1978. *Symbiotic Starters of Thermophilic Bacteria in the Production of Cultured Milk Products* (in Russian). Moscow, Sofia: Pishchevaya Promȳshlennost', Technika.

SNEZHOK (En, Fr, De).

Short Description: USSR; nontraditional product; sweetened whole milk yoghurt with fruit syrup.

Microbiology: The starter culture consists of *Streptococcus thermophilus* and *Lactobacillus delbrueckii* subsp. *bulgaricus.*

References:

FIL-IDF Dictionary. 1983, p. 199.

Koroleva, N. S., and E. V. Melnikova. 1978. *Streptococcus thermophilus* strain 28M, for use in two-strain inocula for preparation of the cultured milk products ryazhenka, varenets, mechnikovskaya, prostokvasha, snezhok and yoghurt (in Russian). *USSR Patent* 603,660. Cited in *Dairy Science Abstracts,* 1979, **41**(1515):170.

SOFT-SERVE ICE CREAM FROM YOGHURT (frozen yoghurt) (En); Creme glacée mou (Fr); Softeiskrem aus Joghurt (De).

Short Description: United States, United Kingdom; nontraditional product; cow's milk and other edible ingredients; pleasant flavor; smooth texture.

Microbiology: The starter culture consists of *Streptococcus thermophilus* and *Lactobacillus delbrueckii* subsp. *bulgaricus.*

Manufacture: The soft ice cream yoghurt is made by mixing refrigerated yoghurt or refrigerated yoghurt diluted with up to an equal amount of water with a heat-treated and cooled sugar-stabilizer mixture (for example, sucrose 174 parts, dextrose 62 parts, alginate 1.5 parts, and a foaming agent 12.5 parts). Flavoring ingredients are added to the mix, which is then frozen in a soft-serve ice cream freezer.

Reference: Rašić, J. Lj., and J. A. Kurmann. 1978, p. 345.

SORAT (En, Fr, De).

Short Description: USSR (Yakut people); traditional product; skimmed cow's milk.

Manufacture: The product is homemade. Milk is boiled for 2-3 h, then inoculated with previously made sorat and fermented at 25-28°C (77-82.4°F) for 4 h in a wooden bowl covered with a fur.

Related Products: Tar prepared in winter, concentrated fermented milk.

Reference: Ränk, G. 1969, p. 9.

SÓSTEJ (En, Fr, De).

Short Description: Hungary, eastern part of historic Hungary, Transylvania, today Rumania; traditional homemade product, whey-drained; ewe's milk, sometimes cow's and buffalo's milk; strong salty taste, dense consistency.

Manufacture: A large quantity of this product was made up until World War II. Milk is treated with homemade rennet extract and allowed to sour spontaneously. After souring most of the whey is drained. Today sóstej is of minor importance. Details of the manufacture are described by Gratz (1929).

Related Products: Fermented concentrated milk, spontaneously soured milk.

References:

Demeter, K. Y. 1941, pp. 686-687.

Gratz, O. 1929. "Tarho" and "Sóstej," fermented milks made by the herdsmen of the Hungarian plain and of Transylvania (in German). *Molkerei-Zeitung* **43:**2575-2578.

This entry was contributed by Dr. F. Ketting, Budapest, Hungary.

SOUPS (En); Soupes (Fr); Suppen (De).

Short Description: Various countries; traditional and nontraditional preparations; mostly cow's, ewe's, and goat's milk.

History: Soups are of very ancient origin and are usually considered meat and emergency foods.

Manufacture: Several types of soup are prepared with fermented fresh milk products: sour-milk soups, usually sweetened or seasoned with herbs and spices, for example, chlodnich (Poland) and buttermilk soups (e.g., buttermilk barley soup).

Reference: Schulz, M. E. 1965, vol. 2, p. 798.

SOUR MILK (En); Lait acide (Fr); Sauermilch (De).

Short Description: Various countries in earlier time; traditional products; mostly cow's, ewe's, goat's milk.

Microbiology: Spontaneously soured milk microorganisms.

Manufacture: There is no fixed definition for sour milk, nor is there any standard or agreement at which level of acidity or lactic acid milk becomes sour. Cow's milk clots (coagulates) when at approximately 18°C (64.4°F) (room temperature) and the acidity reaches 67.5°Th (0.60% titratable acidity).

Products: Spontaneously soured milk.

SPARKLING FERMENTED MILK BEVERAGES (carbonated fermented milk products) (En); Lait fermenté (boisson) mousseux (Fr); Schäumende (gespritzte) Sauermilchgetränke (De).

Short Description: Germany, France (around 1920); nontraditional products; cow's milk.

Manufacture: The addition of carbon dioxide (CO_2) gas to a fermented milk produces an effervescent beverage. In order to prevent casein or whey separation, the coagulated milk must be homogenized at low temperature. The products are also called champagnized fermented milk products.

SPONTANEOUSLY SOURED MILK (En); Lait sûri (Fr); Selbstgesäuerte Sauermilch (De).

Short Description: Various countries; traditional product; mostly cow's ewe's, goat's milk.

History: In ancient times only very few tribes consumed spontaneously soured milk. Gradually it became customary to boil milk and then to inoculate it with a previously soured milk. Under most circumstances, spontaneous souring was induced by the lactic acid bacteria from the wooden container in which the milk was kept.

Manufacture: Spontaneous souring occurs in raw milk that is not refrigerated or boiled. Even after heat treatment souring will occur after bacterial contamination and if kept at room temperature.

Food Value: For some peoples spontaneously soured milk is considered inedible (e.g., Masai, Azerbaijans).

Related Products: Spontaneously soured fresh milk products are categorized into milk products (sóstej, clabbered milk, Dickmilch, caudiaux) and cream products (tsutsugi).

STONE (FERMENTED) MILK (En); "Lait en forme de pierres" (Fr); Steinmilch (De).

Short Description: Tibet (China); traditional product; mare's milk; residue of the distillation of kumiss is called stone milk.

Products: Tschürra (Tibet), surtschick (Siberia).

Reference: Demeter, K. J. 1941, p. 712.

STREPTOCOCCUS LACTIS MILK (En); Lait fermenté au Streptococcus lactis (Fr); Streptococcus lactis-Sauermilch (De).

Short Description: Various countries (Germany, Switzerland); nontraditional fermented milk; cow's milk; baby foods.

Microbiology: The starter culture consists of *Streptococcus lactis,* a common bacterial species associated with milk and dairy products. Mesophilic bacteria with optimum growth at around 21°C (70°F) (room temperature).

Food Value: *S. lactis* milk is mainly used for preparing baby foods (*see* Baby foods, fermented).

Products: Eledon, pelargon, baby foods, fermented.

Reference: Orla-Jesen, S. 1942. *The Lactic Acid Bacteria,* 2nd ed. Copenhagen: Hos Ejnar Munksgaard.

STREPTOCOCCUS THERMOPHILUS FERMENTED MILK (thermophilus milk) (En); Lait fermenté au Streptococcus thermophilus (Fr); Streptococcus thermophilus Sauermilch (De).

Short Description: Various countries; nontraditional fermented milk; mostly cow's milk.

Microbiology: The starter culture consists of *Streptococcus thermophilus* (proposed *S. salivarius* subsp. *thermophilus*), which is a well-known bacterial species, especially in dairying. It is an important part of the microflora of yoghurt and several other fermented fresh milk products.

Manufacture: Characteristics of *S. thermophilus* for the production of fermented milks are rapid acid development at ±40°C (±104°F) and an insipid, flat and not pleasant taste (to overcome this other microorganisms are admixed).

Food Value: Health-related properties of *S. thermophilus* fermented milk are claimed but not well substantiated (see the reference dealing with cholesterol level and antimicrobial activity).

References:

Pulusani, S. R., and D. R. Rao. 1983. Whole body, liver and plasma cholesterol levels in rats fed thermophilus, bulgaricus and acidophilus milks. *Journal of Food Science* **48:**280-281.

Rao, D. R., B. M. Reddy, G. R. Sunki, and S. R. Pulusani. 1981. Influence of antimicrobial compound(s) extracted from milk fermented by Streptococcus thermophilus on keeping quality of meat and milk. *Journal of Food Quality* **4:**247-258.

SUNDAE-STYLE YOGHURT (En); Yaourt ferme aux fruits (Fr); Stichfester Frucht-Joghurt (De).

Short Description: United States; nontraditional product; cow's milk; yoghurt containing fruit preserve either on the top of the product or in the bottom of the container (*see* Yoghurt).

SUSA (En, Fr, De).

Short Description: Kenya; traditional product; camel's milk; fermented milk.

Reference: Farah, Z., I. Streiff, and M. R. Bachmann. 1990. Preparation and consumer acceptability tests of fermented camel milk in Kenya. *Journal of Dairy Research* **57**(2):281-282. Cited in *Dairy Science Abstracts,* 1990, **52:**488.

SWEET ACIDOPHILUS BIFIDUS MILK (En); Lait acidophilus-bifidus, non fermenté (Fr); Ungesäuerte Acidophilus-Bifidus-Milch (De).

Short Description: Japan; nontraditional product; cow's milk; sweet milk product, nonfermented.

Microbiology: The starter culture consists of *Lactobacillus acidophilus* and *Bifidobacterium longum.*

Manufacture: The product is made by adding concentrated suspensions of *L. acidophilus* and of *B. longum* to cold pasteurized milk in a surge tank. After mixing, the milk is packaged and refrigerated. The flavor of the finished product is similar to that of the freshly pasteurized milk used.

Reference: Rašić, J. Lj., and J. A. Kurmann. 1983, p. 131.

SWEET ACIDOPHILUS MILK (En); Lait acidophilus, non fermenté (Fr); Ungesäuerte Acidophilus-Milch (De).

Short Description: United States; nontraditional product; cow's milk; taste is the same as regular milk (only a suspension of acidophilus bacteria has been added); sweet milk product, nonfermented.

History: An acidophilus product developed in the United States following studies that involved the addition of a frozen concentrate of *Lactobacillus acidophilus* to cold, pasteurized milk (Duggan et al. 1959). The milk is not fermented; it contains viable *L. acidophilus* bacteria at a concentration of several million per ml. The commercial manufacture and sale of "sweet" acidophilus milk in the United States began in 1975, followed by rapid expansion of the market.

Microbiology: The concentrated culture consists of selected *L. acidophilus* strains characterized by their resistance to bile and antagonism to pathogens such as salmonellae and staphylococci.

Manufacture: Sweet acidophilus milk is made from low-fat milk or whole milk. A suspension of concentrated viable cells of the above-mentioned culture is added to cold pasteurized, homogenized milk with thorough mixing. The inoculated milk is then packaged and kept in cold storage. The taste of sweet acidophilus milk is similar to that of freshly pasteurized milk. No flavor or off-flavor is contributed by the culture. The added acidophilus bacteria do not multiply at refrigeration temperatures.

Several variants of sweet acidophilus milk are produced in North America and they are usually sold under specific names. For example, one product, called "Di-gest" is a homogenized, pasteurized low-fat milk that contains 8×10^6 *L. acidophilus* cells per ml and is fortified with vitamins A and D.

Food Value: Regular consumption of sweet acidophilus milk ensures a continued supply of acidophilus bacteria to the intestinal tract where they help to maintain a favorable balance among the indigenous gut microorganisms to the best advantage of the user. Frequent observations are that the acidophilus organisms eliminate unpleasant odor in feces.

Related Product: Sweet acidophilus-bifidus milk.

Reference: Duggan, D. E., A. W. Anderson, and P. R. Elliker. 1959. A frozen concentrate of *Lactobacillus acidophilus* for the preparation of a palatable acidophilus milk. *Food Technology* **13:**465-469.

SWEET BIFIDUS MILK (En); Lait bifidus non fermenté (Fr); Ungesäuerte Bifidus-Milch (De).

Short Description: Germany, Japan; nontraditional product; cow's milk; sweet milk product, not fermented.

Manufacture: The product is made by inoculating dried sweet milk or sterilized milk with cultures or freeze-dried powder containing bifidobacteria.

Reference: Rašić, J. Lj., and J. A. Kurmann. 1983, p. 131.

SWISS-STYLE YOGHURT (En); Yaourt brassé (Fr); Gerührter Joghurt (De).

Short Description: United States; nontraditional product; cow's milk; yoghurt containing uniformly distributed fruit preserves or flavor preparations. This type is also called stirred, stirred-curd, or California-style yoghurt (*see* Yoghurt).

SYUZMA (En, Fr, De); Siuzma, Sjuzma.

Short Description: USSR (Azerbaijan); traditional national product; cow's milk; pasty texture; direct consumption; clean acid taste; smooth, spreadable consistency; agreeable flavor.

Microbiology: The starter culture consisting of *Streptococcus thermophilus* and *Lactobacillus delbrueckii* subsp. *bulgaricus* (1:1 ratio) is mixed in at a rate of 5%.

Manufacture: The product is made from milk standardized to 3.3% fat, pasteurized at 80-85°C (176 -185°F) for 10 min, cooled to 40-45°C (104-113°F) and inoculated with the above culture. The milk is incubated until an acidity of 80-85°Th (0.72-0.76% titratable acidity) is reached, then the coagulum is cut and allowed to stand for 10-20 min for partial separation of the whey. After whey drainage, the curd is put in cloth bags and pressed until a moisture level of 70% is obtained. Packaging is customarily into briquettes of 100, 200, and 500 g.

Food Value: The finished product contains not less than 15% fat, not more than 70% moisture and the acidity is not above 200°Th (1.8% titratable acidity).

Related Products: Chakka, chekize.

Reference: Azimov, A. M. 1982. Syuzma. *Proceedings 21st International Dairy Congress* **1:**(2)619. Moscow: Mir Publisher.

T

TAGAR (En, Fr, De).

Short Description: USSR (Siberia); nontraditional product; fermented milk.

Reference: Khamagacheva, I. S., N. I. Khamnaeva, and A. K. Kulikova. 1987. Stereospecificity of lactic acid in cultured milk products (in Russian). *Pishchevaya Tekhnologiya* **3**:67-68. Cited in *Dairy Science Abstracts,* 1988, **50**(5305):602.

TAIRU (En, Fr, De).

Short Description: Malaysia; traditional product; cow's milk; Malaysians of Indian origin drink it almost daily diluted with water; used for cooking rice dishes; can also be eaten directly with rice or bread; texture and consistency is that of sour cream; the odor is rather strong and cowlike.

Microbiology: No details about microorganisms have been reported.

Manufacture: Similar to that of yoghurt. Homogenize and pasteurize whole milk, cool to 45°C (113°F), inoculate with 2.5-3% starter and incubate at 45°C (113°F) for 3-4 hours or until the acidity reaches 77.5-90°Th (0.7-0.8% titratable acidity).

Food Value: Per 100 g the product contains 90.5 g water, 9.5 g total solids, 2.9 g fat, 2.0 g protein, 2.9 g carbohydrates, 0.7 g ash, 280 mg calcium,

220 mg phosphorus, 50 kcal. When used in cooking, tairu provides a creamy texture and subtle flavor desirable in gravies. It is also used in the cooking of some spicy rice dishes. Tairu provides a better flavor in dishes than the conventional combination of coconut milk and tamarind. It is consumed by people of all socioeconomic levels at a rate of approximately 30-60 g per person per day (Ahmad 1977).

Related Product: Malaysian fermented soybean extract tairu (soy-extract tairu).

References:

Ahmad, I. H. 1977. Malaysian tairu. *Symposium on Indigenous Foods.* Bangkok, Thailand. Cited by Steinkraus, H. 1983, pp. 260-266.

Reddy, N. R., M. D. Pierson, and D. K. Salunke. 1986. *Legume-Based Fermented Foods,* pp. 129-131. Boca Raton, Fl.: CRC Press, Inc.

TALKUNA (En, Fr, De); talkuna.

Short Description: Finland (Häme region); traditional product; fermented milk-cereal preparation; mesophilic fermented milk is add to barley flour.

Reference: Forsén, R. 1966, p. 11.

TAMAROGGTT (En, Fr, De); tamaroggt; signifies the presence of tamr (date) and oggtt (cultured skim milk).

Short Description: Saudi Arabia; nontraditional product; oggtt preparation; heat-treated after culturing.

Microbiology: The starter culture consists of *Streptococcus thermophilus* and *Lactobacillus delbrueckii* subsp. *bulgaricus.*

Manufacture: Manufacture of tamaroggtt involves the proper incorporation of dates in the cultured skim milk product called oggtt. Reconstituted skim milk powder is cultured at 42°C (107.6°F) for 5 h with the above culture and then heated under continuous stirring to form a paste. This is blended with (1) chopped dates, (2) chopped dates plus anise seed and sesame seed, or (3) chopped dates plus cocoa powder. Samples may be shaped and are then air- and oven-dried prior to being packaged in polyethylene bags.

Food Value: The composition of (1), (2) and (3), respectively, is: 10.25, 9.75, and 10.1% moisture; 25.71, 32.35, and 33.50% protein; 59.10, 48.95, and 49.90% total sugars; and 2.79, 2.48, and 3.25% ash.

Reference: Ruquaie, I. M., and H. E. Al Nakhal. 1987. Tamaroggt-new product from dates and oggtt. *Journal of Food Science and Technology India* **24:**230-232. Cited in *Dairy Science Abstracts,* 1988, **50**(2176):241.

TAN (En, Fr, De); than.

Short Description: USSR (Armenia); traditional product; goat's ewe's or mixed milk; concentrated fermented milk product; fermentation in cloth bags with a mixed microflora similar to that of yoghurt.

Reference: Oberman, H. 1985. Fermented milks. In *Microbiology of Fermented Foods.* ed. B. J. B. Wood, p. 175. London: Elsevier.

TAR (En, Fr, De); root word for many fermented milk products of central Asia.

Short Description: USSR (Yakut region); traditional fermented food; emergency foodstuff, probably of very old origin (Ränk 1969); key product.

Microbiology: Mesophilic fermented milk microflora.

Manufacture: It is similar to that of sorat (*see* sorat) prepared in winter time. Tar is manufactured during the summer at warm temperatures causing the acidification and decomposition of the milk mixtures to become very advanced during storage. Tar is a concentrated fermented milk because of its high solids content.

Food Value: A relatively complete emergency or reserve foodstuff for the long winter period. High nutritive value because of concentration and availability of nutrients as a result of the fermentation, including calcium from bones, dietary fiber from plant materials and other substances depending on the ingredients of the mixture. The Yakuts mix into the sour milk "tar" such items as stems and leaves of *Lilium spectabile,* the roots and leaves of *Potentilla anserina, Angelica sylvestris, Rumex acetose,* and various bark meals and berries, chiefly lingonberries. They also add fish, cartilage, bones, and bread crusts. The fish and bones are disintegrated by the lactic acid (Edlitz 1969).

References:

Edlitz, K. 1969. Food and emergency food in the circumpolar area. *Studia Ethnographica Uppsaliensia* **22:**76.
Ränk, G. 1969, p. 9.

TARAG (En, Fr, De).

Short Description: Mongolia; traditional product; ewe's, goat's, cow's, and yak milk; set type, without gas production; viscous consistency; fresh, acidic.

Microbiology: A new "enriched tarag" cultured preparation has been developed based on a mixed starter culture consisting of *Bifidobacterium* spp., *Lactobacillus delbrueckii* subsp. *bulgaricus,* and kefir grains in a 1:0.5:0.5 proportion. This culture has been shown to be inhibitory to a strain of *Escherichia coli* and a strain of *Salmonella sonnei* (Khamnaeva et al. 1985).

Related Products: Clabbered milk, tar.

References:

Accolas, J. P., J. P. Defontaines, and F. Aubin. 1975. Rural activities in the Democratic Mongol Republic (in French). *Etudes Mongol (Paris)* **6:**7-98.
Khamnaeva, N. I., I. S. Khamagaeva, V. F. Tovarov, and V. I. Sharobalko. 1985. Antibiotic properties of a mixed starter (in Russian). *Molochnaya Promȳshlennost'.* **3:**31-32. Cited *Dairy Science Abstracts,* 1986, **48**(3372):398.
Ränk, G. 1969, p. 10.

TARASSUN (En, Fr, De); taraszun.

Short Description: USSR (Burjiat people, Irkutsk region, Siberia); traditional product; popular beverage, but gradually disappearing because a substitute is increasingly made from rye.

Related Product: Kourunga.

Reference: Melnikov, N. 1899. The Burjiats of Irkutsk (in German). *Internationales Archiv für Ethnographie* **12:**206-207. Cited by Maurizio, A. 1933, p. 89.

TARATOR (En, Fr, De).

Short Description: Bulgaria; nontraditional product; cow's, ewe's, or mixed milks; a frozen yoghurt containing 40% cucumber matter and small quantities of dill, parsley, and common salt, as well as 0.2% pectin; used as hors d'oeuvre (not as a soup); spicy, piquant flavor.

Related Products: Yoghurt preparations.

Reference: Rašić, J. Lj., and J. A. Kurmann. 1978, p. 343.

TARHÓ (En, Fr, De).

Short Description: Hungary; traditional product, national type of yoghurt; ewe's, cow's milk; sometimes, when no tarhó is available for inoculation, the stomach contents from a young slaughtered lamb are used instead; originally prepared from ewe's and later from cow's milk.

Related Products: Hungarian fermented milk products, yoghurt.

References:
FIL-IDF Dictionary. 1983, p. 213.
Gratz, O. 1929, *Molkerei-Zeitung* **43:**2575-2578. Cited by Demeter, K. J. 1941, p. 699.

TARYK (En, Fr, De); root word tar.

Short Description: Mongolia; a type of sorat (*see* sorat).

Reference: Seroševskij, V. L. 1896. *Jakuty,* p. 313, Remark 5. St. Petersburg. Cited by Ränk, G. 1969, p. 10.

TÄTTE (En, Fr, De); tettemjølk.

Short Description: Scandinavia (Norway, Sweden); traditional, mesophilic ropy milk.

Related Products: Long milk, langfil.

References:

Bergsaker, J. 1982. Butterwort and other plants in the dairying of the past (in Norwegian). *Meieriposten* **71**:482, 490. Cited in *Dairy Science Abstracts,* 1983, **45**(1213):144.

Nilsson, C. 1952. The microflora of Swedish taette-milk (in Swedish). *Svenska Mejeritidningen* **42**(38):411-416. Cited in *Dairy Science Abstracts,* 1952, **14**:284.

TEAR GRASS-MILK FERMENTED PRODUCT (En); Lait fermenté avec extract d'herbe (Fr); Fermentierte Milch mit Grassaft (De).

Short Description: Japan; nontraditional beverage; an aqueous extract of tear grass and milk or liquid milk product; fermented product.

Microbiology: The starter culture consists of *Lactobacillus delbrueckii* subsp. *bulgaricus.*

Manufacture: The product is made by mixing an aqueous extract of tear grass (*Coix lachrymajobi* Linne var. *mayhen* Stapf) with milk, whey, or any other liquid milk product and fermenting with the above culture to form a lactic beverage.

Reference: Hagiwara, Y. 1981. Tear grass fermentation product and process. *United States Patent* 4,298,620. Cited in *Dairy Science Abstracts,* 1982, **44**(4563):514.

THARA (En, Fr, De).

Short Description: Nepal; traditional product; yoghurt buttermilk.

Reference: Tokita, F., A. Hosono, F. Takahashi, T. Ishida, and H. Otani. 1981. Animal products in Nepal. III. The Sherpas and their livestock farming (in Japanese). *Japanese Journal of Dairy and Food Science* **30**:55-60. Cited in *Dairy Science Abstracts,* 1982, **44**(6616):728.

TIBETAN FERMENTED FRESH MILK PRODUCTS (En); Laits fermentés de la région Tibet (Fr); Tibetische Sauermilchprodukte (De).

Short Description: Tibet and adjacent Chinese area; traditional products.

History: Tibet and the adjoining Chinese area is an important source of old traditional fermented milks of which very little is known elsewhere.

Food Value: Tibetan fermented milks are mainly used in medicine for combating digestive and liver problems.

Products: Kumiss, kheran, da-ra (buttermilk).

T.-M.-MILK (En); Lait T.-M. (Fr); T-M.-Milch (De); denotes *Thermobacterium mobile* milk.

Short Description: Germany; nontraditional product; cow's milk.

Microbiology: The starter culture consists of *Zymomonas mobilis* (formerly named *Pseudomonas lindneri* or *Thermobacterium mobile*). Apart from the production of some lactic acid, the fermentation observed with this organism resembles that of yeast alcoholic fermentation. Some strains may produce as much as 10% alcohol.

Manufacture: Milk with 2% added glucose or sucrose is inoculated with the above organism isolated from parts of an amaryllis plant of the genus *Agave americana.* The result is mild acid development and some alcohol production (max. 0.5%) and continuing postproduction souring.

Food Value: Observations are that the organism suppresses putrefactive intestinal digestion.

Reference: Linder, P. 1929. *Funke-Festschrift* (in German), p. 53-59. Verlag Deutsche Molkereizeitung. Cited by Demeter, K. J. 1941, pp. 712-713.

TONED MILK, FERMENTED FRESH MILK PRODUCTS (En); Toned milk (lait de bufflesse additionné de lait écremé) (Fr); Toned milk (Büffelmilch mit wiederaufgelöster Magermilch gemischt) (De).

Short Description: India mostly; nontraditional product; buffalo's milk.

History: Well-known in India, toned milk is a mixture of buffalo's milk and reconstituted skim milk (from powder). This combination brings the high solids and fat percentage of the buffalo's milk to levels close to those of market milk found in Europe and elsewhere and also makes available a milk that is more affordable than buffalo's milk. The term *toned milk* is not used much any longer, except in some areas.

Manufacture: There are two kinds of toned milk, regular and double-toned. The first is standardized to 3% fat and 8.5-9.0% nonfat solids by blending equal amounts of buffalo's milk and reconstituted skim milk; and the second is standardized to about 1.5% fat and 10% nonfat solids by mixing buffalo's milk with reconstituted skim milk in a ratio of 1:2. In some families, toned milk is used to manufacture fermented fresh milk products.

Reference: *FIL-IDF Dictionary.* 1983, p. 216.

TORBA YOGHURT (En); Yaourt torba (Fr); Torba Joghurt (De); denotes sack or bag yoghurt.

Short Description: Turkey; traditional product; cow's, ewe's, or mixed milks; concentrated yoghurt; produced throughout Turkey.

Microbiology: The starter culture consists of *Streptococcus thermophilus* and *Lactobacillus delbrueckii* subsp. *bulgaricus.*

Manufacture: Torba yoghurt is obtained when whey is allowed to drain from fresh yoghurt or ayran that is filled into sacks or bags. When adding salt, torba yoghurt can be stored for one month at room temperature. In many villages milk producers make this kind of yoghurt from surplus milk or yoghurt. The small yoghurt manufacturing plants use unsold yoghurt to make torba yoghurt, whereas farmers make it from buttermilk. Normally families use torba yoghurt when freshly made or to prepare ayran, cacik (yoghurt salad), and various other yoghurt-derived products.

Food Value: There is no standard method for making this product so its composition varies. The following composition has been reported by one researcher (in %): 18.83 (13.86-24.04) total solids, 5.1 (2.4-8.8) fat, 9.36 (7.23-11.32) protein, 0.21 (0.14-1.11) salt, 2.02 (1.33-3.50) ash, and 104.5°SH (25-135°SH) acidity.

This entry was contributed by Prof. Dr. H. Yaigin,
Bornova-Izmir, Turkey.

TRAHANA, GREEK (En); Trahane gricque (Fr); Griechischer Trahana (De); also known as kapestoes or zamplaricos.

Short Description: Greece and Greek part of Cyprus; traditional product; ewe's milk; made in wheat-producing areas where sheep are kept; related to Trahana, Turkish; used as first solid food when weaning babies, also fed to young children and incorporated in meals (breakfast, soups), food for winter time; heat-treated after culturing.

Microbiology: The starter culture consists of yoghurt-related microorganisms.

Manufacture: Trahana is made by mixing flour with fresh or fermented ewe's milk, then drying and grating the dry product (a flow diagram is given by Economidou and Steinkraus 1977).

Food Value: The product contains (approximately) 11.1% moisture, 4.35% total lipids, 14.05% protein, 75.7% starch, and has an acidity of 200°Th (1.80% titratable acidity).

Related Products: Fermented milk-cereal preparations; trahana, Turkish; Kishk, Egyptian.

References:

Economidou, P. L., and K. H. Steinkraus. 1977. Greek Trahana. *Symposium on Indigenous Foods,* Bangkok, Thailand. Cited by Steinkraus, K. H. 1983, p. 273.

Stephanopoulos, O., and N. Tzanetakis. 1977. The microbial flora of acid Trahan. I. Preliminary studies (in French). *Industries Alimentaires et Agricoles* **94:**1279-1280. Cited in *Dairy Science Abstracts,* 1978, **40**(3610):401.

TRAHANA, TURKISH (En); Trahana turc (Fr); Türkischer Trahana (De).

Short Description: Turkey; traditional product, key product.

Microbiology: The starter culture consists of yoghurt-related microorganisms.

Manufacture: The product is prepared from parboiled wheat flour and yoghurt in a proportion of 2:1. To this mixture vegetables are added, and the mass is allowed to ferment for several days after which it is sun-dried. Annual consumption is estimated at 3 kg/person.

Related Product: Trahana, Greek.

References:
Cadena, M. A., and R. K. Robinson. 1978. Factors affecting the quality of fermented milk-wheat mixtures. *Proceedings 20th International Congress,* vol. E, pp. 994-995. Brussels: International Dairy Federation.
Platt, B. S. 1964. Biological ennoblement: Improvement of the nutritive value of foods and dietary regimens by biological agencies. *Food Technology* **18**(5):662-667, 669-670.
Steinkraus, K. M. 1983, pp. 271-274.

TSCHIGAN (En, Fr, De); čegen, čigen, čigen-arik.

Short Description: Mongolia; traditional product; mare's milk.

History: Whereas kumiss is made by the Turkic peoples, tschigan similar to kumiss is made by Mongols and those strongly influenced by Mongolian culture.

Microbiology: The starter culture consists of kumisslike microorganisms.

Related Product: Kumiss.

Reference: Ränk, G. 1969, p. 12.

TSCHURRA (En, Fr, De).

Short Description: Tibet (China); the residue of kumiss after distillation; dried to hardness, therefore called stone milk; directly consumed as small pieces that swell in the mouth and are eaten instead of bread.

Reference: Demeter, K. J. 1941, p. 712.

TSUTSUGI (En, Fr, De).

Short Description: Mongolia; traditional product; tsutsugi is naturally fermented cream, which is frozen and eaten like ice cream.

Microbiology: Spontaneously soured milk microorganisms.

Reference: Miaki, T. 1980. Mongolian nomadic culture and animal production (in Japanese). *Animal Husbandry* **34:**391-394. Cited in *Dairy Science Abstracts,* 1982, **44**(1871):214.

TULUM YOGHURT (En); Yaourt tulum (Fr); Tulum Joghurt (De); denotes skim yoghurt.

Short Description: Turkey, some regions; traditional product, whey-drained; specifically made to preserve the nutritional value of milk and thus to store any excess of milk beyond family requirements.

Microbiology: The starter culture consists of yoghurtlike microorganisms.

Manufacture: This yoghurt product has no definite manufacturing process. Fresh, heated sour milk or yoghurt and some salt are mixed in special goat or sheep skins for manufacturing. When milk is poured into them it coagulates spontaneously. Whey is allowed to drain from pores in the skin. The skin is regularly washed and kept in a clean, cold place. Generally 2-3 months are needed to fill up a skin by daily additions of milk. Whenever a skin is filled with tulum yoghurt, it is made into "butter." Some may be mixed with spices and dried. The resulting buttermilk is used to make various kinds of cheese.

Food Value: The composition of tulum yoghurt is as follows: 35.67% (26.35-44.44%) total solids, 22.5% (13-35%) fat, 10.06% (8.04-13.5%) protein, 3.1% (1.4-4.6%) salt, 5.2% (3.16-6.03%) ash, and 94.5°SH (62-111°SH). Tulum yoghurt is consumed as a bread spread and is used to make pastries.

This entry was contributed by Prof. Dr. H. Yaygin,
Bornova-Izmir, Turkey.

TZATZIKI (En, Fr, De).

Short Description: Greece; traditional product; well-known refreshing Greek culinary food preparation, a summer dish, yoghurt-cucumber mixture.

Manufacture: Mixture of drained-type natural yoghurt and cucumbers, vinegar, black olives, olive oil, salt, pepper, and mint leaves or garlic.

Related Products: Fermented milk-vegetable mixtures.

Reference: Salaman, R. 1983. *Greek Food,* p. 39. London: Fontana Paper Books.

U

UDAN (En, Fr, De).

Short Description: USSR (Yakutsk Republic); traditional fermented milk beverage consumed during the winter months; yak or other milk; very popular.

Manufacture: Soured milk mixed with butter granules (if available) and water.

Reference: Hintze, K. 1934. *Geography and History of Nutrition* (in German), p. 125. Leipzig: Georg Thieme Verlag.

UHT-FERMENTED MILK (En); Lait-UHT fermenté (Fr); Fermentierte UHT-Milch (De).

Short Description: France; nontraditional product; cow's milk; a new type of fermented milk product patented by Ferialdi and Moisan (1982).

Microbiology: Lactic streptococci or lactobacilli suspended in sterilized water.

Manufacture: UHT-sterilized milk is inoculated aseptically by direct injection into the milk stream between the sterilizer and aseptic packaging machine. Fermentation takes place in the packages during storage at 18°C (64.4°F) for 1 wk. The final product is claimed to have excellent keeping quality.

Reference: Ferialdi, R., and R. Moisan. 1982. Method for producing a cultured milk (in French). *French Patent* 2,502,465 Al. Cited in *Dairy Science Abstracts,* 1983, **45**(2668):307.

UMAN (En, Fr, De).

Short Description: USSR (Yakutsk Republic); traditional fermented milk beverage; diluted with water; product of rural areas.

Related Products: Trug, tar.

Reference: Ränk, G. 1970, p. 9.

UMDAA (En); Undaa (Fr); Umdaa (De).

Short Description: Mongolian Republic, Central Asia; traditional cultured milk beverage; prepared from different types of milk other than mare's milk.

Microbiology: Lactic acid and alcoholic fermentation similar to airag.

Related Products: Airag, khoormog.

Reference: Accolas, J. P., J. P. Defontaines, and F. Aubin. 1978. Milk and dairy products in the Mongolian Republic (in French). *Le Lait* **58:**284.

URDA (En, Fr, De); denotes whey cheese. The name was adopted by the peoples of southeastern Europe (Greeks, Albanians, Bulgarians, Serbians, Hungarians, Czechs, Slovaks, Poles, Ukrainians).

Short Description: Ancient region of Hungarian Carpathian Mountains; traditional beverage; made from sheep cheese whey; whey enriched with whey proteins and induced lactic acid and alcoholic fermentations; generally consumed by shepherds and cheesemakers; frothy, pleasant odor, tangy sour taste and salty.

Microbiology: Microscopic studies have shown the presence of *Zooglea,* very short and rod bacteria and yeastlike microorganisms (*Blastomycetes*).

Manufacture: The cheese whey is brought to near boiling to precipitate the whey proteins. The flocculated protein is removed and added to the whey barrel (whey protein enrichment). After several days the liquid undergoes an alcoholic fermentation with gas production.

Food Value: The approximate chemical composition of the beverage is 91% water, 2% fat, 1.9% whey proteins, 1.3% lactose, 1.66% lactic acid, 0.22% volatile acids, 1.33% alcohol (or 1.68 vol %) and 0.5% ash.

Related Product: Skuta.

References:

Note: More recent technical references are not available in the international literature.

Dunare, N. 1969. Dairy products prepared by Rumanian shepherds (in German). In Földes, L. 1969, pp. 631-632.

Laxa, O. 1907. Sheep cheese varieties prepared by the Slav peoples of the Western region (in French). *Revue Générale du Lait,* **6**(22):510.

URGUTNIK (En, Fr, De).

Short Description: Bulgaria, Balkan mountain regions; traditional sour milk; usually prepared from sheep milk; pleasant sour taste, buttermilk-like and yoghurtlike.

Reference: Fleischmann, W., and H. Weigmann. 1932, p. 370.

V

VARENETS (En, Fr, De).

Short Description: USSR (Ukraine); traditional (national) yoghurtlike product; cow's milk; set or stirred type; sour taste and cooked flavor.

Microbiology: The starter consists of *Streptococcus thermophilus* and *Lactobacillus delbrueckii* subsp. *bulgaricus* (ratio of 4:1 to 10:1).

Manufacture: Varenets is made from cow's milk standardized to 3.2% fat and pasteurized at 95°C (203°F) for 2-3 h. The milk is cooled and inoculated with 1-5% of the starter, then filled into containers and incubated at 43-45°C (109.4-113.0°F) for 4-5 h until coagulated and then cooled. When manufacturing the stirred type, the inoculated milk is incubated at 43-45°C (109.4-113.0°F) in the vat until coagulated (4-5 h) and then cooled and packaged. The final product has an acidity of 100-120°Th (0.90-1.08% lactic acid).

Related Products: Molodost, rjazhenka, yoghurt.

Reference: Koroleva, N. S. 1975. *Technical Microbiology of Whole Milk Products* (in Russian). Moscow: Pishchevoĭ Promȳshlennosti.

VIILI (En, Fr, De); viiliä, denotes thick milk (clabbered milk).

Short Description: Finland (western and northern parts); traditional (national) product; cow's milk; set type, very viscous; used at breakfast and as a snack.

It is gummy, but at the same time it can be easily cut with a spoon. It has a mildly sour and aromatic taste. There is a layer of cream on the surface of viili, and its appearance is clearly distinguished by mold growth (*Geotrichum candidum*) on the surface of the product.

Microbiology: The starter is composed of mesophilic lactic acid bacteria and contains slime-forming (capsule-forming) variants of the strains *Streptococcus lactis, S. lactis* subsp. *cremoris* and *diacetylactis, Leuconostoc mesenteroides* subsp. *dextranicum,* and *Geotrichum candidum.*

Manufacture: Viili is prepared from milk that is standardized (2.5 or 3.9% fat), and pasteurized (85°C/185°F, 20 min. or 95°C/133°F, 5 min.) but unhomogenized. Inoculation (at a rate of 4% for viili with 2.5% and 3% for viili with 3.9% fat content) is with above-mentioned starter culture. As a result of development of *Geotrichum candidum,* there will be a velvetlike layer on the surface of the finished product. The starter is the most essential factor in viili manufacture because it is responsible for the characteristic appearance, consistency, and taste of the product. After filling into 200-ml cups, it is incubated at 18-21°C (64.4-69.8°F) 16-18 h at which time a pH of 4.3 is reached. Then the product is cooled to 5-6°C (41.0-42.8°F) and cold-stored. Shelf life is approximately 14 days.

Food Value: An interesting aspect of the nutritional-physiological value of viili is the presence of antigenic proteins produced by slime-forming, encapsulated *S. lactis* subsp. *cremoris* bacteria.

References:

Finnish Cooperative Dairies Association. (Undated). Valio viili original Finnish cultured milk products. Technical paper, available from Research and Development Department of Valio, Division of Product Development, Finnish Cooperative Dairies Association, Meijerite 4, SF-00101 Helsinki 10, Finland.

Niskasaari, K. 1988. Surface components of *Streptococcus lactis* subsp. *cremoris* from Viili. *Acta Universitas Ouluensi,* Series A, Thesis, University, Oulu, Finland.

VIILIPIIMÄ (En, Fr, De); denotes thick (soured, clabbered) milk (viili), which has been fermented (piimä).

Short Description: Finland; traditional beverage; unskimmed cow's milk, an original type of viili.

Reference: Forsén, R. 1966, p. 10.

VIKING (En, Fr, De); trade name.

Short Description: Switzerland; nontraditional product; cow's milk; pleasant, delicate flavor.

Microbiology: Mesophilic lactic acid bacteria starter culture.

Manufacture: Manufacture procedure of mesophilic fermented milk products. Incubation is carried out in cups and at room temperature.

Food Value: 3.5% protein and 8% fat (the composition lies between those of milk and sour cream).

Reference: Swiss Milk Producers Association. Cristallina Viking recipes, sales literature, undated (in French and German). Bern, Switzerland: Swiss Milk Producers Association.

VINEGAR FROM WHEY (En); Vinaigre de lactosérum (Fr); Molkenessig (De); derived from the French "vinaigre," meaning sour wine.

Short Description: Various countries; traditional product, vinegar; whey.

History: The manufacture of vinegar is of ancient origin and the oldest literature provides some details of the methods and appliances used. Vinegar is mentioned in the Old Testament and milk vinegar was prepared by the Scythians and used as a spice (Strabo VII:4,6). Although the term *vinegar* was originally applied to the product obtained by acetification of wine, it lost this original meaning long ago. Acetification is the formation of acetic acid, usually from ethyl alcohol.

Microbiology: Two microbiological processes are essential in the production of vinegar. The first is an alcoholic fermentation by lactic yeast (e.g., *Kluyveromyces marxianus* subp. *marxianus, Saccharomyces cerevisiae, Candida kefir*). The second is an oxidative fermentation of ethyl alcohol to acetic acid by acetic bacteria (*Aectobacter aceti*).

Manufacture: Vinegar may be prepared from almost any aqueous substance that contains sugar and other nutrients to provide an alcoholic fermentation followed by an acetic fermentation. According to the substrate fermented, the vinegar is named wine vinegar, apple cider vinegar, whey vinegar, and so forth. For other details about manufacture of whey vinegar *see* Whey cultured products.

Food Value: Vinegar is primarily a dilute solution of acetic acid obtained through a fermentation process; however, it also contains the unaltered soluble ingredients from which it is made (e.g., whey vinegar contains lactose), as well as many fermentation products other than acetic acid (lactic acid, etc.). There are legal requirements specifying the chemical composition of vinegar (e.g., minimum 5% acetic acid, maximum 15% acid, expressed as acetic acid).

Reference: Pederson, Carl S. 1978. *Microbiology of Food Fermentations,* 2nd ed., pp. 351-355. Westport, Conn.: AVI Publishing.

VITA (En, Fr, De); trade name.

Short Description: Bulgaria; nontraditional product; cow's milk; skim and full-cream "Vita."

Microbiology: *Lactobacillus delbrueckii* subsp. *bulgaricus* starter culture.

Manufacture: Manufacturing process is similar to that of yoghurt making.

Reference: Ilinov, P., and P. Naumova. 1981. Some data on the lipid composition of the new type Bulgarian yoghurt "Vita." *Proceedings, Bulgarian Academy of Science,* pp. 155-159. Cited in *Dairy Science Abstracts,* 1984, **46**(8486):963.

VITALAKT (En, Fr, De); trade name.

Short Description: USSR; nontraditional product; cow's milk; infant food with a propionic acidophilus mixture.

Microbiology: The mixture is composed of *Lactobacillus acidophilus* and propionibacteria.

Food Value: Its use is recommended for artificial and mixed feeding in infants.

Related Products: Propionic acid bacteria-containing fermented milk products, acidophilus baby foods, baby foods.

Reference: Nabukhotnyi, T. K., S. A. Cherevko, V. P. Pavlyuk, and T. N. Zhernovoi. 1985. Some physicochemical and biological properties of the propionic acidophilus mixture Vitalakt (in Russian). *Ratsional'noe Pitanie* **20:**106, 108. Cited in *Dairy Science Abstracts,* 1986, **48:**617.

VITANA (En, Fr, De); trade name; denotes vitaminized yoghurt.

Short Description: Switzerland; nontraditional vitamin-fortified yoghurt; partially skimmed cow's milk.

Microbiology: Yoghurt starter culture microorganisms.

Food Value: The yoghurt is supplemented with nine vitamins (A, B_6, D, tocopherol, thiamin, riboflavin, nicotinic acid, ascorbic acid, and pantothenic acid) and is available in several flavors (strawberry, pineapple, orange, raspberry, Bircher-Muesli (granola) and celery). Vitamin addition is such that a one-cup serving supplies 100% of the daily requirements of each vitamin.

Related Product: Yoghurt.

Reference: Anonymous. 1976. Nine most important vitamins in one yoghurt (in German). *Schweizerische Milchzeitung* **102:**636. Cited in *Dairy Science Abstracts,* 1977, **39**(2282):262.

W

WEIGHT WATCHERS YOGHURT (En); Yaourt selon "Weight Watchers" (Fr); "Weight Watchers" Joghurt (De).

Short Description: United States and wherever diet programs and products are sold by Weight Watchers International Inc., New York; yoghurt from skim milk.

Food Value: Yoghurt from skim milk, protein-enriched, sweetened with aspartame, containing stabilizers and flavors; vitaminized with vitamins A and D; predominantly used in weight loss and weight control diets, endorsed by dieticians and nutritionists.

Reference: Weight Watchers International Inc., New York, N.Y.

WHEY BEER (En); Bière de lactosérum (Fr); Molkenbier (De).

Short Description: Germany; nontraditional product; whey; not produced today; alcoholic beverage.

History: Much research on whey beer was conducted in Germany in the late 1930s and during World War II (Hesse 1948).

Microbiology: Brewer's yeast or *Kluyveromyces marxianus* var. *lactis.*

Manufacture: Whey has many properties that make it suitable for the manufacture of beerlike beverages. Because whey contains substances similar to the colloids of beer wort, it has great capacity for binding carbonic acid. Whey, like beer wort, has a high mineral salt content. Some constituents in whey, after prolonged heating under pressure, develop caramellike flavors similar to the taste and odor of cured malt. Lactose is only slightly sweet, so it does not alter the taste of the finished beverage. For further details about the manufacture see references.

References:

Hesse, A. 1948. Whey beverages (A compilation of some newer studies) (in German). *Zeitschrift für Lebensmittel-Untersuchung und Forschung* **88:**499.

Zunterer, G. K. 1946. Procedure for the preparation of whey beverages (in German). *Brauwelt* **20:**325.

WHEY BEVERAGES, CULTURED, NONALCOHOLIC (En); Boissons fermentées au lactosérum, sans alcool (Fr); Fermentierte Molkengetränke, ohne Alkohol (De).

Short Description: Various countries; nontraditional products; mainly of domestic and artisanal interest.

For the application of whey in whey drinks the most appropriate modifications are directed at change in the composition of whey, for instance by the removal of sugars (lactose), salts (minerals), or fats (Driessen and van den Berg 1991).

Beverages from Whole Whey

Manufacture: The cheapest, most efficient method of preparing a whey-based beverage is to drain the whey from the cheese vat, filter and pasteurize it, deodorize it if desired, and then ferment it with a lactic acid bacteria and/or bifidobacteria starter culture. Sugar and an appropriate flavoring are added, the most compatible ones being orange and lemon flavorings. Carbonation is optional. Traditional milk industry or soft drink industry packaging may be used. A nonalcoholic beverage concentrate can be produced, which, upon dilution with tap water, is then ready for repacking or consumption.

Food Value: When fruit-flavored, whey beverages have a number of beneficial characteristics. They are low in calories, are refreshingly thirst-quenching, and are generally less acidic than fruit juices. They are nutritious, appeal to health-conscious consumers, and, overall, can provide a good profit margin (Prendergast 1985).

References:

Driessen, F. M., and M. G. van den Berg. 1991. New developments in whey drinks. *IDF Bulletin* **250:**11-19.
Prendergast, K. 1985. Whey drinks-technology, processing and marketing. *Journal of Soc. Dairy Technology* **38:**103-105. Cited in *Dairy Science Abstracts,* 1986, **48:**83.

Nonalcoholic Beverages from Deproteinized Whey

Manufacture: Deproteinization of whey is achieved through acidification and heat or by ultrafiltration. Using deproteinized whey has several advantages: The beverages are clear, there is no sediment formation, and they resemble soft drinks. The disadvantage is the removal of the whey protein. For deproteinization procedures of whey see Vinegar from Whey and the references.

References:

Holsinger, V. H., L. P. Posati, and E. D. DeVilbiss. 1974. Whey beverages: A review. *Journal of Dairy Science* **57:**849-851.
Zunterer, G. K. 1946. On the problems of deproteinizing whey for the preparation of whey beverages (in German). *Brauwelt* **17:**266.
Zunterer, G. K. 1946. Procedure for the preparation of whey beverages (in German). *Brauwelt* **20:**325.

WHEY BEVERAGES WITH LESS THAN 1% ALCOHOL (En); Boissons alcooliques à base de lactosérum avec moins de 1% d'alcool (Fr); Alkoholische Molkengetränke mit weniger als 1% Alkohol (De).

Short Description: Various countries; nontraditional product.

Manufacture: A good beverage of this category should be transparent, clear, and preferably carbonated or sparkling. Deproteinization of whey would be especially important in the production of such beverages. Use of lactose-hydrolyzed whey would increase the acid and alcohol production (Miyamoto et al. 1986).

Related Products: Milone; beverages.

References:

Bambha, P. P., P. A. S. Setty, and V. K. N. Nambudriapad. 1972. "Whevit"—nourishing soft drink. *Indian Dairyman* **24**(7):153.
Holsinger, V. H., L. P. Posati, and E. D. DeVilbiss. 1974. Whey beverages: A review. *Journal of Dairy Science* **57:**849-851.

Miyamoto, R., S. Iwamura, H. Komatsu, T. Yoneya, K. Kataoka, and R. Nakae. 1986. Studies on the production of alcohol-containing fermented beverages using lactose. Hydrolyzed milk (in Japanese). *Japanese Journal of Dairy and Food Science* **35:**A143-A150. Cited in *Dairy Science Abstracts,* 1987, **49:**205.
Schulz, M. E., and K. Fackelmeier. 1948. Fermented beverages from plant extracts and whey (in German). *Milchwissenschaft* **3:**165.

WHEY "CHAMPAGNE" (En); "Champagne" de lactosérum (Fr); Molken-"Champagner" (De); inappropriate term.

Short Description: Poland; nontraditional beverage; cheese whey; low alcoholic content.

Microbiology: Alcoholic fermentation by baker's yeast.

Manufacture: The product is made from fresh whey, either undiluted or with 12% water added. About 7% sugar is used in its manufacture, which uses baker's yeast. Raisins or extracts may be used as flavoring agents. Caramel added at a rate of 1% provides color and taste.

Related Product: Alcoholic beverages from whey.

References:

Anonymous. 1971. Whey "champagne" (in Polish). *Przegląd Mleczarski* **20**(1):13-14. Cited in *Dairy Science Abstracts,* 1971, **33**(3941):594.
Bodmershof, W. 1959. Method for the preparation of sparkling milk beverage of good keeping quality. *Austrian patent* 205,328. Cited in *Dairy Science Abstracts,* 1960, **22:**2468.

WHEY CULTURED PRODUCTS (En); Produits fermentés à base de lactosérum (Fr); Fermentierte Molkenprodukte (De).

Short Description: Various countries; traditional and nontraditional products; different milk types.

Manufacture: In dairy technology, whey is a by-product of cheese making. It is the residue from milk after removal of the casein and most of the fat. Acid whey is obtained in the manufacture of fresh cheeses, such as cottage cheese and quarg. Sweet whey results from the production of practically all other natural cheeses that are made from rennet-coagulated curd.

Food Value: Sweet whey contains more total solids (6.5 versus 5.2%), more lactose (4.9 versus 4.3%), more protein (0.8 versus 0.6%), and has a higher ash content (0.56 versus 0.46%) than acid whey. However, acid whey contains more lactic acid (0.7-0.8 versus 0.1-0.2%). Significant amounts of water-soluble vitamins are also present.

Whey composition varies greatly and depends on milk composition, which is never constant; cheese making procedure, and subsequent treatment (e.g., ultrafiltration, reverse osmosis, demineralization, lactose hydrolysis, and skimming). From a nutritional standpoint, the following can be said about whey:

- All whey products and whey itself have high nutritional value.
- Among food proteins whey protein ranks extremely high in biological value.
- The lacose in whey is special in that it is a sugar that produces conditions favoring mineral absorption during digestion.
- Even though one-half of the lactose molecule is glucose, the passage of glucose into the blood is very slow.
- Whey contains a large mixture of mineral elements making it comparable to mineral water.
- In general, whey is a rich source of potassium.
- Whey is claimed to be a thirst quencher.

Related Products: The following cultured whey products are not pasteurized and are therefore mainly of domestic and artisanal interest: whey beverages, cultured; alcoholic beverages from whey with less than 1% alcohol; whey beer; whey wine; whey "champagne."

Other uses of whey are whey vinegar, whey fermented for animal feeds, and whey fermented for silage production.

WHEY FERMENTED FOR ANIMAL FEEDS (En); Lactosérum fermenté pour l'affourragement animal (Fr); Fermentierte Molke für die Tierfütterung (De).

Short Description: Various countries; nontraditional product; mainly cow's milk whey.

Microbiology: All cheese whey contains cheese starter bacteria. The whey is soured to various degrees depending on uncontrolled fermentation conditions. Rarely is cheese whey fermented under controlled conditions, such as with yoghurt bacteria or specifically selected mesophilic starter cultures.

Food Value: Cheese whey has been fed to animals, especially hogs, probably since the beginnings of the cheese industry. Cultured whey, for example, with yoghurt starter culture or with mesophilic starter cultures, has been shown to be advantageous for young animals (wholesome, nutritious, growth-stimulating, pathogen-suppressing).

Related Products: Animal feeds, whey protein concentrate fermented, yeast-whey products for animal feeds.

WHEY FERMENTED FOR SILAGE PREPARATION (En); Lactosérum fermenté pour des préparations d'ensilages (Fr); Fermentierte Molke für die Silagezubereitung (De).

Short Description: Various countries; nontraditional preparation; cow's milk; sometimes the cultures employed in silage preparations are propagated on whey as the nutritive medium (*see* Silage preparations).

Reference: Marth, E. H. 1973. Fermentation products from whey. In *By-Products from Milk,* ed. B. H. Webb and E. O. Whittier, p. 71. Westport, Conn.: AVI Publishing.

WHEY KVAS. *See* Kvas from Whey, Polish; Kvas from whey, Russian; and Kvas, New Milk Beverage.

WHEY PROTEIN CONCENTRATE, FERMENTED (En); Concentré de protéines du lactosérum fermenté (Fr); Fermentiertes Molkenprotein-Konzentrat (De).

Short Description: Canada; nontraditional product; fermented whey protein concentrate.

Food Value: Fed to weaned piglets at a rate of 33.9% on a total solid basis in isonitrogenous and isocaloric maize-soybean meal diets.

Related Products: Animal feeds and bacterial concentrates.

Reference: Cinq-Mars, D., G. Bélanger, B. Lachance, and G. J. Brisson. 1986. Fermented whey protein concentrate fed to weaned piglets. *Canadian Journal of Animal Science* **66:**1117-1123. Cited in *Dairy Science Abstracts,* 1987, **49**(5822):661.

WHEY VINEGAR (En); Vinaigre de lactosérum (Fr), Molkenessig (De).

Short Description: United States and USSR; nontraditional product; whey.

Microbiology: The first fermentation produces alcohol (yeasts) and the second fermentation converts alcohol to acetic acid (*Acetobacter*).

Manufacture: Whey is first deproteinized by heat-processing; however, a minimum of heat is applied to reduce flavor defects in the finished product. Deproteinization can also be achieved by: (1) precipitation of protein with tannin; (2) precipitation with ethanol added for subsequent acidification; (3) evaporation of whey to dryness and resuspension in water; lactose and mineral salts will then dissolve, but the protein remains insoluble; and (4) removal by ultrafiltration.

Acidification is accomplished in a vinegar generator and with the use of wood chips. The finished vinegar is filtered before bottling. It is also possible to make vinegar from whey by fortifying it with alcohol and without emphasis on protein removal. Another technique consists of first allowing the whey to undergo an alcoholic fermentation, which is then followed by acetifying with the submerged fermentation procedure.

Food Value: The resulting product would be recommended for persons whose digestive systems are sensitive to conventional vinegar.

Related Product: Vinegar.

References:

Hadorn, H., and K. Zürcher. 1973. Manufacture, analysis and evaluation of whey vinegar (in German). *Mitteilungen aus dem Gebiete der Lebensmitteluntersuchung und Hygiene* **64:**480-503.

Marth, E. H. 1970. Fermentation products from whey. In *By-Products from Milk,* ed. B. H. Webb and E. O. Whittier, pp. 69-74. Westport, Conn.: AVI Publishing.

WHEY WINE (En); Vin de lactosérum (Fr); Molkenwein (De).

Short Description: United States; nontraditional product; cheese whey, acid whey powder; whey wine is aimed at the younger consumer.

Microbiology: Alcoholic fermentation by yeast.

Manufacture: The alcohol content is low and the wines are usually fruit-flavored.

Food Value: Whey wines would seem to fit in with the trend toward natural beverages, and some promising products have been developed with the young and "natural" market in mind.

References:

Holsinger, V. H., L. P. Posati, and E. D. DeVilbiss. 1974. Whey beverages: A review. *Journal of Dairy Science* **57**:849-851.

Kosikowski, F. V., and W. Wzorek. 1976. Whey wine from concentrates of reconstituted acid whey powder. *Journal of Dairy Science* **60**:1982-1986. Cited in *Dairy Science Abstracts,* 1978, **40**:267.

Mann, E. J. 1980. Alcohols from whey. *Dairy Industry International* **45**:47-48. Cited in *Dairy Science Abstracts,* 1980, **42**(4239):496.

WINE. Inappropriate but frequently used term, for example, milk wine. *See* Kumiss; Whey Wine.

X

XURUD (En, Fr, De).

Short Description: Mongolian republic; traditional fermented milk.

References:

Hosono, A. 1986. Cultured milks of the Orient (in Japanese). *New Food Industry* **28**(11):65-76. Cited in *Dairy Science Abstracts,* 1987, **49**(2708):317.

Kambe, M. 1986. Traditional cultured milks of the world (in Japanese). *New Food Industry* **28**(10):39-50. Cited in *Dairy Science Abstracts,* 1987, **49**(3434):398.

Y

YAK MILK PRODUCTS, FERMENTED (En); Laits fermentés à partir de lait de Yak (Fr); Yak-Sauermilchprodukte (De).

Short description: USSR, Tibet region; traditional products; yak milk.

History: According to archeological findings in Tibet, yaks were domesticated around 2500 B.C. This long-haired mammal (*Bos grunniens*) is now found in Tibet and adjacent elevated parts of central Asia.

Food Value: Yak milk composition has been reported to be 17.3% total solids, 6.5% fat, 5.8% protein, 4.8% lactose, and 0.9% ash.

Related Products: Fermented yak milk products are airan (Kirghizian people), zho.

References:

Jenness, R., and R. E. Sloan. 1970. The composition of milks of various species: a review. *Dairy Science Abstracts* **32:**611.

Zeuner, E. 1967, pp. 216-217.

YAKULT (En, Fr, De); trade name.

Short Description: Japan; nontraditional product; skim cow's milk; flavored fermented beverage; liquid consistency; agreeable taste and pleasant flavor.

History: The manufacturing procedure of yakult was developed by Dr. Shirota following his studies in 1930. The present name "Yakult" was registered in 1938. Yakult is now commercially produced in Japan, Hong Kong, Taiwan, Brazil, South Korea, and Thailand.

Microbiology: An intestinal isolate of *Lactobacillus casei* is used. This organism shows optimum growth at 37°C (98.6°F) and resistance to gastric acid and bile salts.

Manufacture: Yakult is made from reconstituted skim milk powder, which, after sterilization at 140°C (284°F)/3-4 sec tempering, is fermented with a culture of *Lactobacillus casei* (strain Shirota) at 37°C (98.6°F) for 4 days. It is then cooled and mixed with a syrup of glucose and sucrose containing natural flavors. The concentrated product is transported to bottling plants, diluted with sterilized water to the desired concentration, and then packaged in polystyrene containers and distributed under refrigeration.

Food Value: The finished product contains more than 10^8/ml viable cells of *L. casei*; the approximate composition is 1.2% protein, 0.1% fat, 0.3% ash, 16.5% carbohydrate, and 81.9% water. Caloric value per 100 g is 292.6 kJ (70 kcal). According to studies conducted in Japan, regular human consumption of yakult is beneficial in maintaining the health of the gastrointestinal system.

References:

Anonymous. 1979. Yakult. Available from Yakult-Honsha Co. Ltd., Tokyo.
Rašić, J. Lj. 1987. Other products. *Proceedings 23rd International Dairy Congress*, pp. 673-682. Dordrecht: D. Reidel.

YAKUT PEOPLES FERMENTED MILK PRODUCTS (En); Lait fermentés du peuple Yakut (Fr); Sauermilchprodukte des Yakut-Volkes (De).

Short Description: USSR (Yakutsk Republic); traditional products; mare's, yak's, reindeer milk.

History: Yakuts are Turkic people of northeastern Siberia along the Lena river (now Yakut or Yakutsk Republic of the Soviet Union). They are breeders of horses, cattle, and reindeer and are known for celebrating the kumiss festival.

Products: Sorat, tar, udan, uman, trug, kumiss.

References:

Jochelson, W. 1933. The Yakut. *Anthropological Papers of the American Museum of Natural History* **33**(part 2, Kumiss Festivals): 197-205.
Ränk, G. 1969, pp. 9-10.

YAZMA (En, Fr, De); Jazma, Jazmia.

Short Description: USSR (Tartar people); traditional fermented milk beverage obtained after stirring of liquid yoghurt or katych (dried yoghurt) into water.

Reference: Fleischmann, W., and H. Weigmann. 1932, p. 370.

YLETTE (En, Fr, De); trade name.

Short Description: Denmark; nontraditional product; cow's milk.

Microbiology: A culture of *Streptococcus lactis* subsp. *cremoris* (75%), *S. lactis* subsp. *diacetilactis* (20%), and *Leuconostoc mesenteroides* subsp. *cremoris* (5%).

Reference: Larsen, J. B. 1982. The content of carbohydrates and lactic acid in cultured liquid milk products. *Proceedings 21st International Dairy Congress* **1**:299-300. Moscow: Mir Publishers.

YMER (En, Fr, De); trade name; from Ymir, Scandinavian mythology.

Short Description: Denmark; traditional product; cow's milk; cultured milk product with increased total solids; firm consistency and high viscosity; pleasant taste and aroma.

Microbiology: A cream culture consisting of *Streptococcus lactis* subsp. *lactis, S. lactis* subsp. *cremoris, S. lactis* subsp. *diacetilactis,* and *Leuconostoc mesenteroides* subsp. *cremoris.*

Manufacture: The conventional procedure of manufacturing ymer involves culturing of heat-treated skim milk with a cream culture at 20-22°C (69.0-71.6°F) to a pH of 4.6. This is followed by cutting the coagulum and warming indirectly with water to remove about 50% of the whey. The fat content is standardized with cream and the final product is homogenized at a low pressure, cooled, and packaged. The modern manufacturing procedure of ymer is based on ultrafiltration to a protein content of 6%; standardization; homogenization (200 kg/cm^2 at 65°C); pasteurization (83°C for 5 min); cooling to 20-22°C (68.0-71.6°F); culturing for 20 h; stirring and cooling to

5°C (41°F); standing at 5°C (41°F) for 24 h; stirring at 5°C (41°F), and packaging. The final product contains more whey proteins and calcium than that made by the conventional procedure. As a result, its consistency is gel-like and the nutritive value of the product is improved.

Food Value: Ymer contains 6% protein, 11% nonfat solids, and a minimum of 3.5% fat.

Related Product: Skyr.

Reference: Samuelsson, E. G., and P. Ulrich. 1982. Processing of "Ymer" based on ultrafiltration. *Proceedings 21st International Dairy Congress* **1**:288–289. Moscow: Mir Publishers.

YOGHURT (En); Yaourt (Fr); Joghurt (De); Yogurt.

Short Description: Asia, Near and Middle East, Balkan area; traditional product; milk of cows, ewes, goats, or buffalo or mixed milk; direct consumption, sometimes used for making drinks, salad dressings, or desserts and in confectionery or bakery products; plain or flavored, set-type or stirred; firm consistency in set yoghurt and creamy viscosity in the stirred type; acidic, refreshing taste and pleasant, characteristic aroma.

History: The history of yoghurt dates to biblical times. It is recorded that when Abraham entertained the three angels, he put before them soured or sweet milk (Genesis 18:8). One legend tells that an angel brought down from heaven the pot that contained the first yoghurt or leben. According to some sources yoghurt originated in Asia, where the ancient Turks lived as nomads. The first Turkish name appeared in the eighth century as *Yogurut,* which by the eleventh century had changed to the present name. In the eighteenth century, Kaempfer on his travels to the East described the palace of the Emperor of Persia and a special room in it designated as "Yoghurt choneh." Other sources mention leben of the ancient Arabs, and still others claim that yoghurt originated in the Balkan area in southeastern Europe where the inhabitants of Thrace used to make soured milk called *Prokish,* later called yoghurt. As areas in India also lay claim to the origin of yoghurt, its true beginnings remain disputed.

Originally yoghurt was homemade from ewe's milk or buffalo milk and partly from goat's milk and cow's milk in containers of wood or clay pots. Propagation was carried out by using a small quantity of previously coagulated milk to seed the next batch of boiled milk.

Yoghurt was used mainly for direct consumption, but also for cooking and baking. Ancient physicians of the Middle East prescribed yoghurt as a cure for gastrointestinal disorders. In the early part of the twentieth century, Metchnikoff in his theory of longevity noted the beneficial effect of yoghurt in the human diet. His theory considerably influenced the popularity of yoghurt and stimulated its consumption in many countries. Government regulations, where elaborated, do not allow any health claims for yoghurt, and scientific evidence on that subject has not been clearly forthcoming. Yoghurt is produced in many countries as a popular food. The considerable growth in yoghurt consumption over the years has been due to its agreeable organoleptic properties, the availability of a large variety of flavored products, and to its public image as a "healthy" food.

In order to prevent changes in the traditional identity of yoghurt, definitions have been devised. Accordingly, FAO/WHO (1977) (cited in Rašić and Kurmann 1978) states that

> Yoghurt is a coagulated milk product obtained by lactic acid fermentation through the action of *Lactobacillus bulgaricus* and *Streptococcus thermophilus* from milk and milk products (pasteurized milk or concentrated milk), with or without additions (milk powder, skim milk powder, etc.). The microorganisms in the final product must be viable and abundant.

Microbiology: The original microflora of homemade yoghurt consisted of such essential microorganisms as *Streptococcus thermophilus* and *lactobacillus delbrueckii* subsp. *bulgaricus,* smaller proportions of other nonessential lactic acid bacteria, and contaminants such as yeasts, milk molds, and coliforms (Rašić and Kurmann 1978).

Today yoghurt is produced via large-scale manufacture involving the use of uniformly processed milk, sanitary equipment, and pure cultures consisting of *Streptococcus thermophilus* and *Lactobacillus delbrueckii* subsp. *bulgaricus* (1:1 ratio). When *S. thermophilus* and *L. delbrueckii* subsp. *bulgaricus* are associated, the coagulation time of milk is shorter than in single-culture growth. It has been established that these bacteria act symbiotically. *L. delbrueckii* subsp. *bulgaricus* stimulates the growth of *S. thermophilus* by supplying it with amino acids and peptides liberated from the casein. Then the growth of *S. thermophilus* is slowed due to the adverse effect of gradually accumulating lactic acid. On the other hand, *S. thermophilus* stimulates the growth of *L. delbrueckii* subsp. *bulgaricus* through the formation of formic acid.

Mother culture is transferred every 2-3 days to ensure maximum bacterial activity. Heat-treated skim milk is inoculated with 2% of yoghurt culture and incubated at 41-42°C (105.8-107.6°F) for 2-3 h until coagulation, followed by cooling.

Concentrated cultures are now increasingly used as a bulk starter inoculum to avoid subculturing in the laboratory. A yoghurt culture is regarded as an essential addition in manufacturing yoghurt.

Manufacture: Manufacturing methods may vary and yoghurt may be of the set type (firm, unbroken coagulum) or stirred type (broken, stirred coagulum). From a flavor standpoint, yoghurt may be plain (natural, no additions) or flavored, but flavorings such as fruit and sugar should not be in excess of 30% by weight of the product. Depending on the fat content, in the United States yoghurt may be classified as yoghurt with a minimum fat content of 3.0%; partially skimmed yoghurt with a fat content between 0.5 and 3.0%; and skim milk yoghurt with a maximum fat content of 0.5%. Other specific categories, such as low-fat or cream yoghurt, also exist.

Pasteurized milk is an essential raw material in the manufacture of yoghurt. Raw milk should be derived from healthy animals and must be absolutely free from antibiotics and other inhibiting substances because yoghurt bacteria are very sensitive to them.

Optional additions may be made to

Increase the solids content of the milk, which improves the consistency of the final yoghurt (milk powder, skim milk powder, etc.)
Produce sweetened yoghurt (sucrose, glucose, fructose)
Enhance the dietetic value of yoghurt or modify organoleptic properties (cultures of suitable lactic acid-producing bacteria in addition to yoghurt bacteria)
Reduce caloric value in the final product (substitution of sugar with an approved nonnutritive sweetener).

The steps in the manufacture of yoghurt are diagrammed in Figure 5.

Flavored yoghurt refers to yoghurt with added flavoring ingredients or flavors, with or without added sugars and/or coloring. Natural flavoring ingredients are considered essential manufacturing raw material ingredients, whereas flavors (extracts, essences, etc.) are additives. The following ingredients are used: fruit (fresh, canned, quick-frozen, powdered), fruit purée, fruit pulp, fruit juice, honey, chocolate, cocoa, vanilla, nuts, coffee, spices, and other acceptable flavorful foods. Fruit is the item most often used; approximately equal parts of fruit and sugar, with some water, are mixed and pasteurized to destroy any yeasts, molds, and undesirable bacteria. There are commercially made special fruit preparations solely intended for yoghurt manufacture. The quality of any such fruit preparations influences to a high degree the aroma, taste, and appearance of the final product.

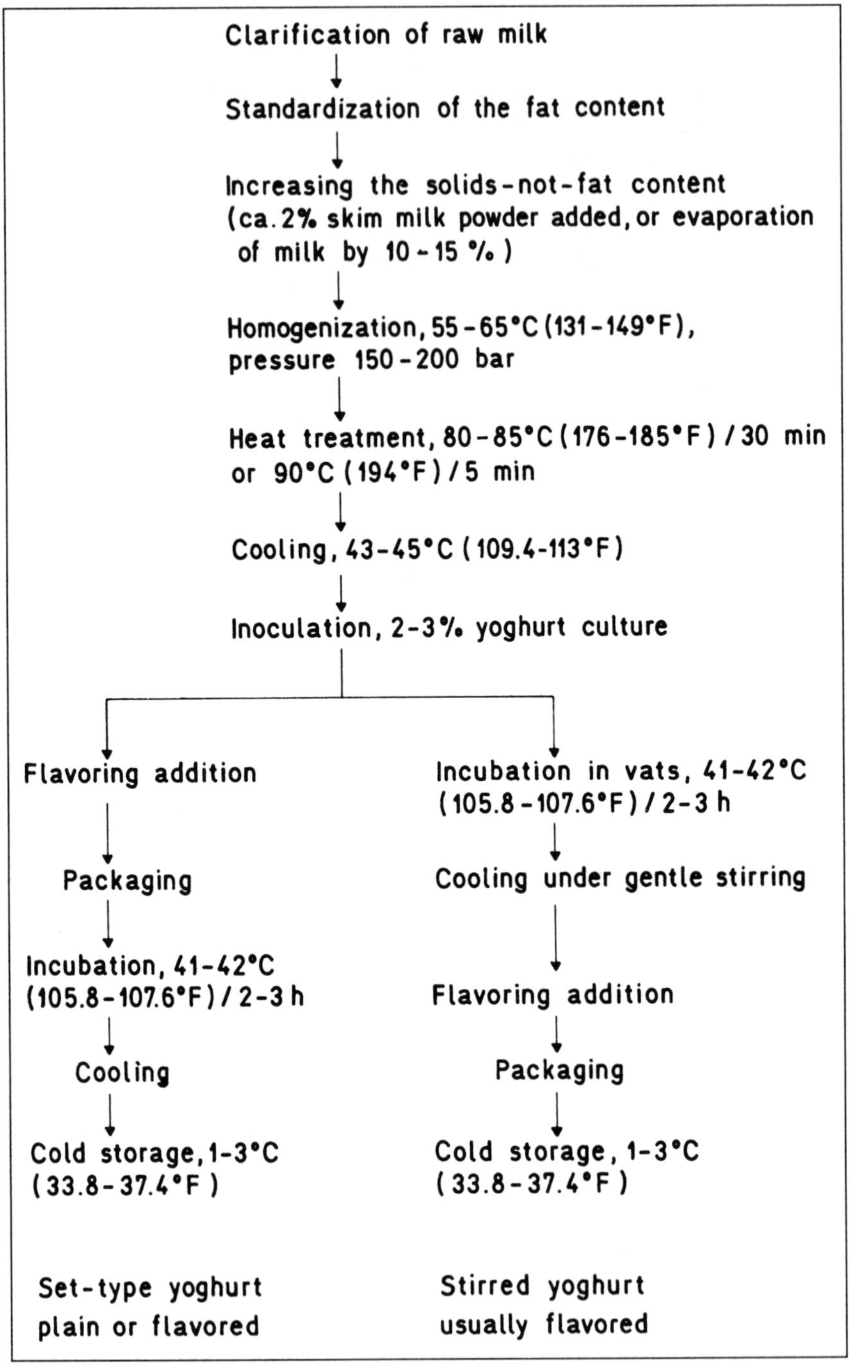

FIGURE 5. Flow diagram of manufacturing yoghurt.

The shelf life of yoghurt under refrigeration ranges between one and four weeks depending on the quality of all ingredients and the sanitary conditions of manufacture and packaging.

Food Value: The composition of yoghurt is similar to that of the milk used for its manufacture. However, yoghurt made from milk fortified with milk powder is obviously richer in protein, lactose, and minerals than the original milk; and, similarly, flavored yoghurt is usually richer in carbohydrates. The manufacture of yoghurt involves modification of some of the milk constituents because of lactic acid fermentation. The content of lactose is reduced by 20-30% or more and that of lactic acid increased from almost zero to 0.8-1.0% or more; 50-70% of the total lactic acid is L(+)-lactic acid and 30-50% is D(−)-lactic acid. In general, for each 0.1% decrease in lactose concentration the titratable acid percentage increases by 0.1%. Contents of amino acids, peptides, and volatile fatty acids are also increased somewhat. Concentrations of some vitamins, for example B_{12} and B_6, are reduced and folic acid and choline are increased. Aroma substances are produced of which acetaldehyde is the principal compound giving plain yoghurt its characteristic flavor. It may reach a concentration of 23-41 ppm at a pH of 4.40-4.00.

Fresh yoghurt contains per gram between 200 million and 1000 million viable yoghurt bacteria. The number declines gradually during storage.

Yoghurt is considered to be more digestible than the milk from which it was made. Lactase-deficient persons tolerate yoghurt better than milk because of the reduced content of lactose and the presence of the bacterial enzyme lactase. The beneficial effect of yoghurt in human nutrition is based on the nutritive value of milk used for the manufacture as well as on the beneficial changes of milk constituents during lactic acid fermentation and the presence of large numbers of viable yoghurt bacteria in the product. Numerous yoghurt products have been made to meet dietary tastes and requirements.

Related Products: Dahi, tarho, zabaday, and others (see Appendix F for Yoghurt, national types).

References:

Kosikowski, F. V. 1977. *Cheese and Fermented Milk Foods,* 2nd ed. Ann Arbor, Mich.: Edwards Brothers, Inc.

Kroger, M. 1976. Quality of yoghurt. *J. Dairy Science* **59:**344-350.

Kurmann, J. A. 1968. Recent insights and problems in yoghurt manufacture (in German). *Schweizerische Milchzeitung,* No. 46.

Kurmann, J. A. 1984. Aspects of the production of fermented milks and consumption statistics. *IDF Bulletin 179* (Fermented Milks), pp. 8-26. Brussels: International Dairy Federation.

Rašić, J. Lj. 1984. The future developments of fermented milks. *IDF Bulletin 179* (Fermented Milks), pp. 27-32. Brussels: International Dairy Federation.
Rašić, J. Lj. 1987. Nutritive value of yoghurt. *Cultured Dairy Products Journal,* August, pp. 6-9.
Rašić, J. Lj., and J. A. Kurmann. 1978. *Yoghurt. Scientific Grounds, Technology, Manufacture and Preparations.* Copenhagen: Technical Publishing House.
Tamine, A. Y., and R. K. Robinson. 1985. *Yoghurt, Science and Technology.* Oxford: Pergamon Press.

YOGHURT AS INGREDIENT IN OTHER FOODS (En); Yaourt comme ingrédient dans d'autre denrées alimentaires (Fr); Joghurt als Zusatzstoff in anderen Lebensmitteln (De). *See also* Ingredients: fermented milk products in other foods.

Yoghurt Bread

Short Description: Switzerland; nontraditional product; bread made with added yoghurt; considered to possess better digestibility than bread made with added milk powder or whey powder.

Reference: Blanc, B. 1973. The value of fermented milk products in modern nutrition (in German). *Schweizerische Milchzeitung,* No. 60/61.

Yoghurt Chocolate

Short Description: Germany (Federal Republic); nontraditional product; cow's milk; a dessert made from yoghurt (1.5% milk fat) and 13% chocolate matter, with the addition of sugar and mint flavor.

Reference: Rašić, J. Lj., and J. A. Kurmann. 1978, p. 346.

Yoghurt Mayonnaise

Short Description: Czechoslovakia; nontraditional product; cow's milk; yoghurt with added sour cream, lemon juice, horseradish and sugar mixed and stirred to a foamy consistency; formulation: 0.80 liter plain yoghurt, 0.15 liter sour cream, 30 g lemon juice, 30 g grated horseradish, and sugar as desired.

Reference: Rašić, J. Lj., and J. A. Kurmann. 1978, p. 359.

YOGHURT PREPARATIONS (En); Préparations de yaourt (Fr); Joghurt-Zubereitungen (De).

Short Description: Near East and Middle East, Europe, United States; traditional and nontraditional products; cow's milk or other kinds of milk; mainly direct consumption; a variety of products with different moisture contents and physical properties (liquid, concentrated, dried, pasty, frozen, jellied) or with modified composition to meet various consumer tastes and requirements; plain or flavored; consistency and flavor depend on the type of product.

Products: Numerous yoghurt preparations are used as drinks, desserts, dressings, in confectionery and bakery products, or in the manufacture of other foods. Several typical products should be mentioned.

Yoghurt drinks are products of liquid consistency (*see* Beverages). They are made by stirring the coagulum of skimmed or partially skimmed yoghurt with a high-speed agitator, then cooling and flavoring; or by diluting yoghurt with milk (1:1 ratio), water (1:0.1-1 ratio) or fruit juice. In the manufacture of long-life drinks, skimmed yoghurt containing sugar, fruit flavoring, and stabilizer is heat-treated after fermentation and then aseptically packaged. Since the product does not contain viable yoghurt bacteria, the name yoghurt is usually replaced by another suitable name.

Concentrated yoghurt is a product containing 22-24% total solids (*see* labneh, Kurut), whereas dried yoghurt, produced by either freeze-drying or spray-drying, contains about 96% total solids. Dried yoghurt is used in bakery products, confectionery items, diet or low-caloric foods, and so forth (*see* Dried yoghurt).

Frozen yoghurt is a plain or flavored product of relatively long storage life (*see* Frozen yoghurt). It is related to other frozen dairy desserts and may resemble ice cream or sherbet. The plain frozen yoghurt is made by adding a pasteurized sugar-stabilizer mixture and flavor to refrigerated yoghurt with subsequent mixing and freezing (soft-serve ice cream from yoghurt). The flavored frozen product contains yoghurt mixed with fruit juice and sugar (sherbet). Other desserts include fruit-jellied yoghurt, cocktail yoghurt, yoghurt creme, yoghurt pudding, whipped yoghurt, and others (*see* Dessert products).

Products used as hors d'oeuvre or appetizers include *apero yoghurt* and *tarator.*

Finally, a growing assortment of yoghurt-containing dietetic products exists, such as protein and/or vitamin-fortified preparations, diabetic yoghurt, low-lactose yoghurt, calcium-fortified yoghurt, low-sodium yoghurt, dietary fiber-fortified yoghurt, and other products.

See also Ewe's milk fermented products; Goat milk firm yoghurt; Goat milk liquid yoghurt; Honey-yoghurt; Reformed yoghurt; Silivri yoghurt; Long-life yoghurt.

Reference: Rašić J. Lj., and J. A. Kurmann. 1978, pp. 325-362.

YRKIT (En, Fr, De).

Short Description: Mongolia; traditional product; a long-life fermented milk.

Manufacture: Yrkit is prepared from katyk or airan, which are stored 5-15 days with periodic addition of fresh or boiled milk. Usually made during the spring or summer months.

Reference: Rudenko, S. I. 1969. Studies about nomads. In Földes, L. 1969, p. 29.

YUGOSLAVIAN FERMENTED MILK PRODUCTS (En); Laits fermentés yougoslaves (Fr); Jugoslavische fermentierte Milchprodukte (De).

Short Description: Yugoslavia; traditional and nontraditional products; cow's, ewe's, goat's, and mixed milks.

Products:

Yoghurt, stirred. Made from cow's milk (2.8% and 3.2% fat). Industrial manufacture.

Yoghurt, set-type. Also called kiselo mleko, made from cow's milk (2.8%, 3.2% and 6.0% fat). Homemade, small-scale, and industrial manufacture.

Yoghurt, set-type. Made from ewe's milk (at the end of the lactation period) with subsequent drainage of whey. Homemade in some mountain villages in the eastern part of the country (zimne, brano milk).

Grusevina. A sour milk made from fresh milk by spontaneous souring or addition of uncontrolled starter culture. Homemade in some regions of the western part of the country.

Yoghurt, fruit-flavored. Stirred-type made from cow's milk. Industrial manufacture.

Kefir, plain. Made from cow's milk using kefir grains. Industrial manufacture.

Sour cream. Made from pasteurized cream (20% and 12% fat). Industrial manufacture (cultured cream).

Kaymak (skorup). Made from cow's, ewe's, or mixed milks. Homemade.

Many local fermented milks made under various names: kisela varenika, basa, samakisselis.

Reference: Rašić, J. Lj. 1982. Special products. Contributing paper. *Proceedings 21st International Dairy Congress* **2:**151 (Discussion, manuscript). Brussels: International Dairy Federation.

YUZHNYI (En, Fr, De); yuznii, yuzhnaya.

Short Description: USSR; traditional product; designation for stirred, natural liquid yoghurt.

Microbiology: Made with selected strains of *Streptococcus thermophilus* and *Lactobacillus delbrueckii* subsp. *bulgaricus,* which are mixed such that the ratio of streptococci and lactobacilli is 4:1 (similar to conventional yoghurt starter culture).

Manufacture: Milk with 3.2% fat content, homogenized and pasteurized, is inoculated with 5% of the starter culture, incubated at 42-45°C (107.6-113.0°F) for 2.5-4 h, and then stirred, cooled, and filled into retail containers. The final product has an acidity of 90-120°Th (0.81-1.08% titratable acidity).

Related Product: Yoghurt.

Reference: Koroleva, N. S. 1975. *Technical Microbiology of Whole Milk Products* (in Russian). Moscow: Pishchevoĭ Promy̆shlennosti.

Z

ZABADAY (En, Fr, De); Zabadi, Laban Zabaday; original name means "thickened milk, like butter."

Short Description: Egypt (major Egyptian fermented milk), Sudan; old traditional product; buffalo's milk, cow's milk, or a mixture of the two; Egyptian type of yoghurt; smooth, porcelainlike surface; bright white when made from buffalo's milk and yellowish white when made from cow's milk; characteristic taste and aroma, full, pleasant, mildly sour; firm consistency without whey separation, custardlike, nearly sliceable.

Microbiology: *Streptococcus thermophilus* and *Lactobacillus delbrueckii* subsp. *bulgaricus* are the predominant microflora. Other organisms, but nonessential among the zabaday microflora, are streptococci (*S. faecalis*) and lactobacilli (*L. casei, L. fermenti, L. helveticus, L. viridescens*). The starter culture, whether for household, small-scale, or large-scale industrial manufacture is the same as for yoghurt.

Manufacture: Kitchen preparations of zabaday use additions of fruits, honey, and kishk; one specific preparation is zabaday salad made with lettuce and mayonnaise. Factory-made products are sold with the addition of fruits.

Food Value: Yoghurt.

Related Product: Yoghurt.

References:

Abou-Donia, S. A. 1984. Egyptian fresh fermented milk products. *New Zealand Journal of Dairy Science and Technology* **19:**7-18.

El-Samragy, Y. A. 1988. The manufacture of zabady from goat milk. *Milchwissenschaft* **43:**92-94.

This entry was contributed by S. A. Abou-Donia, Alexandria, Egypt.

ZDOROVE (En, Fr, De); a term denoting healthy.

Short Description: USSR; nontraditional whey-drained product; cow's milk; concentrated fermented milk; direct consumption; pasty consistency; sour taste and agreeable aroma.

Microbiology: The starter culture is composed of *Streptococcus thermophilus* and mesophilic lactic acid streptococci (ratio 1:1).

Manufacture: Zdorove is made from heat-treated skim milk inoculated with 5% of the starter culture. This is incubated at 36-38°C (96.8-100.4°F) for 10-12 h until firmly coagulated and an acidity of 80-85°Th is obtained (0.72-0.77% titratable acidity). The coagulum is cut into 2 cm cubes and held for 40-50 min to bring about partial whey separation. After whey drainage the curd is pressed to reduce the moisture content to at least 85%; it is then passed through a colloid mill, and cream and other optional ingredients (sugar, fruit, syrup, vitamin C, salt) are incorporated. The final product has good spreadability.

Related Products: Fermented milk protein paste, acidophilus paste, laktofil, ymer.

References:

Bogdanova, G. I., and K. Novoseleva. 1962. Milk protein paste (in Russian). *Molochnaya Promȳshlennost'* **23:**16-17.

Bogdanova, G. I., and E. A. Bogdanova. 1974. *New Whole Milk Products of Improved Quality* (in Russian). Moscow: Pishchevoĭ Promȳshlennosti.

ZEBU CATTLE MILK-BASED FERMENTED MILK PRODUCTS (En); Laits fermentés à partir de lait Zebu (Fr); Fermentierte Zebu-Milch (De).

Short Description: India, China, and Southeast Asia; traditional and nontraditional products; zebu milk.

History: Zebu cattle predominate in India, China, and Southeast Asia. They are characterized by a prominent fleshy hump over the shoulders, short horns, large drooping ears, very large dewlap, large folds under the belly, short stiff hair, and loose body build—all features evolved to facilitate heat loss. Zebu cattle are quite tolerant of insects, resistant to tropical diseases, and to living in areas with long periods of dryness. Numerous crosses exist between zebu cattle and other dairy cattle breeds. The zebu was domesticated as early as other dairy cows. Drawings of zebu cattle have been found in northern Mesopotamia that date from 4500 B.C.

Food Value: The average chemical composition of zebu milk in India resembles that of cows' milk: 13.5% total solids; 4.7% fat; 2.6% casein; 0.6% whey protein; 4.9% lactose; 0.7% minerals (Basu et al. 1962), but in general, the total solids, fat, and mineral percentage are slightly higher. A zebu cow produces approximately 1000 kg (2200 lbs) milk per 7-month lactation period. Crossbreeds produce much larger quantities.

Reference: Basu, K. P., T. M. Paul, N. B. Shroff, and M. A. Rahman. 1962. *Indian Council of Agriculture Research Report* Series 8. Cited in *Dairy Science Abstracts,* 1970, **32**(10):611.

ZEN (En, Fr, De); a term of contemplative Buddhism meaning life, vitality; trade name.

Short Description: France; nontraditional products; cow's milk; fermented milk.

Microbiology: Bifidus and acidophilus cultures.

Manufacture: Two preparations exist: (1) Milk is fermented with the culture and fruit preserves are then added; also added are specific amounts of vitamins A, B_1, B_2, B_6 and niacin; and (2) A similarly made product is based on a vegetable-milk mixture that is enriched with magnesium (*see* Bircher-Muesli).

Reference: Anonymous. 1988. *Revue Laitière Française,* No. 476.

ZHEMAJCJU (En, Fr, De).

Short Description: USSR; nontraditional beverage; fermented sweet buttermilk; pleasant taste.

Microbiology: A cream starter culture consisting of *Streptococcus lactis, S. lactis* subsp. *cremoris, S. lactis* subsp. *diacetylactis,* and/or *Leuconostoc mesenteroides* subsp. *cremoris.*

Manufacture: Sweet buttermilk with an acidity of not more than 20°Th (0.18% titratable acidity) is mixed with cream to raise the fat content 1% and with nonfat milk solids (skim milk powder) to raise the total solids to 10.5%. Then the mixture is pasteurized at 85-87°C (185.0-188.6°F), homogenized, and cooled to 22-25°C (71.6-77.0°F). After addition of 2-5% of the starter culture and incubation for 12-16 h to an acidity of 70-80°Th (0.63-0.72% titratable acidity), the product is stirred, refrigerated to 6-8°C (42.8-46.4°F) and filled into packages.

Related Products: Buttermilk, natural (fermented).

Reference: Berzinskas, G. G., and S. K. Urbene. 1978. Development of a new fermented milk beverage based on buttermilk (in German). *Milchforschung-Milchpraxis* **20**(2):35. Cited in *Milchwissenschaft,* 1978, **33:**593.

ZHENTITSA (En, Fr, De).

Short Description: USSR (eastern Carpathian Mountains); traditional product; sheep milk.

Microbiology: *See* Brano milk.

Manufacture: Fermented at 35-40°C (95-104°F), followed by aging at low temperatures. Preparation of zhentitsa is similar to that of brano milk.

Related Product: Brano milk.

Reference: Matuszewski, T., and J. Supinska-Jakubowska. 1949. *Mikrobiologia Mleczarska.* Warsaw: PIWR.

ZHO (En, Fr, De).

Short Description: Tibet; traditional product; yak milk; pleasant tasting fermented milk; in the border territory the Chinese people prepare zho from cow's milk; it is relatively tasteless.

Food Value: Zho is believed to prolong human life and to be beneficial to the gastrointestinal tract.

Reference: Hermann, M. 1949. *The Nomads of Tibet. The Socio-Economic Foundations of Nomadic Cultures in Amdo and Central Asia—Origin and Development of Animal Husbandry* (in German). Vienna, Austria. Cited by Takamya, T. 1978, p. 102.

ZIMNE (En, Fr, De); means winter-milk.

Short Description: Yugoslavia; traditional concentrated, whey-drained fermented milk; sheep milk; pleasant taste, typical odor, reminiscent of young walnuts; consistency of the product is similar to that of viscous cream.

Microbiology: The culture is composed of *Streptococcus thermophilus* and *Lactobacillus delbrueckii* subsp. *bulgaricus;* typical lactic acid bacteria present in the product.

Manufacture: Immediately after milking, sheep milk is filtered, boiled in an open kettle, allowed to cool to room temperature, poured into wooden vessels (5-10 liter capacity), inoculated with 2% culture at a temperature of 30-40°C (86-104°F), and allowed to incubate (ferment) in the covered containers. The next day a further quantity of similarly treated milk is carefully added, but without excessive mixing. Further portions of milk may be added on up to four subsequent days, and the full vessels are left for about 7 days during which period the milk coagulates and whey begins to separate along the walls. Further treatment includes removal of the separated whey and protection of the product against contamination with molds, yeast, and other organisms. Under local conditions the storage temperature decreases from 18-20°C (64.4-68.0°F) in October to 8°C (46.4°F) in December.

Food Value: In tests 4 months after manufacture, samples contained approximately 30% total solids and 15% fat; the acidity was 175°Th (1.57% titratable acidity).

Related Products: Concentrated fermented milks.

Reference: Zivkovic, Z. 1969. Manufacture of soured milk for winter use (in Slovakian). *In Proceedings of the 1st Conference on Processing of Ewes' Milk. Part.I. Zbornik referatov,* Zilina, pp. 151-159. Cited in *Dairy Science Abstracts,* 1970, **32:**475.

ŽINČICA (En, Fr, De); Zincisa, Shintschitza; means sheep milk whey.

Short Description: Czechoslovakia; traditional kefirlike beverage for shepherds; sweet whey from sheep milk.

Microbiology: A lactic acid fermentation by *Lactobacillus casei, L. plantarum, L. lactis,* and *Leuconostoc mesenteroides* subsp. *dextranicum* (Prekoppova and Prekopp 1973).

Manufacture: Whey from Bryndza cheese is heated to precipitate the whey proteins. The coagulum is removed and transferred, with some of the whey, into a barrel. Every day this addition is repeated. The slurried coagulated mass undergoes a lactic acid fermentation. After addition of salt, the fermented whey and whey protein mixture is consumed as a daily snack food.

Related Products: Whey protein-based fermented milk products, azi.

References:

FIL-IDF Dictionary. 1983, p. 236.

Prekopp, I. 1970. Bryndza, Ostic pok and Parenica-Slovak ewes' milk cheeses (in Slovakian). *Výživa Lidu* **25**(8):127-128. Cited in *Dairy Science Abstracts,* 1975, **37**:89.

Prekoppova, J., and I. Prekopp. 1973. Isolation and identification of microorganisms from sheep milk and sheep milk products (in Slovakian). *Veda a Vyskum v Potravinarskom Priemysle* **25**:93-102. Cited in *Dairy Science Abstracts,* 1975, **37**:89.

ZIVDA (En, Fr, De).

Short Description: Israel; nontraditional product; cow's milk; coagulated with a culture of *Streptococcus lactis.*

References:

FIL-IDF Dictionary, 1983, p. 236.

Keller P., D. Sklan, and S. Gordin. 1974. Effect of diluents on bacterial counts in milk and milk products. *Journal of Dairy Science* **57**:128-128. Cited in *Dairy Science Abstracts,* 1974, **36**:421.

Appendixes

Appendix A
Conversion Tables

Conversion of Titrable Acidity

°SH	°TH	°D	% Lactic Acid	°SH	°TH	°D	% Lactic Acid
1	2.5	2.25	0.0225	51	127.5	114.75	1.1475
2	5.0	4.50	0.0450	52	130.0	117.00	1.1700
3	7.5	6.75	0.0675	53	132.5	119.25	1.1925
4	10.0	9.00	0.0900	54	135.0	121.50	1.2150
5	12.5	11.25	0.1125	55	137.5	123.75	1.2375
6	15.0	13.50	0.1350	56	140.0	126.00	1.2600
7	17.5	15.75	0.1575	57	142.5	128.25	1.2825
8	20.0	18.00	0.1800	58	145.0	130.50	1.3050
9	22.5	20.25	0.2025	59	147.5	132.75	1.3275
10	25.0	22.50	0.2250	60	150.0	135.00	1.3500
11	27.5	24.75	0.2475	61	152.5	137.25	1.3725
12	30.0	27.00	0.2700	62	155.0	139.50	1.3950
13	32.5	29.25	0.2925	63	157.5	141.75	1.4175
14	35.0	31.50	0.3150	64	160.0	144.00	1.4400
15	37.5	33.75	0.3375	65	162.5	146.25	1.4625
16	40.0	36.00	0.3600	66	165.0	148.50	1.4850
17	42.5	38.25	0.3825	67	167.5	150.75	1.5075
18	45.0	40.50	0.4050	68	170.0	153.00	1.5300
19	47.5	42.75	0.4275	69	172.5	155.25	1.5525
20	50.0	45.00	0.4500	70	175.0	157.50	1.5750
21	52.5	47.25	0.4725	71	177.5	159.75	1.5975
22	55.0	49.50	0.4950	72	180.0	162.00	1.6200
23	57.5	51.75	0.5175	73	182.5	164.25	1.6425
24	60.0	54.00	0.5400	74	185.0	166.50	1.6650
25	62.5	56.25	0.5625	75	187.5	168.75	1.6875
26	65.0	58.50	0.5850	76	190.0	171.00	1.7100
27	67.5	60.75	0.6075	77	192.5	173.25	1.7325
28	70.0	63.00	0.6300	78	195.0	175.50	1.7550
29	72.5	65.25	0.6525	79	197.5	177.75	1.7775
30	75.0	67.50	0.6750	80	200.0	180.00	1.8000
31	77.5	69.75	0.6975	81	202.5	182.25	1.8225
32	80.0	72.00	0.7200	82	205.0	184.50	1.8450
33	82.5	74.25	0.7425	83	207.5	186.75	1.8675
34	85.0	76.50	0.7650	84	210.0	189.00	1.8900
35	87.5	78.75	0.7875	85	212.5	191.25	1.9125
36	90.0	81.00	0.8100	86	215.0	193.50	1.9350
37	92.5	83.25	0.8325	87	217.5	195.75	1.9575
38	95.0	85.50	0.8550	88	220.0	198.00	1.9800
39	97.5	87.75	0.8775	89	222.5	200.25	2.0025
40	100.0	90.00	0.9000	90	225.0	202.50	2.0250
41	102.5	92.25	0.9225	91	227.5	204.75	2.0475
42	105.0	94.50	0.9450	92	230.0	207.00	2.0700
43	107.5	96.75	0.9675	93	232.5	209.25	2.0925
44	110.0	99.00	0.9900	94	235.0	211.50	2.1150
45	112.5	101.25	1.0125	95	237.5	213.75	2.1375
46	115.0	103.50	1.0350	96	240.0	216.00	2.1600
47	117.5	105.75	1.0575	97	242.5	218.25	2.1825
48	120.0	108.00	1.0800	98	245.0	220.50	2.2050
49	122.5	110.25	1.1025	99	247.5	222.75	2.2275
50	125.0	112.50	1.1250	100	250.0	225.00	2.2500

Conversion of Temperature Degree

°C	°F	°C	°F	°C	°F	°C	°F
−20	−4	23	73.4	66	150.8	109	228.2
−19	−2.2	24	75.2	67	152.6	110	230.0
−18	−0.4	25	77.0	68	154.4	111	231.8
−17	1.4	26	78.8	69	156.2	112	233.6
−16	3.2	27	80.6	70	158.0	113	235.4
−15	5.0	28	82.4	71	159.8	114	237.2
−14	6.8	29	84.2	72	161.6	115	239.0
−13	8.6	30	86.0	73	163.4	116	240.8
−12	10.4	31	87.8	74	165.2	117	242.6
−11	12.2	32	89.6	75	167.0	118	244.4
−10	14.0	33	91.4	76	168.8	119	246.2
− 9	15.8	34	93.2	77	170.6	120	248.0
− 8	17.6	35	95.0	78	172.4	121	249.8
− 7	19.4	36	96.8	79	174.2	122	251.6
− 6	21.2	37	98.6	80	176.0	123	253.4
− 5	23.0	38	100.4	81	177.8	124	255.2
− 4	24.8	39	102.2	82	179.6	125	257.0
− 3	26.6	40	104.0	83	181.4	126	258.8
− 2	28.4	41	105.8	84	183.2	127	260.6
− 1	30.2	42	107.6	85	185.0	128	262.4
0	32.0	43	109.4	86	186.8	129	264.2
1	33.8	44	111.2	87	188.6	130	266.0
2	35.6	45	113.0	88	190.4	131	267.8
3	37.4	46	114.8	89	192.2	132	269.6
4	39.2	47	116.6	90	194.0	133	271.4
5	41.0	48	118.4	91	195.8	134	273.2
6	42.8	49	120.2	92	197.6	135	275.0
7	44.6	50	122.0	93	199.4	136	276.8
8	46.4	51	123.8	94	201.2	137	278.6
9	48.2	52	125.6	95	203.0	138	280.4
10	50.0	53	127.4	96	204.8	139	282.2
11	51.8	54	129.2	97	206.6	140	284.0
12	53.6	55	131.0	98	208.4	141	285.8
13	55.4	56	132.8	99	210.2	142	287.6
14	57.2	57	134.6	100	212.0	143	289.4
15	59.0	58	136.4	101	213.8	144	291.2
16	60.8	59	138.2	102	215.6	145	293.0
17	62.6	60	140.0	103	217.4	146	294.8
18	64.4	61	141.8	104	219.2	147	296.6
19	66.2	62	143.6	105	221.0	148	298.4
20	68.0	63	145.4	106	222.8	149	300.2
21	69.8	64	147.2	107	224.6	150	302.0
22	71.6	65	149.0	108	226.4		

Fahrenheit conversion: °F = 9/5 (°C + 32)
Celsius conversion: °C = 5/9 (°F − 32)

Appendix B
Products by Regions

This appendix is intended as a compilation of known products by specific category. Items listed do not necessarily appear in the text.

Africa

Algeria: Aoules; Tammart
Chad: Roba, *see* **Rob;** Yoghurt
Congo Republic: Yoghurt
Egypt: Kishk; Kishk seiamy (vegetable mixture); Laban hamid; Laban kerbah; Laban khad; Laban matrad; Laban rayeb; Laban zabady; Laban zeer; Labneh
Ethiopia (using amaharic terms): Aghwat (whey beverage); Arera (buttermilklike); Ayib (whey-drained buttermilk heat-treated); Ergo; Hitu (sour milk)
Kenya: Mala; Maziwa lala
Nigeria: Nono; Samco's
South Africa: Cultured buttermilk; Cultured cream; Cultured milk; Maas; Sour milk in clay pots with natural strains of streptococci (homemade); Yoghurt (set, stirred, plain fruit, and flavored types)
Sudan (similiar to southern parts of Egypt): Goubasha; Laban rayeb; Zabady
Tunesia: Yoghurt (from reconstituted and recombined or fresh milk)

North America

Canada: Bifidus milk products; Biogarde products; Buttermilk; Cultured buttermilk; Kefirlike products; Yoghurt (cow's and goat's milk)
Hawaii, United States: Cultured mold milk

United States: Acidophilus milk; Bifidus milk; Bifidus yoghurt; Cultured cream; Cultured lowfat milk; Cultured milk; Cultured skim milk; Frozen yoghurt; Kefir; Yoghurt, Sunday style (fruit on the bottom) and Swiss style (fruit blended in yoghurt)

South America

Argentina: Kefir; Yoghurt
Brazil: Biogarde products; Yakult; Yoghurt (flavored)
Chile: Progurt; Yoghurt
Colombia: Yoghurt
Mexico: Yakult; Yoghurt

Asia, Central*

Mongolia: Acidophilin; Arak; Bjaslag; Kefir; Khoormog; Mongolian fermented fresh milk products; Prostokvasha; Smetanka; Tarag; Tarassun (Kurunga); Tsutsugi; Umdaa; Xurud

Asia, Eastern

China: Tibetan fermented fresh milk products; Yoghurt (insignificant)
Japan: Bifidus products; Joie; Mil-Mil; Yakult; Yoghurt
South Korea: Kumiss type; Yoghurt
Taiwan: Acidophilus milk; Cultured buttermilk; Cultured milk drinks; Kefir; Kumiss; Skyr; Yakult; Yoghurt
Tibet: Da-ra (buttermilk); Kheran®; Kumiss; Tibetan fermented fresh milk products

Asia, Southern

India: Acidophilus milk; Bulgarian buttermilk; Buttermilk, natural; Chakka, Indian; Chhach; Chhana, Channa; Cultured buttermilk; Dahi; Jalebi; Kefir; Kumiss; Misti dahi; Payyodhi; Shrikand; Yoghurt
Nepal: Churpi; Dahi; Sho (yoghurt); Shomar (whey of buttermilk cheese mixed with milk, fermented); Shosim; Thara (yoghurt buttermilk from churned yoghurt)
Malaysia: Tairu, Taire

---

*USSR (Russia) is included under Europe.

Asia, Southeast

Indonesia: Dahi, dadih
Philippines: Yoghurt

Asia, Southwest

Afghanistan: Chackka; Dough; Kurut
Cyprus: Trahana, Greek; Yoghurt
Egypt, see Africa
Iran: Dough; Kashk; Mast; Yoghurt
Iraq: Leben, Laban; Kishk; Yoghurt
Israel: Gil; Israeli fermented fresh milk products; Kefir; Kosher milk products; Leben (Zivdah); Lebenié; Lebenit; Revion; Shamenet; Yoghurt; Zabad (see Zabadaỳ)
Jordan: Laban, Leban; Labneh
Lebanon: Jub-Jub; Kishk; Laban; Labneh; Labneh anbaris
Saudi Arabia: homemade Laban (from goat's, sheep's, or a mixture of the two milks) and Madeer (oggt); industrially made Laban (plain liquid yoghurt), Laban munakkah (flavored set yoghurt), Labneh (concentrated plain yoghurt), and Zabadi, Zabaday (plain set yoghurt)
Syria: homemade Airan, Buttermilk, Chanklich, Karicha, and Laban; industrially made Laban and Labneh
Turkey: Ayran; Kefir; Kumiss; Kurut; Peskütan; Trahana; Yoghurt types = Kis yoghurt, Konserve yoghurt, Silivri yoghurt, Torba yoghurt, Tulum yoghurt, Vakum yoghurt, Yoghurt

Australia and New Zealand

Australia: Acidophilus yoghurt; Bifidus products; Cultured buttermilk; Sour cream; Yoghurt and Frozen yoghurt
New Zealand: Buttermilk; Cultured buttermilk; Kefir; Yoghurt

Europe

Albania: Kos
Austria: Acidophilus milk; Fru-Fru; Kefir; Sour cream; Yoghurt
Balkans: Balkan fermented fresh milk products
Belgium: B*A®; Bioghurt®; Buttermilk; Kefir; Yoghurt
Bulgaria: Bulgaricus milk; Brano mljako; Kiselo mljako; Yoghurt (plain and fruit)

Czechoslovakia: Acidophilus products; Baby milk; Bifidus products; Biokys; Biolactis®; Femilactis; Kefir; Smetanka; Sostej; Sour milk; Yoghurt
Denmark: A-37 (Acidophilus milk); AB-fermented milks; Cultura®; Cultura® drinks; Cultured buttermilk; Cultured cream; Junket (Tykmaelk); Ylette (plain); Ymer; Yoghurt (plain, natural)
England: Acidophilus milk; Bifidus products; Buttermilk; Cultured buttermilk; Cultured cream; Kefir; Yoghurt
Finland: Acidophilus piimä; Harmaa; Jamakka; Kesävelli; Kirnupiimä (buttermilk); Kokkeli; Piimä; Pitkäpimä; Taikkuna; Viili; Viilipiimä
France: A*B products; Bifidus products; Bio; Biolait (bifidus beverage); Caudiaux; Freelin; Gweden; Ofilus; Yoghurt (plain, fruit); Zen
Germany: Bifidus products; Bifighurt; Biogarde; Bioghurt; Buttermilk; Cultured buttermilk; Cultured cream; Dickmilch (sour milk); Kefir; Klotzmilch; Kosana; Kumiss; Sour cream; Yoghurt
Greece: Sakoulas (domestic strained yoghurt); Trahana, Greek; Tzatziki; Yoghurt = Cows's, ewe's, goat's with creamy layer on the top and strained (whey-drained)
Hungary: Aludtthej; Hungarian fermented fresh milk products; Iro; Kefir; Sostej; sour cream; Tarho; Teijföl; Yoghurt
Iceland: Sheep's milk products; Skyr; Yoghurt
Ireland: Clabber; Cultured buttermilk; Yoghurt
Italy: Bifidus products; Gioddu; Miciuratu; Sour cream; Yoghurt
Netherlands: Bifidus products; Biogarde products; Buttermilk, natural; Crème fraîche; Cultured buttermilk; Sour cream; Yoghurt (very high yoghurt consumption)
Norway: Buttermilk; Cultured milk (Kulturmelk); Cultured skim milk; Kefir; Sour cream and low fat cream; Taette, *see* **Ropy milk;** Yoghurt
Poland: Acidophilus products; Felisowka; Kefir; Kumisslike products; Kyas from whey, Polish; Lactorol; Milk "champagne"; Polkrem; Whey "champagne"; Yoghurt
Rumania: Buttermilk; Fermented milks with long keeping quality; Numerous local products; Oxygala (ancient product); Yoghurt
Spain: Acidophilus yoghurt; Bifidus products; Sour cream; Yoghurt (cow's, ewe's, goat's milk)
Sweden: Acidophilus milk (low-fat); Arla acidophilus milk; Bifidus milk; Buttermilk (Sur kärnmjölk); Crème fraîche; Filbunke (soured whole milk); Filmjölk (Cultured buttermilk); Grädfil (Cultured cream); Kefir; Laktofil; Langfil (Swedish ropy milk); Lättfil (Low-fat buttermilk); Yoghurt
Switzerland: Bifidus milk products; Buttermilk; Cultured cream; Kefir; Mesophilic fermented milk; Viking (Nordic acidified milk); Yoghurt

(high consumption level) beverage, cream (dessert product), flan, flavored, frozen, fruit, and plain

Yugoslavia: Basa; Brano milk; Grushevina; Kaymak (Skorup); Kefir; Kisela varenika; Kiselo mleko (Set yoghurt); Local products, numerous; Samakisselis; Sour cream; Yoghurt, set type; Yugoslavian fermented milk products; Zimne

USSR: Acidophilus milk products; Bifidus milk products; Chakka, Russian; Chal; Huslanka; Jushnaya; Katyk; Kefir; Kumiss; Local products, numerous; Matsoni; Metchnikov Prostokvasha (Yoghurt); Prostokvasha; Rjazhenka; Shubat; Syuzma; Varenets; Zhentitsa

Appendix C
Products by Milk Types, Cow's Milk Excepted

This appendix is intended as a compilation of known products by specific category. Items listed do not necessarily appear in the text.

Asses' Milk

Asses' milk, fermented; Kumiss

Buffalo Milk

Beverages: Katyk (kumiss type); Laban zeer (buttermilk type); Rob (kefir type)
Buffalo milk fermented fresh products
Milks: Dahi; Kiselo mljako; Leben; Lo; Matzoon; Prostokvasha; Silivri yoghurt; Snezhanka; Yoghurt; Zabaday
Milks, churned: Laban zeer
Mixtures: Lo (of milk types); Miltone (with peanut-milk); Toned milk (with skim milk)
Sweets: Shrikhand
Various products: Khoa; Paneer
Whey-drained, concentrated milks: Chanklich (dried); Sostej

Camel's Milk

Camel's milk fermented products
Churned milk, dried: Oggt, cooked
Cream: Aragan
Milk based beverages with lactic and alcoholic fermentations: Airan; Boza; Chal, cal; Irek-mai; Kefir; Khoormog; Kumiss; Kurunga; Shubat; Tujo

Donkey Milk

See Asses' milk.

Ewe's Milk

Brandy: Arak; Arsa
Beverages, milk based, lactic and alcoholic fermentations: Airan; Gioddu; Khoormog; Kojurtnak; Kvas from milk (only lactic fermentation); Skuta
Beverages, whey-based, lactic and alcoholic fermentations: Urda; Žinčica
Buttermilks from churned milks: Laban zeer; Madeer (heat-treated)
Cream: Kajmak
Dried milks: Bulgaricus milk; Chanklich (whey-drained); Gibneh-labneh (whey-drained); Klila; Kurut
Ewe's milk fermented products
Milks, concentrated: Preconcentrated = Huslanka and Lo; Whey-drained = Basa, Chanklich (sun-dried), Gibneh-Labneh (dried), Kis yoghurt (heat-treated), Savanyutej, Sostej
Milks, set or stirred: Ewe's milk yoghurt with a creamy skin layer on top; Kisela varenika; Kiselo mleko; Kiselo mleko, slano; Kiselo mljako; Kos; Leben; Mast; Matzoon; Silivri yoghurt; Snezhanka; Tan; Tarho; Urgutnik; Yoghurt; Zhentitsa
Various products: Kihslo
Whey-based products: Azi; *see also* Beverages

Filled Milk

Filled milk fermented fresh milk products; Yoghurt (Tchad)

Goat's Milk

Beverages, milk based, lactic fermentation
Beverages, goat milk liquid yoghurt, lactic and alcoholic fermentations: Airan; Gioddu; Kefir; Kojurtnak; Kumiss; Rob; Skuta
Brandy: Arak
Buttermilks, from churned milks: Laban zeer (concentrated); Madeer (heat-treated, sun-dried)
Concentrated products: Kis yoghurt (whey-drained or heat-treated); Laban zeer; Oggt, cooked (dried buttermilk)
Dried products: Bulgaricus milk; Gibneh-Labaneh (dried, whey-drained); Kurut; Madeer
Goat milk products
Milks, set type or stirred: Goat milk firm yoghurt; Goat milk, mesophilic

fermented milks; Laban munakkah; Leben; Mast; Matzoon; Nono; Silivri yoghurt; Tan; Tarag; Yoghurt; Zabaday
Whey-based products: Azi; Beverages

Human Milk

Human milk, fermented fresh products; Sour milk beverages

Jhopa Milk

Dahi; Jhopa milk fermented products

Llama Milk

Llama milk fermented fresh products; Undefined products

Mare's Milk

Beverages, milk based, lactic and alcoholic fermentations: Airag; Kumiss; Tschigan
Beverages Brandy based: Köörik (diluted with milk for infants)
Brandy: Airag; Airak; Arsa; Karakosmos; Köörik (diluted with milk, beverage)
Mare's milk fermented milk products
Milk formula products for babies, beverages: Bifiline; Femilact; Lactana B; Malyutka; Pelargon
Milks: Bässrik (mixed with cow's milk); Lo (viscous)

Recombined Milk

Acidophilus milk; Cultured buttermilk; Labneh; Recombined milk, fermented; Yoghurt

Reconstituted Milk

Acidophilus milk; Reconstituted milk, fermented; Yoghurt beverage; Yoghurtlike products

Reindeer Milk

Kumisslike products; Reindeer milk-based fermented fresh milk products; Ropy fermented milk

Sheep Milk

See Ewe's milk

Toned Milk

Toned milk, fermented fresh milk products; Not defined products in India

Yak Milk

Beverages, milk-based: Khoormog (lactic and alcoholic fermentation); Udan
Milks: Lo (viscous); Tarag; Zho
Yak milk products, fermented

Zebu Cattle Milk

Various products; Zebu cattle milk-based fermented milk products

Appendix D
Products by Starter Culture Microorganisms

This appendix is intended as a compilation of known products by specific category. Items listed do not necessarily appear in the text. Note that (P) indicates pharmaceutical preparations.

Starter Culture Microorganisms

Acetic Acid Cultures

Acetobacter aceti (oxidative fermentation of ethyl alcohol) Azi; Dnepryanskii; Kefir; Kumiss; Smetanka; Vinegar; Whey vinegar

Acidophilus (Lactobacillus Acidophilus) Cultures, Pure or Combined With Other Culture Microorganisms (Types 1-12)

Lb. acidophilus, pure (Type 1): Acidophilus baby foods—proteolytic, antibiotic properties; Acidophilus cream; Acidophilus drinks; Acidophilus ice cream; Acidophilus milk—antibiotic, survival gastrointestinal tract, polysaccharide-producting strains; Acidophilus milk, dietetic; Acidophilus natural buttermilk (Czechoslovakia), supplemented polysaccharide-producing strains; Acidophilus albumin paste; Acidophilus paste; Arla acidophilus milk; Biolakt; Enpac (P)—antibiotic-resistant, against side effects of antibiotics; Gefilus® (P)—survival in gastrointestinal tract, adhesion ability, against intestinal disorders; Laccillia (P)—against intestinal disorders; Legume-containing fermented fresh milk products; Megadophilus (P)—against gastrointestinal disorders; Moskowski, supplemented polysaccharide-producing strains; Novaya; Raibi; Ribolac® (P)—resistant antibiotics and sulfa drugs, against intestinal disorders; Sweet acidophilus milk

Lb. acidophilus, bifidobacteria (Type 2): Biomild; Cultura®—human intestinal strains, concentrated cultures, *B. bifidum* and *B. longum;*

Cultura® drink—human intestinal strains, concentrated cultures, *B. bifidum* and *B. longum;* Diphilus milk—*Lb. acidophilus, B. bifidum,* antibacterial properties; Mil-Mil—*Lb. acidophilus, B. bifidum, B. breve*

Lb. acidophilus, Streptococcus thermophilus (Type 3): Bioghurt®—*Lb. acidophilus,* human intestinal strains; Infloran Berna (P)—against intestinal disorders; Parag, whey beverage

Lb. acidophilus, bifidobacteria *Str. thermophilus* (Type 4): Biogarde®-fermented milk—*Lb. acidophilus, B. bifidum,* human intestinal strains; Biogarde® ice cream—*Lb. acidophilus, B. bifidum,* human intestinal strains; Mild (low-acid) cultured fresh milk products; Ofilus

Lb. acidophilus, bifidobacteria *Pediococcus acidilactici* (Type 5): Bifidus baby foods (Femilact); Czechoslovakia, *B. bifidum;* Biokys, *B. bifidum*

Lb. acidophilus, yoghurt cultures (Type 6a): Acidophilus yoghurt; Aco-yoghurt)—*Lb. acidophilus* intestinal strains, antagonistic against pathogens

Lb. acidophilus, bifidobacteria, yoghurt cultures (Type 6b): Acidophilus bifidus yoghurt—*Lb. acidophilus, B. bifidum,* human intestinal strains; Lünebest

Lb. acidophilus, mesophilic lactic acid cultures (Type 7a): A-fil milk; Acidophilin—*Lb. acidophilus, Str. lactis* subsp. *lactis,* kefir culture, antibacterial activity of *Lb. acidophilus;* Acidophilus cream cultured, USSR; Acidophilus cultured buttermilk—*Lb. acidophilus,* human intestinal strains; Acidophilus natural buttermilk, USSR; *Lb. acidophilus,* polysaccharide-producing strains

Lb. acidophilus, bifidobacteria, mesophilic lactic acid cultures (Type 7b): Ofilus double douceur, *see* **Ofilus;** Progurt; Slavyanka, LD-type

Lb. acidophilus, Lb. delbrueckii subsp. *bulgaricus* (Type 8): Lactinex (P)—against oral infections

Lb. acidophilus, Lb. casei (Type 9): Lactobacillus milk for biotherapy of infantile diarrhea

Lb. acidophilus, propionibacteria (Type 10): Vitalakt, formula feeding babies

Lb. acidophilus, Lb. delbrueckii subsp. *bulgaricus, Saccharomyces lactis* (Type 11): Kumiss from cow's milk

Lb. acidophilus, yeast culture (Type 12): Acidophilus yeast beverage, lactose-fermenting yeasts; Acidophilus yeast milk—*Lb. acidophilus* and lactose-fermenting yeasts, antagonistic to intestinal pathogens

Animal Feed Cultures

Lactic acid bacteria and bifidobacteria: Animal feeds and bacterial concentrates—*Lb. acidophilus* or *Lb. delbrueckii* subsp. *bulgaricus,* or *B. thermophilum,* animal strains antagonistic to pathogens; Silage preparations—*Lb. plantarum, Pediococcus acidilactici*

Bifidobacteria Cultures, Pure or Combined with Other Culture Microorganisms (Types 1-9)

Bifidobacteria, pure (Type 1): Bifider® (P)—antibacterial activity; Bifidus baby foods = Bifiline and Lactana B; Bifidus milk—*B. bifidum* or *B. longum,* human intestinal strains, against imbalance of the gut flora and intestinal disorders; Bifidus milk with yoghurt flavor, *B. bifidum* or *B. longum;* Eugalan Töpfer Forte (P)—management of liver cirrhosis; Euga-lein Töpfer (P)—against chronic constipation; Lactopriv (P)—against side effects of antibiotic therapy; Life Start Original (P)—*B. infantis,* against intestinal disorders; Life Start Two (P)—*B. bifidum,* against intestinal disorders; Liobif (P)—*B. bifidum,* against enteric infections of babies and children; Lyobifidus (P)—against diarrhea; Sweet bifidus milk

Bifidobacteria *Lb. acidophilus* (Type 2): see *Lb. acidophilus* culture, Type 2

Bifidobacteria, *Str. thermophilus* (Type 3): Bifighurt®—*B. longum,* slime-forming variant

Bifidobacteria, *Lb. acidophilus,* Str. thermophilus (Type 4): see *Lb. acidophilus* culture, Type 4

Bifidobacteria, *Lb. acidophilus, Pediococcus acidilactici* (Type 5): see *Lb. acidophilus* culture, Type 5

Bifidobacteria, yoghurt cultures (Type 6a): Bifidus yoghurt, *B. bifidum* or *B. longum*

Bifidobacteria, *Lb. acidophilus,* yoghurt culture (Type 6b): see *Lb. acidophilus* culture, Type 6b

Bifidobacteria, mesophilic lactic acid cultures (Type 7a)

Bifidobacteria, *Lb. acidophilus,* mesophilic lactic acid cultures (Type 7b): see *Lb. acidophilus* culture, Type 7b

Bifidobacteria, *lactobacilli* (Type 8): Bifilakt

Bifidobacteria, *Lb. delbrueckii* subsp. *bulgaricus,* kefir grains (Type 9): Tarag

Enterococcus Cultures

Enterococcus, pure: Bioflorine®(P)—*Enterococcus faecium* SF 68; Paraghurt (P)

Mixed: Enterococci-containing fermented fresh milk products = products *E. faecalis, E. faecium;* Jalebi = *E. faecalis;* Labneh

Kefir and Kefirlike Cultures

Kefir grains: Kefir (see original composition); Tarag

Kefir cultures (originating from kefir milk): Acidophilin, mixed culture; Baby kefir; Elvit, mixed culture

Kefirlike cultures (originating from kefir): Kefir cultured milk; Kefirlike products

Kumiss and Kumisslike Cultures

Kumiss cultures, original: Kumiss
Kumisslike cultures: Kumiss from cow's milk

Lactobacilli, Except Lb. Acidophilus

Lb. brevis: Laban zeer, mixed culture
Lb. buchneri: Jalebi, mixed culture
Lb. casei: Aerin, mixed culture; Biolactis®, mixed culture (P); Laban rayeb, mixed culture; Laban zeer, mixed culture; Lactobacillus milk for biotherapy of infantile diarrhea, mixed culture; Yakult, pure; Žinčica, mixed culture
Lb. delbrueckii subsp. *bulgaricus:* Animal feeds and bacterial concentrates used in mixed culture; Azi used in mixed culture; Biolactin, pure culture (P); Bulgarian rod concentrate, pure (P); Bulgaricum LB-51, pure (P); Bulgaricum cultured buttermilk, mixed culture; Bulgaricum tablets, pure (P); Bulgaricus milk, pure; Huslanka, mixed culture, Katyk, mixed culture; Kheran, mixed culture; Kolomenski, mixed culture; Korot, mixed culture; Kumiss, mixed culture; Kumiss from cow's milk, mixed culture; Kurt, freeze-dried, mixed culture; Kurt, mixed culture; Kurunga, from pasteurized milk, mixed culture; Lactinex, mixed culture (P); Magou, pure; Multicurdled fermented milk products, mixed; Prohlada, mixed culture; Reformed yoghurt, mixed cultured; Tarag, mixed culture; Tear grass-milk fermented product, pure; Vita, pure
Lb. fermentum: Jalebi, mixed culture
Lb. helveticus: Azi, mixed culture; Katyk, mixed culture; Kurt, freeze-dried, mixed culture
Lb. lactis: Žinčica, mixed culture
Nonspecified species: Bifilakt, mixed cultured
Lb. plantarum: Laban zeer, mixed culture; Silage preparations, mixed culture; Žinčica, mixed culture

Mesophilic Lactic Acid Cultures, by Types D, L, LD, and O

Str. lactis subsp. *lactis, Str. lactis* subsp. *cremoris, Str. lactis,* subsp. *diacetilactis* (Type D): Carbonated dahi; Cream cultured fermented milk, low acetaldehyde-producing strains of *Str. lactis* subsp. *diacetilactis;* Crème fraîche, fermented; Crowdies; Lactorol; Prostokvasha
Str. lactis subsp. *lactis, Str. lactis* subsp. *cremoris, Leuconostoc mesenteroides*

subsp. *cremoris* (Type L): Cultured buttermilk, low-acetaldehyde-producing strains of *Str. lactis* subsp. *diacetilactis;* Cultured cream, sour cream—low acetaldehyde-producing strains of *Str. lactis* subsp. *diacetilactis;* Dahi, mild acid; Dnepryanski, polysaccharide-producing strains; Fru-fru; Ropy milk, polysaccharide-producing strains; Saya®; Schemaitschju; Zhemajcju

Str. lactis subsp. *lactis, Str. lactis* subsp. *cremoris, Str. lactis* subsp. *diacetilactis, Leuconostoc mesenteroides* subsp. *cremoris* (Type LD): Cultured buttermilk; Cultured cream, sour cream; Dahi, mild acid; Filmjölk; Fru-fru; Goat milk mesophilic fermented milk; Junket; Lactofil; Lactorol; Lättfil; Mala; Schemaitschju; Ylette; Ymer; Zhemajcju

Streptococcus lactis subsp. *lactis, Str. lactis* subsp. *cremoris* (Type O): Mesophilic fermented fresh milk products

Mesophilic Lactic acid Cultures, by Various Compositions

Culture type L and *Lb. acidophilus:* Acidophilus (cultured) buttermilk

Culture type L, *Lb. acidophilus* or *Lb. casei* and *Acetobacter lactis:* Smetanka, cultured cream

Culture type L, *Leuconostoc mesenteroides* subsp. *dextranicum* and *Lb. casei:* Aerin

Culture type L, yoghurt culture, *Lb. acidophilus* and *B. bifidum:* Multicurdled fermented milk products

Culture type L or D or LD, propionibacteria and/or kefir culture or *Lb. acidophilus:* Elvit

Culture type L or LD and baker's yeast: Felisowka; Kefir-cultured milk

Culture type LD and *Lb. acidophilus:* A-fil milk; Acidophilus (cultured) buttermilk; Slavyanka

Culture type LD and *Geotrichum candidum:* Viili, capsula-forming strains of *Str. lactis* subsp. *cremoris/lactis*

Str. lactis subsp. *diacetilactis* and *Lb. acidophilus:* Acidophilus cream, cultured; Acidophilus (natural) buttermilk

Str. lactis subsp. *diacetilactis* and *Lb. delbrueckii* subsp. *bulgaricus:* Kolomenski

Str. lactis subsp. *diacetilactis, Str. lactis* subsp. *lactis* and *Str. thermophilus:* Lyubitelski

Str. lactis subsp. *lactis:* Baby foods, fermented; Chakka, Indian; Colostrum milk, fermented; Demeter (fermented fresh milk) products; Eledon; Pelargon; Shrikand; Streptococcus lactis milk; Zivda

Str. lactis subsp. *lactis, Lb. acidophilus* and *kefir culture:* Acidophilin

Str. lactis var. longi (ropy strain) and *Leuconostoc mesenteroides* subsp. *cremoris:* Long milk

Mold Cultures

Mixed: Gua-nai—*Amylomyces rouxii, Rhizopus oryzea, Aspergillus oryzea;* Viili, *Geotrichum candidum*

Pure: Cultured mold milk, *Saccharomycopsis* spp. and *Rhizopus* spp.; Cultured mold-containing fermented milks—*Aspergillus oryzea, Amylomyces rouxii, Geotrichum candidum, Pleurotus ostreatus, Rhizopus oryzea, Rhizopus javanicus*

Pediococci Cultures

Pediococcus acidilactici: Bifidus baby foods, mixed culture; Biokys, mixed culture; Silage preparations

Propionibacteria Cultures

Mixed: Animal feeds and bacterial concentrates; Elvit, *Propionibacterium freudenreichii* subsp. *shermanii;* Vitalakt, species not mentioned

Thermophilic Lactic Streptococci

Mixed: Bifighurt®; Biogarde® fermented milk; Biogarde® ice cream; Bioghurt; Chakka, Russian; Chal; Hydrolyzed-lactose fermented fresh milk products; Katyk; Lyubitelskii; Multicurdled fermented milk products; Ofilus; Parag; Zdorove

Pure: Cream turo—slime-producing, heat-resistant strains; Kheran, traditional product; Kheran®, nontraditional product; Lapte-akru; Russkiǐ; Streptococcus thermophilus fermented milk

Yeast Cultures

Mixed: Acidophilus yeast beverage, lactose-fermenting yeasts; Acidophilus yeast milk, lactose-fermenting yeasts; Busa—*Saccharomyces busa* asiaticae (species not mentioned in Manual of Yeast), mixed culture; Cellarmilk—*Torula* yeast, mixed culture; Felisowka, baker's yeast; Gua-nai—*Endomycopsis burtonic,* mixed culture; Huslanka—lactose-fermenting yeasts, mixed culture; Hydrolyzed-lactose fermented fresh milk products—*Saccharomyces fragilis,* mixed culture; Jalebi, *Saccharomyces* spp.; Katyk—*Torula* spp., mixed culture; Kjälder mjölk, *Torula* spp.; Kuban—*Mycoderma, Torula lactis,* other undefined species; Kumiss from cow's milk—*Saccharomyces lactis,* mixed culture; Kurunga—*Candida* spp., *Torula* spp.; Kvas from whey, Polish; Kvas, new milk beverage, baker's yeast; Laban rayeb, *Kluyveromyces marxianus* subsp. *marxianus;*Leben—*Kluyveromyces marxianus* subsp. *marxianus, Saccharomyces cerevisiae,* mixed culture; Prohlada, lactose-fermenting yeasts; Vinegar from whey—*Kluyveromyces marxianus* subsp. *marxianus, Saccharomyces cerevisiae, Candida kefir;* Whey vinegar, yeasts not specified

Pure: Galazyme, champagne or baker's yeast; Kvas from whey, baker's yeast; Salomat—*Saccharomyces lactis, Saccharomyces carlbergensis;* Whey beer, brewer's yeast or *Kluyveromyces marxianus* subsp. *marxianus;* Whey "champagne," baker's yeast

Yoghurt Cultures

Mixed: Acidophilus bifidus yoghurt; Acidophilus yoghurt; Aco-yoghurt; B*A®; Bifidus yoghurt; Biolakton: Dahi; Lünebest; Mild (low-acid) cultured fresh milk products

Pure: Airan, Bulgarian; Arera; Bio-yoghurt; Brano milk; Chanklich; Demeter (fermented fresh milk) products; Dessert products; Diabetic yoghurt; Dietetic fermented milks; Dried yoghurt; Egg milk product, cultured; Ewe's milk fermented products; Ewe's milk yoghurt with a creamy skin layer on top; Frozen yoghurt; Gioddu; Goat milk firm yoghurt; Goat milk liquid yoghurt; Honey-yoghurt; Kashk; Kiselo mleko; Kiselo mleko, slano; Kiselo mljako; Kurut, Turkish; Laben; Labneh; lactosat; Legume-containing fermented fresh milk products; Metchnikov prostokvasha; Molodost; Niyoghurt; Rjazhenka; Skyr; Snezhok; Soft-serve ice cream from yoghurt; Syuzma; Tamaroggtt; Torba yoghurt; Varenets; Vitana; Yoghurt; Yuzhnyi; Zabady; Zimne

Special Strain Culture Microorganisms

Acetaldehyde Producing, Low

Str. lactis subsp. *diacetylactis:* Cultured buttermilk; Cultured cream; Cream culture fermented products

Antibacterial Activity

Bif. bifidum: Diphilus milk

Lb. acidophilus: Acidophilin; Aco-yoghurt; Biolakt; Gefilus®; Lactobacillus milk for biotherapy of infantile diarrhea

Lb. casei: Lactobacillus milk for biotherapy of infantile diarrhea

See also Appendix E, Intestinal pathogens

Antibiotic Resistant, Some

Lb. acidophilus: Acidophilus baby foods; Enpac (P); Ribolac®

Anticholesteremic Effect

Lb. delbrueckii subsp. *bulgaricus:* Bulgaricus milk
Streptococcus thermophilus: Streptococcus thermophilus fermented milk

Antitumor Activity

Lb. delbrueckii subsp. *bulgaricus:* Bulgaricum L.B. (P)

Bile Resistance

Yakult (Shirota strain)

Fecal Enzyme Activity, Depressing

Lb. acidophilus: Gefilus®

Gastric Acid Resistance

Lb. casei: Yakult

Gastrointestinal Transit, Ability

Adhesion ability: *Bif. bifidum*—Lyobifidus; *Lb. acidophilus*—Gefilus® (GG strain); *Lb. casei*—Yakult (Shirota strain); *Lb. delbrueckii* subsp. *bulgaricus*—Bulgarian rod concentrate

Heat-Resistant Strains

Str. thermophilus: Cream turo; Yoghurt strains, Dried yoghurt

Human Strains, Declared as Such

Bifidobacteria: *B. bifidum* = Acidophilus bifidus yoghurt, Bifidus milk, Biogarde®-fermented milk, Biokys, Cultura®, Diphilus milk; *B. longum* = B*A®

Lactobacillus: *Lb. acidophilus* = Acidophilus baby foods, Acidophilus bifidus yoghurt, Acidophilus milk, Acidophilus yoghurt, Biogarde®-fermented milk, Bioghurt®, Biokys, Biolakt, Cultura®, Diphilus milk, Enpac (P), Malysh, Malyutka; *Lb. casei* = Aerin, Yakult (Strain Shirota); *Str. thermophilus* = Kheran

Intestinal Bacteria, Selected Products

See Appendix F

Phenol Resistance

Lb. acidophilus: Acidophilus baby foods

Proteolytic Strains

Lb. acidophilus: Acidophilus baby foods

Ropy Strains

Bifidobacteria: *B. longum* = Bifighurt®
Lactobacillus: *Lb. acidophilus* = Acidophilus milk, Acidophilus (natural) buttermilk, Moskowski; *Lb. delbrueckii* subsp. *bulgaricus* = Bulgaricus milk
Leuconostoc spp.: Ropy milk
Streptococcus: *Str. lactis* subsp. *cremoris* = Ropy milk, Viili; *Str. lactis* var. *longi* = Long milk; *Str. thermophilus* = Cream turo
See also Appendix F

Slow Acid Variant

Lb. delbrueckii subsp. *bulgaricum:* Bulgaricum milk; Kheran

Transit, Survival

Lb. acidophilus: Acidophilus milk; Gefilus® (GG strain)

Vitamin-Producing Strains, Utilization

Vitamin B group: Propionic acid bacteria fermented milk products (vit. B_{12}); Yeasts = Acidophilus yeast milk, Kwas

Yeast, Lactose-Fermenting

Acidophilus yeast beverage; Acidophilus yeast milk; Hydrolyzed-lactose fermented fresh milk products

Original Microflora in Traditional Products

Airan: not the same everywhere; Bulgaria = yoghurtlike bacteria; Russia = lactic acid bacteria and yeasts; Turkey-yoghurtlike bacteria and two yeast species.
Buttermilk: culture types = O and/or L, D or LD; various combinations and proportions.

Clabber: composition of microflora not well specified; spontaneously soured mainly with mesophilic lactic acid bacteria (streptococci and lactobacilli).

Dahi: mesophilic lactic streptococci with or without *Lb. delbrueckii* subsp. *bulgaricus* or *Str. thermophilus* or both.

Kefir grains: mesophilic lactic streptococci, leuconostocs, lactobacilli, yeasts and acetic acid bacteria; *Str. lactis* subsp. *lactis*/subsp. *cremoris, Leuconostoc* spp., *Lb. kefir, Lb. casei,* sometimes *Lb. acidophilus, Candida kefir, Kluyveromyces marxianus* subsp. *marxianus, Acetobacter aceti.*

Kheran: *Str. thermophilus* is an important component; microflora insufficiently investigated.

Kumiss: *Lb. delbrueckii* subsp. *bulgaricus, Saccharomyces lactis* (lactose-fermenting yeast), *Saccharomyces cartilaginosus* (lactose-nonfermenting yeast), *Mycoderma* (carbohydrate nonfermenting).

Kurunga: *Str. lactis* subsp. *lactis*/subsp. *cremoris, Str. lactis* subsp. *diacetilactis, Str. thermophilus, Lb. delbrueckii* subsp. *bulgaricus, Lb. acidophilus, Candida* spp., *Torulopsis* spp.

Laban, leben: predominantly yoghurt bacteria and small proportions of mesophilic lactic streptococci, mesophilic and thermophilic lactobacilli and yeasts; *Str. lactis* subsp. *lactis, Lb. casei, Lb. plantarum, Lb. lactis, Lb. helveticus.*

Yoghurt: predominantly *Str. thermophilus* and *Lb. delbrueckii* subsp. *bulgaricus;* possible presence of *Lb. lactis, Lb. helveticus,* sometimes *Lb. casei, Lb. plantarum, Str. lactis* subsp. *lactis,* yeasts and molds, (most often *Geotrichum candidum*).

Appendix E
Products by Food Value and Health Claims

This appendix is intended as a compilation of known products by specific category. Items listed do not necessarily appear in the text. Note that (P) indicates pharmaceutical preparations.

Animal, Probiotic Effect (Young Animals)

Alternative to antibiotics: Korolac; Lactosat
Animal stress: Animal feeds and bacterial concentrate (unsanitary conditions)
Growth stimulating: Whey fermented for animal feeds
Isocaloric and isonitrous diet: Whey protein concentrate, fermented
Pathogens, suppressing: Whey fermented for animals
Reestablishment and maintenance of normal balance of bacteria in digestive tract: Silage preparations (calves, pigs, other farm animals)

Diarrhea, Aids in Therapy

Adults: Acidophilus milk; Bifider®; Bifidogène® (P); Bifidus milk; Bioflorine® (P); Bulgaricus milk; Bulgaricum tablets; Diphilus milk; Enpac; Infloran Berna (P); Kefir; Kheran; Kumiss; Laccillia (P); Lactobacilli-containing fermented fresh milk; Life Start Original (P); Life Start Two (P); Lyobifidus (P); Megadophilus (P); Omniflora (P); Ribolac® (P)
Babies: Acidified baby milk formula; Baby foods, fermented; Bifidus baby foods; Bulgaricus milk; Quarg milk
Children: Acidophilus milk formula (artificial); Bifilakt; Bulgaricum tablets; Bulgaricus milk; Gefilus®; Lactobacillus milk; Lactobacillus milk for biotherapy of infantile diarrhea; Malyutka (acute disease)
Chronic diarrhea: Infloran Berna (P)
Travelers' diarrhea: Gefilus®

Dietetic Foods

Demineralized: Demineralized long-life beverages; Low-(content) fermented milk products; Low-sodium yoghurt
Diabetic: Diabetic yoghurt; Fermented milk-vegetable mixtures
Dietary fiber, enriched: Bircher-Muesli; Fermented milk-vegetable mixtures; Plant admixtures, *see* Appendix F
Dietetic, requirements: Dietetic fermented milk
Fortified calcium: Bifidus yoghurt; Tar; Yoghurt
Fortified fat: Zen
Fortified protein: Acidophilus-albumin paste
Fortified vitamin: *see* Appendix F
Hydrolyzed, lactose: Hydrolyzed-lactose fermented fresh milk products; Multicurdled fermented milk products; Whey beverages with less than 1% alcohol
Low-calorie and low-carbohydate: Low-(content) fermented milk products
Low-fat: Acidophilus paste; Low-(content) fermented milk products
Low-sodium: Low-sodium yoghurt
Older persons, diet: Gerolakt
Weight loss and control: Weight watchers yoghurt

Digestion

Digestibility, improved: Acidophilus milk; Buttermilk, natural; Dessert products; Goat milk liquid yoghurt; Rabadi; Tibetan fermented fresh milk products; Yoghurt; Yoghurt as ingredients in other foods
Digestive system: fine casein flocculation and reduction of gastric pH = Acidified baby milk formula; peristaltic function, improved = Bulgaricum L.B. (P); sensitivity to vinegar = Whey vinegar
Predigestion of milk components: lactose digestion, (low lactose content, β-galactosidase enriched) = Hydrolyzed-lactose fermented fresh milk products, Multicurdled fermented milk products, Rabadi, Recombined milk, fermented (yoghurt-based), Reconstituted milk, fermented (yoghurt-based), Yoghurt
Proteins: Rabadi
Resorption, minerals: elderly people, improved gastric acidity, low buffering capacity = Bulgaricus milk (acid), Dahi (acid), Yoghurt (acid); improved = Rabadi, Whey cultured products (sugar, flavors)
Putrefactive digestion, elimination of unpleasant odor in feces: Lacto-Bacilline (P); Sweet acidophilus milk (for typical meat eaters); T.M.-milk

Diseases, Prophylaxis, Aids in Therapy

Allergic reactions: Fermented milk-vegetable mixtures
Anticholesterolemic effect: Acidophilus milk; Bifidus yoghurt; Bulgaricus milk (for babies); Cultured mold-containing fermented milks; Streptococcus thermophilus fermented milk
Antitumor properties: Bulgaricum L.B. (P) (glycopeptides in cell wall); Cultured mold-containing fermented milks
Atrophic infants: Acidified baby milk formula
Cardiovascular diseases: Low-sodium yoghurt
Constipation: *see* Stress factors
Cosmetic purposes: Molkosan (skin)
Diarrhea: *see* Diarrhea, aids in therapy
Gastrointestinal disorders: *see* Diarrhea, aids in therapy
Growth of babies and children with subclinical infection with growth-depressing microorganisms (improved): Acidified baby milk formula (artificial); Acidophilus-albumin paste; Acidophilus milk; Acidophilus yoghurt; Aerin; Bifidus milk; Lactulose-enriched fermented fresh milk products
High blood pressure: Low-sodium yoghurt (P)
Immune system, stimulation, nonspecific immunity: Bulgaricum L.B. (P); Bulgaricus milk; Lactobacillus milk for biotherapy of infantile diarrhea; Ropy milk; Viili
Kidney diseases: Low-sodium yoghurt
Leucocytosis: Bulgaricus milk
Liver diseases, chronic: Bifidus milk; Eugalan Töpfer Forte; Kheran; Life Start Two (P); Tibetan fermented fresh milk products
Oral infections: Lactinex (P); Lactobacilli-containing fermented fresh milks (P); Lactopriv
Prolongation of human life (aids in control of intestinal putrefaction): Yoghurt; Zho
Respiratory diseases (preschool children): Elvit
Skin diseases: Lactobacilli-containing fermented fresh milks (P) (burns, eczemas, lesions, purulent inflammations)
Strength, eroded by fatigue: Kheran
Therapeutic dose: Kumiss; Yoghurt

Intestinal Pathogens, Aids in Inhibition

Antimicrobial activity: Acidophilin; Acidophilus yeast milk; Aco-yoghurt (children); Aerin (children); Biolakt; Bulgarian rod concentrate; Life

Start Original (P); Ribolac® (P); Streptococcus thermophilus fermented milk
E. coli: Tarag (in vitro)
Moniliasis: Enpac
Mycobacterium tuberculosis: Airag; Kumiss; Mare's milk fermented milk products (pulmonary); Shubat (pulmonary)
Salmonella: Sweet acidophilus milk
Staphylococcus: Enpac®

Stress Factors for Intestinal Flora, Easing of Symptoms

Aging: Gerolakt
Antibiotic therapy: Bifidus baby foods; Enpac (P) (antibiotic resistance); Eugalan Töpfer Forte; Gefilus®; Infloran Berna (P); Lactopriv (P); Ribolac® (P)
Constipation, chronic: Bifidus milk; Eugalan Töpfer Forte; Euga-Lein Töpfer (gluten-free, for children); Gefilus® (P); Infloran Berna (P)
Emotional stress: Acidophilus products; Bifidus products; Yoghurt

Appendix F
Products by General Subject

This appendix is intended as a compilation of known products by specific category. Unless they appear in **bold** type, items listed do not necessarily appear in the text. Note that (P) indicates pharmaceutical preparations.

Acetic acid-containing products. *See* Appendix D.
Acidified products
  Baby milks: Acidified baby milk formula; Marriott's acidified baby milk formula
  Milks: Acidified milk products; Acilmilk
  Sweets: Chhana; Rasgolla
Acidophilus products
  Baby and infant foods. *See herein:* Baby foods, foods for small children
  Beverages: Acidophilin; Acidophilus drinks, pure; Acidophilus yeast beverage; Cultura drink
  Buttermilk: Acidophilus (cultured) buttermilk; Acidophilus (natural) buttermilk
  Cultured cream: Acidophilus cream, cultured
  Ice cream: Acidophilus ice cream
  Milks: AB-fermented milk and milk products; Acidophilus milk, dietetic; Acidophilus yeast milk; A-fil milk; Arla acidophilus milk; Bioghurt®; Biokys; Cultura®
  Paste: Acidophilus-albumin paste; Acidophilus paste
  Pharmaceutical preparations: Acidophilus Zyma; Enpac; Infloran Berna; Laccillia; Megadophilus
  Starter cultures, types. *See* Appendix D: Acidophilus cultures, types 1-12
  Sweet milks: Sweet acidophilus bifidus milk; Sweet acidophilus milk
  Yoghurt: Acidophilus bifidus yoghurt; Acidophilus yoghurt; Aco-yoghurt
Admixtures. *See herein:* Products by food types and special preparations
Alcohols. *See* **Milk brandy**

Alternative manufactured fermented milk products: Demeter (fermented fresh milk) products

Ancient products (in chronological order)

Before Christ (B.C.): Burghul (Near East); Buttermilk, natural or traditional (India); Dahi (300 B.C. to 75 A.D., India); Kumiss (ancient Scythia); Lac concretum (ancient Scythia); Opus lactarum (ancient Roman Empire); Oxygala (ancient Roman Empire); Schiston (ancient Roman Empire); Shrikhand (800 to 300 B.C., ancient India); Smy (ancient Egypt);

after Christ (0 to A.D. 1000), first mention: Acor jucundes. *See* **Caudiaux;** Dickmilch (German peoples. *See* **Clabber;** Yoghurt (Persian Emperor)

period A.D. 1000 to 1500, first mention: Airan; Balkan and fermented milk products; Clabber (Ireland); Karakosmos (Mongolia); Kefir (Caucasus region); Kheran (Sibera, Tartar people); Tar (Yakut region); Tibetan fermented fresh milk products; Yoghurt, national types. *See herein:* Yoghurt, national names

Animal feeds

Animal feed supplement: Animal feeds and bacterial concentrates; Colostrum milk, fermented; Lactosat

Green fodder preservation: Animal feeds and bacterial concentrates; Silage preparations

Probiotics: Animal feed and bacterial concentrates; Animal probiotic effect. *See* Appendix E

Silage: Whey cultured products; Whey fermented for animal feeds; Whey fermented for silage preparation; Whey protein concentrate, fermented

Veterinary preparations. *See* Appendix E

Baby foods, foods for small children

Acidified baby milk formula

Acidophilus products: Acidophilus baby foods; Acidophilus cream, cultured; Baldyrgan; Malyutka

Aerin

Alcohol-milk beverages: Arak; Köörzik

Baby foods, fermented: half-skimmed milk formula, Eledon; whole milk formula, Pelargon

Baby kefir

Bifidus baby foods

dried formula product: Femilact; Lactana B

liquid formula products: Bifiline; Malyutka

Bifilakt

Biolakt

Biolakton

Fermentations, lactic and alcoholic: Acidophilus yeast milk; Airan; Chal; Galazyme; Gioddu; Katyk; Kefir; Kefirlike products; Koormog; Kuban; Kumiss; Kurunga; Kvas; Ma-tung; Micriuratu; Shubat; Tarassun; T. M. milk; Tschigan
Flavorings, flavors
aromas: Belguss (vanilla)
fruit juice: Acidophilus drink pure
fruit pulp: Aerin; Jovokoktejl; Shakterskii; Yoghurt
heat-treated after culturing: Frulati (buttermilk-based); Jogi-drink (yoghurt-based); Rivella (whey-based)
lactose, hydrolyzed: Whey beverages with less than 1% alcohol
salt: Dough; Korot; Liban
spices, not specified: Dough; Jub-Jub; Korot
sucrose: Galazyme; Kolomenski; Korot; Liban; Maĭskiĭ Shakterski
sweeteners, glucose or sucrose: Mil-Mil; T. M. milk; Yakult
syrups: Kolomenski; Russkiĭ; Shakterskii; Yakult
white-wine: Buttermilk champagne
Bifidobacteria-containing products
Bifidus baby foods
Cultures and products. *See* Appendix D: Bifidobacteria cultures, types 1-9
Pharmaceutical preparations: Bifidogène; Eugalan Töpfer forte; Euga-Lein Töpfer forte; Infloran Berna; Lactopriv; Life-Start Original; Lyobifidus
Biogarde (Sanofi)
Bifighurt
Biogarde
Biogarde clotted milk
Biogarde ice cream
Biogarde sour milk
Bioghurt
Bio-products
Alternative manufacture fermented fresh milk products
Bio-prefix: Biodynamic milk yoghurt; Biogarde; Bioghurt; Biokys; Biolactis; Biolakt; Biolakton; Biomax Si; Biomild
Demeter (fermented fresh milk) products
Brandy
geographical origin: Kirgisia, Chorsa; Mongolia, Karakosmos; Siberia, Arak; Tibet, Aker; Turcic (Arabian, Mongolian peoples), Arak
Milk brandy
Origin by milk types
cow's, sheep's, goat's milk: Arak

Microorganisms
cream starter cultures: Cream cultured fermented products; Cream fraîche, fermented
kefir starter culture: Kefir sour cream
Lb. acidophilus: Acidophilus cream, cultured, Czechoslovakia; Acidophilus cream, cultured, USSR
polysaccharide producing strains: Cultured cream, sour cream
Smetanka cultured cream (special starter composition)
spontaneously fermented: Tsutsugi
Str. thermophilus: Lapte akru
yoghurt cultures: Cultured cream, sour cream
Preparations, products
baby food, children: Acidophilus cream, USSR
bread spread: Kajmak
culinary preparations: Cream fraîche, fermented; Cultured cream, sour cream; Kajmak
dessert: Lapte akru
drinks: Samco's
frozen: Tsutsugi
Cultured buttermilk products (often low fat content)
Acidophilus buttermilk
Bulgarian buttermilk
Cultured buttermilk
Flake buttermilk
Kefir buttermilk
Kosama-milk
Mazi-lala
Scandinavian buttermilks: Filmjölk; Lättfil
Yoghurt buttermilk
Cultured cream. *See herein:* Cream, cultured
Cultured egg-milk. *See* **Egg-milk product, cultured**
Cultured molds milk
Cultured mold-containing fermented milks
Cultured natural buttermilk
Cultures. *See* Appendix D
Cultures, wild, empirical
Maja (for yoghurt, Serbia, Turkey, Yugoslavia)
Podkvasa (for yoghurt, Bulgaria)
Roba (for Leben)
Stimulation of fermentation, stomach additions: Airan (sheep); Kojurtnak (ram or wether); Tarho (young lamb)
Curds and curd derivates. *See herein:* Pasty (texture) fermented milk products

Kumiss
  Domestic preparations
    camel's milk: Chal, Cal; Shubat, Shuvat, Suvat; Tujo (Kazakhstan, USSR)
    donkey's milk, kumiss
    mare's milk (original type of kumiss)
Kumiss-related beverages
  Kuban
  Kumiss from cow's milk
  Kurunga
Kvasseno mljako. *See* **Kvas from milk, Bulgarian**
Kwas. *See* **Kvas**

Labneh (various spellings)
Lact-, prefix, indicates milk
  Fermented milks: Lactofil; Lactorol
  Fermented whey beverage: Lactrone
  Infant food: Lactana
  Pharmaceutical preparations: Lactinex; Lacto-Bacilline; Lactopriv
Lactobacillus acidophilus products. *See herein:* Acidophilus products
Lactobacillus bulgaricus products. *See herein:* Bulgaricus products
Lactobacillus casei products. *See* **Lactobacilli-containing fermented fresh milks**
Lactose-hydrolyzed fermented products. *See* **Hydrolyzed-lactose fermented fresh milk products**
Laktofil. *See* **Lactofil**
Langfil. *See* **Long milk**
Langmjölk. *See* **Long milk**
Lio-, Lyo-, prefix (indicates freeze-dried)
  Liobif
  Lyobififidus
Long-life fermented milks
  Long-life fermented milks products
  Long-life yoghurt
Lyophilized products
  Dried yoghurt
  Freeze-dried preparations

Maast. *See* **Mast**
Matsoni. *See* **Matzoon**
Matsun. *See* **Matzoon**

Glossary of Technical Terms

For the general reader, various technical terms are defined here. Many others are explained as they are used throughout the book.

clarification The removal of sediment from milk by means of mechanical force, usually a centrifuge; it does not materially reduce the bacterial content.

coagulation Clotting or curdling is one of the most characteristic properties of the milk protein called casein. When under the influence of acid casein is altered and causes a change in the milk from a fluid to a semisolid or curdled state (formation of a coagulum). Coagulation can be brought about by the addition of acid or the production in the milk of acid by bacteria. Certain enzymes also cause coagulation (*see* rennet).

cream layer formation The rise of fat globules to the surface of milk. Also called creaming ability; this phenomenon does not occur in homogenized milk. The thickness of the cream layer depends on time of standing. Creaming is considered complete after 24 hours.

culture A growth of particular types of microorganisms, e.g., lactic acid bacteria, bifidobacteria, yeasts, and others in a liquid medium or on a solid medium, formed as a result of the previous inoculation and incubation of that medium.

cultured, culturing *See* fermented, fermenting.

curd Clotted protein formed when milk is treated with rennet or formed by means of acid either added or produced in the milk by fermentation of the lactose.

fermented, fermenting The fermentation (conversion) of milk sugar by bacteria or yeasts during the manufacture or incubation of milk into mainly acids but also other minor by-products.

homogenization Breaking up mechanically the fat globules of milk to smaller and approximately equal size to prevent the cream from rising to the surface.

incubation The process of development and growth of microorganisms after inoculation (*see* inoculation). Temperature control is very critical during incubation.

inoculation The introduction of specific microorganisms into a nutrient medium, for example, milk. Inoculation is almost always followed by incubating the microbial culture in the medium at a specific temperature for an appropriate period of time.

inoculum The material containing live microorganisms used to inoculate a nutrient medium, for example, milk.

pasteurization The exposure of milk and milk products to a heat treatment (with temperature and holding time specified) that destroys all the pathogens and nearly all other microorganisms.

rennet A preparation, for example, from the stomach of calves, containing a mixture of enzymes having the power to coagulate the casein of milk. Rennet is used in cheese making. It is not used in the manufacture of fermented fresh milk products.

starter A pure or mixed culture of microorganisms that is added to the raw material, such as heat-treated milk, in order to initiate production of yoghurt, kefir, cultured buttermilk, or other products.

whey The fluid remaining after casein and most of the fat have been removed from milk. The casein may have been coagulated by rennet (as in hard cheese making) or by means of acid, either added or produced in the milk by fermentation of the lactose, as in the making of soft cheeses such as quarg or cottage cheese; also called lacto-serum.